Jellyfish and Polyps

Jellyfish and Polyps

Cnidarians as Sustainable Resources for Biotechnological Applications and Bioprospecting

Editors

Antonella Leone
Gian Luigi Mariottini
Stefano Piraino

MDPI • Basel • Beijing • Wuhan • Barcelona • Belgrade • Manchester • Tokyo • Cluj • Tianjin

Editors
Antonella Leone
National Research Council,
Institute of Sciences of
Food Production
Italy

Gian Luigi Mariottini
Department of Earth,
Environment and Life Sciences,
University of Genova
Italy

Stefano Piraino
Università del Salento,
Dipartimento di
Scienze e Tecnologie
Biologiche ed Ambientali
Italy

Editorial Office
MDPI
St. Alban-Anlage 66
4052 Basel, Switzerland

This is a reprint of articles from the Special Issue published online in the open access journal *Marine Drugs* (ISSN 1660-3397) (available at: https://www.mdpi.com/journal/marinedrugs/special_issues/Jellyfish).

For citation purposes, cite each article independently as indicated on the article page online and as indicated below:

LastName, A.A.; LastName, B.B.; LastName, C.C. Article Title. *Journal Name* **Year**, *Article Number*, *Page Range*.

ISBN 978-3-03943-208-0 (Hbk)
ISBN 978-3-03943-209-7 (PDF)

Cover image courtesy of Michele Solca.

Contents

About the Editors

Antonella Leone is a biologist who holds a PhD in Plant Genetics and who focused her scientific interests and research activities on the study of the biological activities of natural compounds extracted from a diverse array of living eukaryotic organisms, including microalgae, plants and vegetables, as well as marine invertebrates. She is currently working on the development of new processing methodologies for novel food and nutraceutical products from jellyfish within the EU Horizon 2020 project "GoJelly", coupling the sustainability of production through the use of neglected resources (jellyfish biomasses, production waste and byproducts) and the use of innovative processes, such as supercritical CO_2 extraction. She also studies the nutraceutical potential and health-promoting effect of dietary components on human cell culture systems by investigations on some underlying action mechanisms of bioactive molecules, such as the modulation of cell–cell communication mediated by gap junctions (GJIC), a key control of cellular homeostasis, development and differentiation, as well as a target mechanism of tumor promotion.

Gian Luigi Mariottini is a biologist (MSc) and physician (MD), qualified for the practice of both professions. In 1986–1987, he carried out research on neuroblastoma at the Giannina Gaslini Pediatric Hospital of Genova (Italy). From 1987 to 2020, he was employed at the University of Genova, Italy, and retired in June 2020 from the Department of Earth, Environment and Life Sciences. His research activity focused on ecotoxicology, planktonology and environmental sciences; his main research field was the study of cnidarian toxicity and cytotoxicity on cultured cells, in the perspective of the utilization of extracts from tissues and nematocysts for drug discovery or for other purposes. He was involved in several national and international research projects and working groups, and is member of the Italian Society for Experimental Biology (Società Italiana di Biologia Sperimentale—SIBS).

Stefano Piraino is an associate professor of Zoology at the University of Salento. Over the last thirty years, he has authored 120 international scientific (ISI) articles in the field of marine invertebrate zoology, ecology, developmental biology and evolution, with a special focus on marine cnidarians (jellyfish and polyps). His research spans from reproductive biology, taxonomy, and systematics, to symbiotic interactions and trophic ecology, as well as from ontogeny to organogenesis, cell differentiation, and more recently, to bioprospecting and blue growth resource exploitation. He is currently a member of the EuroMarine and JRC-EASIN European networks, as well as the Chairman of the Academic Board in Biology at the University of Salento.

Preface to "Jellyfish and Polyps"

Climate change and other concurrent anthropogenic causes are influencing the frequency and abundance of jellyfish blooms, with large impacts on the structure and functioning of marine plankton ecosystems, as well as on human activities in coastal zones. In parallel, sea anemones, corals and less familiar forms of benthic polypoid cnidarians constitute a major group of suspension feeders governing the energy transfer from water column to seafloor organisms.

Their outstanding ecological importance in worldwide marine ecosystems calls for increased global monitoring of cnidarian ecology and life cycles. At the same time, many cnidarians are now regarded as a potential sustainable resource, calling for new investigations on their chemical and biochemical composition, the physical–chemical features and supramolecular organization of their protein components, the screening and identification of bioactive molecules, the associated microbiota and their sustainable biotechnological exploitation in different fields of applied research.

The apparent vulnerability of their soft bodies, coupled to their limited swimming ability and wide biodiversity with about 13,400 living described species, make cnidarians the top candidate for the development of biochemical strategies for survival (feeding, defense) and reproduction, including symbiosis or other relationships with microbes and other organisms. Venomous compounds occurring in extracts of cnidarians are viewed with particular interest for both aims of the mitigation of their adverse effects and their possible beneficial use for humans. Furthermore, in the pharmacopeia of traditional medicine of Eastern Countries, jellyfish are regarded as a treatment for disorders and diseases and represent a valuable foodstuff with health benefits, suggesting the occurrence of bioactive compounds that could be useful as nutraceuticals. Despite the increasing attention concerning jellyfish blooms, scientific knowledge of their biology, biochemical composition and potential in drug discovery, supporting their possible utilization and exploitation, is still limited.

This Special Issue collects novel research papers and original reviews focusing on bioprospecting marine cnidarians and on the exploitation of their biomasses and derived compounds for biotechnological and biomedical applications, as well as active ingredients for pharmaceutical, nutraceutical, cosmetic and cosmeceutical uses.

Antonella Leone, Gian Luigi Mariottini, Stefano Piraino
Editors

Article

Proteomic Analyses of the Unexplored Sea Anemone *Bunodactis verrucosa*

Dany Domínguez-Pérez [1,2,†], Alexandre Campos [1,2,†], Armando Alexei Rodríguez [3], Maria V. Turkina [4], Tiago Ribeiro [1], Hugo Osorio [5,6,7], Vítor Vasconcelos [1,2] and Agostinho Antunes [1,2,*]

[1] CIIMAR/CIMAR, Interdisciplinary Centre of Marine and Environmental Research, University of Porto, Terminal de Cruzeiros do Porto de Leixões, Av. General Norton de Matos, s/n, 4450-208 Porto, Portugal; danydguezperez@gmail.com (D.D.-P.); amoclclix@gmail.com (A.C.); tiago.amribeiro8@gmail.com (T.R.); vmvascon@fc.up.pt (V.V.)

[2] Biology Department, Faculty of Sciences, University of Porto, Rua do Campo Alegre, s/n, 4169-007 Porto, Portugal

[3] Department of Experimental and Clinical Peptide Chemistry, Hanover Medical School (MHH), Feodor-Lynen-Straße 31, D-30625 Hannover, Germany; aara259@gmail.com

[4] Division of Cell Biology, Department of Clinical and Experimental Medicine, Linköping University, SE-581 85 Linköping, Sweden; maria.turkina@liu.se

[5] Instituto de Investigação e Inovação em Saúde- i3S, Universidade do Porto, Rua Alfredo Allen, 208, 4200-135 Porto, Portugal; hosorio@ipatimup.pt

[6] Ipatimup, Institute of Molecular Pathology and Immunology of the University of Porto, Rua Júlio Amaral de Carvalho, 45, 4200-135 Porto, Portugal

[7] Department of Pathology and Oncology, Faculty of Medicine, University of Porto, Al. Prof. Hernâni Monteiro, 4200-319 Porto, Portugal

[*] Correspondence: aantunes@ciimar.up.pt; Tel.: +353-22-340-1813

[†] These authors contributed equally to this work.

Received: 21 November 2017; Accepted: 15 January 2018; Published: 24 January 2018

Abstract: Cnidarian toxic products, particularly peptide toxins, constitute a promising target for biomedicine research. Indeed, cnidarians are considered as the largest phylum of generally toxic animals. However, research on peptides and toxins of sea anemones is still limited. Moreover, most of the toxins from sea anemones have been discovered by classical purification approaches. Recently, high-throughput methodologies have been used for this purpose but in other Phyla. Hence, the present work was focused on the proteomic analyses of whole-body extract from the unexplored sea anemone *Bunodactis verrucosa*. The proteomic analyses applied were based on two methods: two-dimensional gel electrophoresis combined with MALDI-TOF/TOF and shotgun proteomic approach. In total, 413 proteins were identified, but only eight proteins were identified from gel-based analyses. Such proteins are mainly involved in basal metabolism and biosynthesis of antibiotics as the most relevant pathways. In addition, some putative toxins including metalloproteinases and neurotoxins were also identified. These findings reinforce the significance of the production of antimicrobial compounds and toxins by sea anemones, which play a significant role in defense and feeding. In general, the present study provides the first proteome map of the sea anemone *B. verrucosa* stablishing a reference for future studies in the discovery of new compounds.

Keywords: cnidarian; sea anemone; proteins; toxins; two-dimensional gel electrophoresis; MALDI-TOF/TOF; shotgun proteomic

---

1. Introduction

Cnidarians represent the largest source of bioactive compounds, as candidates for pharmacological tools [1] and even new drugs for therapeutic treatments [2–4]. Unlike toxin from terrestrial animals, cnidarian venoms have not received as much scientific attention [5]. Each one of the around 11,000 living species [6] possess nematocysts [7], which is the organ specialized in the production, discharge and inoculation of toxins [8]. Hence, the toxic feature can be theoretically ascribed to all the members of this Phylum, since nematocysts are the only ones of the three categories of cnidae found in all cnidarians [8]. However, without including components of the venom described at the transcriptomic level, only about 250 compounds have been reported until 2012 [9], although this figure has not increased significantly at the proteomic level in the last five years. The venom of cnidarians is composed mainly by peptides, proteins, enzymes, protease inhibitors and non-proteinaceous substances [9].

Most of the known toxins from cnidarians belong to the Order Actiniaria, Class Anthozoa (sea anemones) [10–36]. Among sea anemones, around 200 non-redundant proteinaceus toxins have been recognized to date, including proteins and peptides [32,37]. In addition, another 69 new toxins were revealed by transcriptomic-based analyses, although an additional set of 627 candidates has been proposed comprising 15 putative neurotoxins [38] and 612 candidate toxin-like transcripts from other venomous taxa [39]. In general, peptide toxins from sea anemones can be classified as cytolysins, protease inhibitors or ion channel toxins (neurotoxins), mainly voltage-gated sodium (Na_V) channel toxins and voltage-gated (K_V) potassium channel toxins [9,35,40–42]. Sea anemones are good candidates as a source of peptide/protein toxins, partly because their toxins are considerably stable compared to other cnidarian toxins (e.g., jellyfish). Only a limited number of sea anemones, however, have been examined for peptide/protein toxins [35], although more than 1000 species have been recorded [43]. Thus, sea anemones represent a relatively unexplored potential source of bioactive/therapeutic compounds.

The *B. verrucosa* is one of the most common species of sea anemones in the intertidal zone (Figure 1) of Portugal coast [44], yet its proteome, including peptide toxins, remains unexplored. The main goal of the present study was to establish a general proteomic analysis of whole-body extracts from the sea anemone *B. verrucose*; a species known to occur in the northeastern Atlantic Ocean, the North Sea and the Mediterranean Sea [45]. The specimens used in this study came from Portugal coast. The combination of shotgun analyses and two-dimensional gel electrophoresis yielded several proteins, including potential toxins. Until now, just a few chemical studies have been reported from this organism. In fact, to the best of our knowledge, this study provides the first proteomic profile of this species. Most of the proteins identified constitute first report for this species.

(a)   (b)

Figure 1. Sampling site at Praia da Memória, Porto, Portugal: (**a**) Picture of tide pools in rocks where inhabits the species of interest *Bunodactis verrucosa*. Note the remained pools at low tide and the relative abundance of mussels in the intertidal community; (**b**) Picture of two individuals of *B. verrucosa* from the sampling site.

2. Results and Discussion

2.1. Two-Dimensional Gel Electrophoresis and MALDI-TOF/TOF Analyses

The gel-based proteome analysis revealed 61 and 36 spots from the soluble fraction (SF) and insoluble fraction (IF), respectively. From the spots analyzed by Matrix-assisted laser desorption/ionization time-of-flight (MALDI-TOF/TOF), 23 peptide sequences belonging to eight proteins were identified in the SF, approximately 38% of the total analyzed (Figure 2, Table 1). Proteins identified in the SF comprised five different enzymes: Superoxide dismutase, Triosephosphate isomerase, Ribonuclease, two Fructose-bisphosphate aldolases and Alpha-enolase. In addition, Peroxiredoxin and two Ferritins were identified. However, three of these proteins matched to "predicted protein" as best hit, but were then further annotated using blastp algorithm in the NCBI with the accession number retrieved from the custom sea anemones databases. Unlike shotgun proteomics, for gel-based analysis were used only two sea anemones databases, since additional search was carried against UniProtKB/Swiss-Prot in the Metazoa section. However, best results corresponded to local analysis. On the other hand, no proteins were identified with statistic confidence from the IF (Figure S1) and in both cases SF and IF, the use of different database like UniProtKB/Swiss-Prot did not improved the identification. The details of blast search and protein identification by MALDI-TOF/TOF mass spectrometry of the protein identified from the 2DE is shown in Table 1. It is noteworthy, that some of the proteins identified have been previously reported in other cnidarians [46–48], but constitute the first report in *B. verrucosa*.

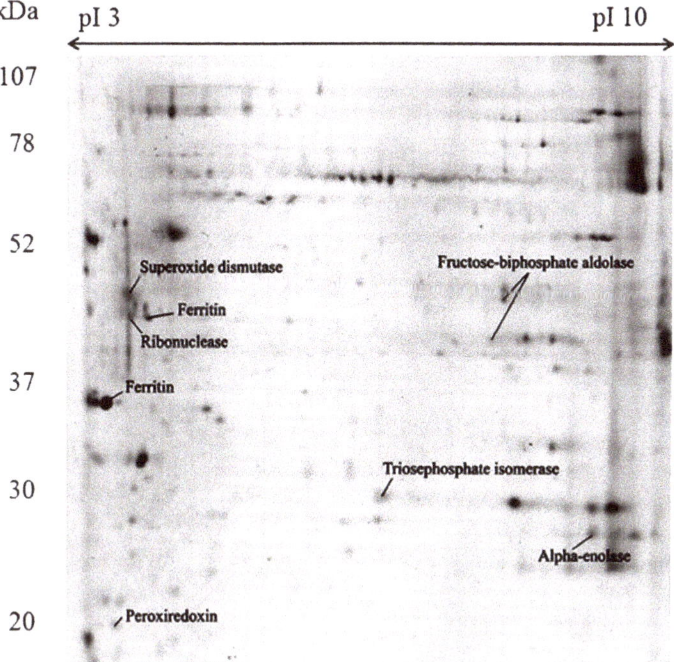

Figure 2. Two-dimensional gel electrophoresis and identification of soluble proteins from the whole-body aqueous extract of *Bunodactis verrucosa*. The first-dimension separation was carried out on 17 cm, pH 3–10 IEF gel strips and the second dimension on 12% sodium dodecyl sulfate-polyacrylamide gel electrophoresis (SDS-PAGE) gels. Gels were stained with colloidal Coomassie blue G-250. Identified proteins are indicated with their most commonly used name.

Table 1. Blast Search summary. Information concerning the proteins identification by Matrix-assisted laser desorption/ionization time-of-flight (MALDI-TOF/TOF) mass spectrometry of the proteins separated in two-dimensional gel electrophoresis.

Protein Name [1]	Species [2]	Protein [3] Score	Accession [4] Number	Ion [5] Score	Peptide Sequence [6]
predicted protein (Peroxiredoxin)	*Nematostella vectensis*	137	XP_001640260.1	15 115	R.LIQAFQFTDK.H K.DYGVLLEDQGVALR.G
Ferritin	*Nematostella vectensis*	124	XP_001632011.1	114	R.QNYHEECEAGINK.Q
Ferritin	*Nematostella vectensis*	117	XP_001627474.1	11 97	K.LMKFQNQR.G R.QNYHEECEAGINK.Q
predicted protein (Ribonuclease)	*Nematostella vectensis*	106	XP_001634183.1	93	R.VEIEAIAIVGEVKDE.
Superoxide dismutase [Mn]	*Exaiptasia pallida*	428	KXJ18609.1	76 67 103	K.DFGSFENFK.X K.KDFGSFENFK. K.AIYDVIDWTNVADR.Y
Triosephosphate isomerase	*Nematostella vectensis*	356	XP_001633516.1	56 22 95 19 121	K.FFVGGNWK.M R.KFFVGGNWK.M K.VIACIGELLSER.E R.NIFGEKDELIGEK.V K.VVIAYEPVWAIGTGK.T
predicted protein/Alpha-enolase	*Nematostella vectensis*	95	XP_001632906.1	10 37 10	K.YNQLLR.I R.AAVPSGASTGIYEALELR.D K.LAMQEFMLLPTGASNFR.E
Fructose-bisphosphate aldolase	*Nematostella vectensis*	151	XP_001629735.1	41 23 28 24	K.LTFSFGR.A R.LLRDQGIIPGIK.V R.LANIGVENTEENRR.L R.LLRDQGIIPGIKVDK.G
Fructose-bisphosphate aldolase	*Nematostella vectensis*	97	XP_001629735.1	28 32	K.LTFSFGR.A R.LANIGVENTEENRR.L

[1] best hit NCBI accession number; [2] the name of the species best hit belongs; [3] Score obtained for the MS ion; [4] NCBI accession number retrieved from the custom database; [5] MASCOT's score for ion peptides; [6] peptides sequences identified with statistical significance.

The identification rates obtained for SF are similar to those reported in previous studies of other marine species, when comparable proteomics protocols were used [49–51]. On the other hand, the absence of identifications in IF is an evidence that our proteomics protocol is likely not optimized for the analysis of the type of proteins present in this fraction. Since IF may be enriched with hydrophobic membrane proteins, the lack of identifications may be related, among other possible causes, to incomplete separation of proteins and to the inefficient digestion of these proteins with trypsin; thus, hindering the generation of a sufficient number of proteolytic peptide fragments for Mass Spectrometry/Mass Spectrometry (MS/MS) sequencing analysis. This limitation of trypsin when cleaving such proteins particularly in the hydrophobic and transmembrane domains can be overcome by combining the activities of other proteases [52,53].

The identified proteins seem to play important roles related with RNA degradation, glycolysis and antioxidant pathways. Moreover, some proteins like alpha aldolase seem to play diverse molecular and physiological roles. In fact, several antibacterial, antiparasitic, antifungical and autoantigen activities have been proposed [54]. Alpha aldolase expression and activity have been associated with the occurrence and metastasis of cancer, as well as with growth, development and reproduction of organisms [54]. Its expression seems to be related to heat shock [55], but it is also probably active under anaerobic condition [54]. In general, some of these proteins act as stress protein against environmental changes by exerting a protective effect on cells.

Ribonucleases, also known as RNases, are common and widely distributed catalytic proteins among animals, involved in the RNA degradation [56]. Three different RNases were detected: triosephosphate isomerase, Fructose-bisphosphate aldolase and Alpha-enolase, which are involved in the glycolytic pathway. Triosephosphate isomerase is a glycolytic enzyme that catalyzes the interconversion of the three-carbon sugars such as dihydroxyacetone phosphate and D-glyceraldehyde 3-phosphate [57]. Aldolases are stereochemistry-specific enzymes acting in a diverse variety of condensation and cleavage reactions [54]. Specifically, fructose-1,6-bisphosphate aldolase is involved in gluconeogenesis and glycolysis, controlling the production of fructose-1,6-bisphosphate from the condensation of dihydroxyacetone phosphate with glyceraldehyde-3-phosphate [58,59]; while Alpha-enolase is a versatile metalloenzyme, that catalyzes the conversion of 2-phosphoglyceric acid to phosphoenolpyruvic acid [54].

On the other hand, Ferritin is one of the most important proteins in iron metabolism, acting as primary iron storage protein or iron transporter, solubilizing iron and thus regulating its homeostasis [60,61]. Peroxiredoxin, also called thioredoxin peroxidase or alkyl hydroperoxide reductase, has been proposed as antioxidant protein [62–64]. Both proteins, seem to play an important role by protecting the cells against reactive oxygen species [65], so they are likely to be natural anti-Ultraviolet (UV) radiation agents [66]. Similarly, superoxide dismutase is another relevant antioxidant protein [65,67]. The high expression of this protein as part of the antioxidant defense system makes sense, since aerobic organisms need to deal with oxygen species produced as a consequence of aerobic respiration and substrate oxidation [67].

2.2. Protein Identification from Shotgun Proteomics Analysis

A methodology based on shotgun analysis was employed to investigate the whole-body proteome of *B. verrucosa*. This methodology has been previously reported as suitable for diverse purposes related to protein identification such as characterization of complex sample, inference of the main enzymatic pathway involved in a tissue, even to reveal venom composition [68–71]. Altogether, 688 peptide sequences were identified among the two replicates of the fractions analyzed (SF and IF), which accounted for 412 groups of non-redundant proteins (), retrieved from custom cnidarians databases. Of all protein detected, 97 were identified from two or more peptides. Only four proteins were detected as potential contaminants in the first search against custom database, while 69 sequences accounted for 35 putative proteins as contaminants against UniProtKB/Swiss-Prot database (Table S2). Of such contaminants, 10 proteins were identified from two or more peptides and were related mostly to human keratin and trypsin. In the case of contaminants, proteolytic fragments from trypsin and keratin were the most commonly found, which are difficult to avoid and thus are ubiquitous in proteomic analysis [72]. The functional annotation of all proteins (except for contaminants) was further addressed

The fact that several IF proteins were identified by this shotgun method shows the increased potential of this method over 2DE/MALDI-TOF/TOF for the analysis of membrane proteins, even when carried out based on the activity of a single protease (trypsin).

All proteins identified from the gel-based analysis were also found among those identified by the shotgun proteomic analysis. As an example, the shotgun analysis allowed the identification of Peroxiredoxin (XP_001640260.1, see Table 1) from two peptides sequences belonging to different organisms (Table S1): one peptide matched Peroxiredoxin-4 (KXJ19217.1) from *E. pallida*, and the second one Peroxiredoxin-4 (KXJ22794.1) from *E. pallida* and peroxiredoxin-like isoform X2 (XP_015769163.1) from *A. digitifera*. In the case of Peroxiredoxin-4 (KXJ22794.1), four peptides were identified for the protein and four for the protein groups (see razor + unique, terms_description in Table S1). However, only nine peptides generated by MALDI-TOF/TOF fragmentation from gel spots, were also detected within peptides resulting from the Orbitrap's approach. Despite the smaller number of protein identified from 2DE gel, this methodology represented a complement for shotgun proteomics analysis, increasing the number of peptides for the reconstruction of each protein. In fact, in 2D-MALDI fingerprint approach the number of peptides matching some proteins such as superoxide dismutase (KXJ18609.1), alpha-enolase (XP_001632906.1), triosephosphate isomerase (XP_001633516.1) and both fructose-bisphosphate aldolase (XP_001629735.1; XP_001629735.1), were identified with higher confidence in gel-based analyses than in the shotgun methodology.

2.3. Protein Gene Ontology Annotation

The proteomics identification pipeline using the Maquant software and 4 sequence databases, retrieved mostly "predicted" protein products. Therefore, these sequences were further blasted and mapped using the Blast2Go software (version 2.4.4) [73], (Figure 3). From a total of the 412 proteins identified with Maxquant software, 408 were successfully mapped using the Blast2Go software (Figure 3). The remaining four proteins, which were not submitted to further analysis, corresponded to potential contaminants. Out of the total number of proteins analyzed (408), 149 proteins were successfully annotated, representing the 36.5%. Thus, 259 proteins remained without Gene Ontology (GO) annotation, of which only four proteins were blasted without hits, 36 were mapped and 219 yielded positive hits. In total, 223 proteins were not included into the GO annotation considering the level 2 of protein classification, likely due to the absent of similar protein sequences in the protein databases. Moreover, most of these proteins retrieved as hits from cnidarian databases were "predicted". This result confirms the limited information known about sea anemones and cnidarians products.

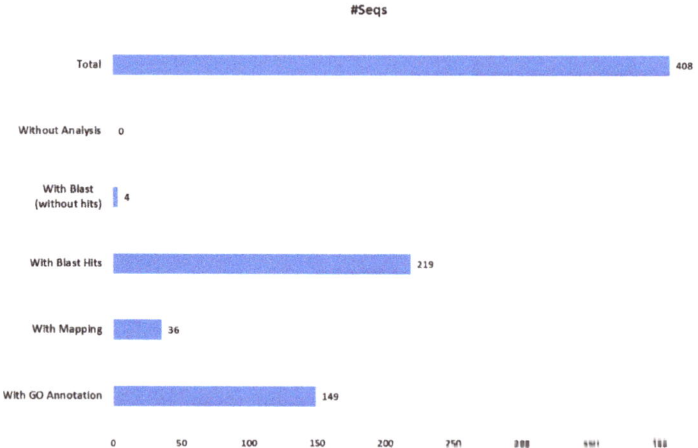

Figure 3. Blast2Go data distribution chart. The number of sequences (#Seqs) analyzed and annotated with Blast2Go software from the four custom cnidarian databases used.

Among the four databases analyzed, most hits corresponded to the species *E. pallida*, followed by *N. vectensis* (Figure 4), as expected according to its relative phylogenetic position [74], although *E. pallida* has the largest number of proteins among the databases used. Afterwards, the proteins identified as positive hits were functionally annotated per the GO nomenclature. Then, GO terms were assigned to each contig and annotated per GO Distribution by Level (2), regarding the three major GO categories: Biological Process (BP), Molecular Function (MF) and Cellular Components (CC).

The groups of proteins obtained from high-throughput analyses were classified per Blast2Go software, considering the GO Distribution by Level (2) (Figure 5). The most represented GO terms in the category of BP were metabolic process (GO:0008152), followed by cellular process (GO:0009987) and single-organism process (GO:0044699). In the case of MF, the most matched GO terms were binding (GO:0005488), catalytic activity (GO:0003824) and structural molecule activity (GO:0005198), in this order; whereas in the category of CC the most significant were cell part (GO:0044464), cell (GO:0005623) and organelle (GO:0043226). It is noteworthy that some proteins can be included in more than one GO term, since each protein could play diverse roles. Thus, some ambiguities can be found in the proteins reported for each category; and also, the total number of protein may apparently be overestimated. Details of GO annotation and protein accession number can be found on Figure S2 and Table S3.

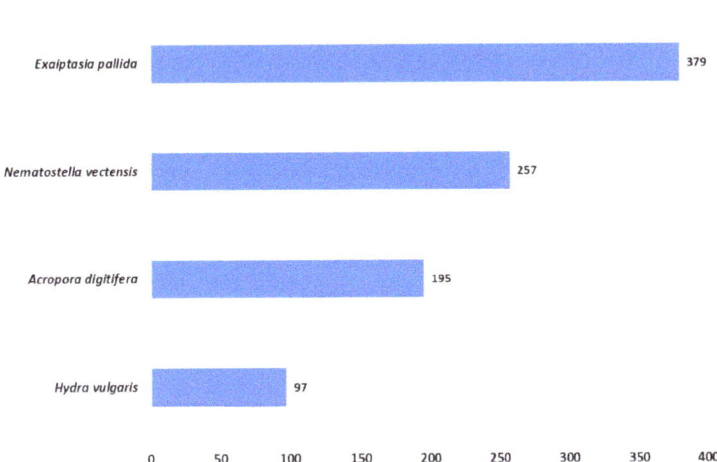

Figure 4. Blast2Go Species distribution chart. Number of blast hits (#BLAST Hits) retrieved are shown from the four cnidarian databases analyzed.

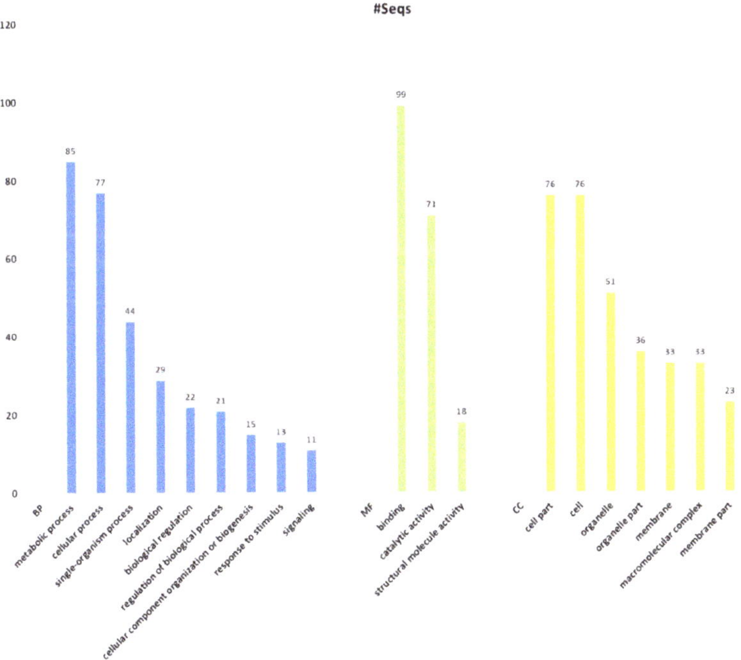

Figure 5. Blast2Go hits Gene Ontology (GO) annotation. Number of sequences (#Seqs) corresponding to blast hits annotation are based on the three major GO Categories of GO Distribution by Level (2): Biological Process (PB) in blue, Molecular Function (MF) in green and Cellular Components (CC) in yellow.

Among the 111 proteins matching to the GO term BP, 85 proteins (76.56%) classified as metabolic process, 77 (69.37%) for cellular process and 44 (39.64%) as single-organism process. In this group, in the GO level 3, 64 proteins were related with the GO name of "primary metabolic process" and "organic substance metabolic process", both belonging to "metabolic process" as parent. Besides, 52 proteins were associated with "cellular metabolic process", which were involved in both metabolic process and cellular process as parents (for details of GO annotation see Figure S2, Table S3).

In total, 86 proteins were included in the category of the CC. Among them, 76 proteins (88.37%) matched for "cell part" and "cell". However, this is an ambiguity, since all sequences detected as "cell part" are part of the "cell" category (Figure S2). Although, other proteins represented by the sublevels, related to cytoplasmic elements as part of intracellular components, were also subcategories of the "cell". The GO "intracellular" was more represented with 73 proteins (84.88%) in level 3 than those "organelle" and "membrane" in the superior level 2, with 51 proteins (59.3%) and 33 proteins (38.37%), respectively.

In addition, 135 proteins were grouped into the MF category. Among them, "binding" with 99 proteins (73.3%) was the most significant one. In this group, a total of 66 (48.89%) proteins were involved in "ion binding", whereas both "heterocyclic compound binding" and "organic cyclic compound binding" hit 62 proteins (45.93%). The second most significant GO term "catalytic activity" comprised 71 proteins (52.59%), of which the most remarkable function was "hydrolase activity", accounting for 37 proteins (27,41%) acting mainly on acid anhydrides, in phosphorus-containing anhydrides. Moreover, 18 of these enzymes were involved in pyrophosphatase activity, of which 17 are associated with nucleoside-triphosphatase activity (Figure S2).

2.4. Top KEGG Pathways

On the other hand, the Kyoto Encyclopedia of Genes and Genomes (KEGG) analyses revealed 28 enzymes involved in 41 different pathways. The accession number of the protein involved in each pathway and other details can be found in Table S4. Considering the number of proteins matched, the most relevant pathways were Purine and Thiamine metabolism, with 18 and 17 proteins matched, respectively (Table 2). In addition, three enzymes: adenylpyrophosphatase, phosphatase and RNA polymerase were found to be involved in the Purine metabolism pathway, whereas only a phosphatase resulted in the Thiamine metabolism. The Purine metabolism pathway is close related to the metabolism of nucleotide [75], since purine constitutes subunits of nucleic acids and precursors for the synthesis of nucleotide cofactors, whereas Thiamine metabolism pathway is fundamental in the metabolism of carbohydrates [76].

Interestingly, one of the most significant among the top twenty pathways was the biosynthesis of antibiotics. In that pathway, a total of 14 proteins, accounted for 13 enzymes grouped into five major families: dehydrogenase, transaminase, carboxykinase (GTP), hydratase, isomerase and aldolase. Most proteins matched in this pathway belong to the larval stage of *N. vectensis*. This result is particularly interesting, because of the abundance of proteins involved in defenses against pathogens, during the most vulnerable stage in the animal life cycle. Thus, this finding supports that sea anemones may be considered as a promising source of antibiotic compounds [77–79]. Other relevant pathways were glycolysis/gluconeogenesis and carbon fixation in photosynthetic organisms, both involved in the production of energy. The presence of proteins associated with carbon fixation in photosynthetic organisms is likely due to symbionts such as zooxanthellae, considering to be present in sea anemones [80,81].

The isomerase detected in the biosynthesis of antibiotics pathway, was the same to that identified in the gel-based analyses as triosephosphate isomerase from *N. vectensis* (XP_001633516.1). This one is also involved in other pathways such as glycolysis/gluconeogenesis, carbon fixation in photosynthetic organisms, fructose and mannose metabolism and inositol phosphate metabolism. The predicted protein (XP_001632906.1), homologue to alpha-enolase, and the fructose-bisphosphate aldolase (XP_001629735.1) from *N. vectensis*, were both involved in the pathways of biosynthesis of antibiotics and glycolysis/gluconeogenesis. In addition, the mentioned predicted protein was also found in the methane metabolism pathway, while the fructose-bisphosphate aldolase also occurred in some pathways such as carbon fixation in photosynthetic organisms, methane metabolism, pentose phosphate pathway and fructose and mannose metabolism. In general, these analyses support the diverse roles of some of the proteins identified, given additional information related to its biological function.

Table 2. Top twenty Kyoto Encyclopedia of Genes and Genomes (KEGG) pathways.

Pathway	#Proteins in the Pathway	#Enzymes in Pathway
Purine metabolism	18	3
Thiamine metabolism	17	1
Biosynthesis of antibiotics	14	13
Glycolysis/Gluconeogenesis	9	6
Carbon fixation in photosynthetic organisms	9	6
Amino sugar and nucleotide sugar metabolism	6	3
Methane metabolism	6	3
Pyruvate metabolism	5	4
Cysteine and methionine metabolism	4	5
Citrate cycle (TCA cycle)	4	3
Fructose and mannose metabolism	4	2
Various types of N-glycan biosynthesis	4	1
Glycosphingolipid biosynthesis—ganglio series	4	1
Glycosaminoglycan degradation	4	1
Glycosphingolipid biosynthesis—globo and isoglobo series	4	1
Other glycan degradation	4	1
Glyoxylate and dicarboxylate metabolism	3	3
Carbon fixation pathways in prokaryotes	3	2
Pentose phosphate pathway	3	2
Histidine metabolism	2	2

2.5. Detection of Potential Toxins

Among all peptides detected, 63 sequences matched for 58 potential toxins (Table S2), but only five toxins with more than one peptide (Table 3). Specifically, the five proteins matched as potential toxins were retrieved from different species other than cnidarians and each was reconstructed from two peptide sequences. Besides, these peptides were not redundant to those proteins reconstructed from the previous analyses with the four cnidarians database. In fact, the origin of such peptides by fragmentation of the protein matched as potential toxin (Table 3), which represents a better explanation for our results. Therefore, it is unlikely a false-positive assumption that the peptides were generated from proteins related to potential toxins.

Table 3. Potential toxins from the sea anemone *Bunodactis verrucosa*. Potential toxins identified by MaxQuant software against the venom section of UniProtKB/Swiss-Prot database.

Protein [1] Name	Species [2]	Score [3]	Accession [4] Number	Ion [5] Score	Peptide Sequence [6]	Fraction [7] (Rep.)
SE-cephalotoxin	*Sepia esculenta*	11.47	CTX_SEPES	62.7	AGYIMGNR	IF (1)
				42.8	LDQINDKLDK	IF (1)
Basic phospholipase A2 vurtoxin	*Vipera renardi*	12.06	PA2B_VIPRE	2.9	CCFVHDCCYGNLPDCNPKIDR	SF (1)
				18.3	NGAIVCGK	IF (1)
Alpha-latroinsectotoxin-Lt1a	*Latrodectus tredecimguttatus*	11.73	LITA_LATTR	22.7	EMGRKLDK	IF (2)
				3.01	NSCMHNDKGCCFPWSCVCWSQTVSR	SF (1)
Zinc metalloproteinase/disintegrin	*Deinagkistrodon acutus*	11.48	VM2M2_DEIAC	27.4	FPYQGSSIILESGNVNDYEVVYPRK	SF (1)
				31.7	NTLESFGEWRAR	IF (1)
Neprilysin-1	*Trittame loki*	11.49	NEP_TRILK	28.4	LAHETNPR	IF (1)
				71.3	LEAMINK	SF (2)

[1] UniProtKB/Swiss-Prot name of the protein identified as potential toxin; [2] name of the species best hit belongs; [3] Protein score which is derived from peptide posterior error probabilities; [4] UniProtKB/Swiss-Prot hit accession number; [5] Andromeda score for the best associated MS/MS spectrum; [6] UniProtKB/Swiss-Prot accession number; [7] fraction (IF: Insoluble fraction; SF: Soluble fraction) where a peptide was detected and replicates they occurred.

The proteins identified as potential toxins comprise several previously reported toxins and other non-reported in cnidarians. Herein, we found two proteins related to metalloproteinases, one zinc metalloproteinase/disintegrin (VM2M2_DEIAC) of the snake *Deinagkistrodon acutus* [82] and another one called neprilysin-1 (NEP_TRILK) from brush-footed trapdoor spider *Trittame loki* [83]. Both proteins represent two of the three classes of metalloproteinases found in the hydra genome: astacin class, matrix metalloproteinase class, and neprilysin [84]. Metalloproteinases have been subsequently reported in hydra [85,86], jellyfish [9,87,88], but less in sea anemones [89]. Their structure and function seem relatively conserved among metazoans [87], since they can play a broad range of roles in biological process related to hydrolytic functions and development [84]. However, the peptides obtained matched specifically to proteins, which have been proposed as venom components [82,83]. In general, the most significant role of these protein (zinc metalloproteinase/disintegrin, neprilysin-1), must be related to its capacity of breakdown the extracellular matrix [84]. Moreover, this protein displays gelatinolytic and fibrinolytic activities, as previously reported from the venoms of four Scyphozoan jellyfishes [88].

Another protein detected matched to a phospholipase A2 (PLA$_2$) called vurtoxin (PA2B_VIPRE) from the steppe viper *Vipera renardi* [90]. Phospholipases A2 are commonly found in the venom of the most toxic animals like cnidarians, cephalopods, insects, arachnids, and reptiles [91]. Specifically, vurtoxin showed homology with the neurotoxic PLA$_2$ ammodytoxins [90]. However, it is not clear if this toxin can act as neurotoxin in this species, since vurtoxin occurred as a minor component in the venom of *V. renardi* [90]. In general, the biological role of PLA2$_s$ could be diverse. PLA$_2$s can act in the arachidonic pathway or in the calcium-dependent hydrolysis of the 2-acyl groups in 3-sn-phosphoglycerides, showing a preference for phosphatidylglycerol over phosphatidylcholine [92,93]. This biological role prevails in cnidarians, showing a significant phylogenetic distance to higher metazoans PLA$_2$s, been proposed as the ancestors [94]. On the other hand, the biological activity of PLA$_2$s in reptiles has been revealed most as antiplatelet, myotoxic, and neurotoxic [93,95].

In addition, two putative neurotoxins named as alpha-latroinsectotoxin-Lt1a (LITA_LATTR) and SE-cephalotoxin (CTX_SEPES) were identified. The first one, also known as alpha-LIT, was purified from venom glands of the Mediterranean black widow spider *Latrodectus mactans tredecimguttatus* [87]. The proposed mechanism of toxicity involved presynaptic effects, acting selectively only for insects [87]. On the other hand, SE-cephalotoxin has been characterized from the salivary gland of cuttlefish *Sepia esculenta* [88]. The lethality of this toxin was very high to crab, seemingly by neurotoxic mechanism, since the symptoms caused loss of movement, flaccid paralysis and even death [88]. However, SE-cephalotoxin has been considered as a new class of proteinaceous toxin, due to the lack of homology with any other toxins, even those cephalotoxins from octoposes [88]. Therefore, the evidences of a potential SE-cephalotoxin from *B. verrucosa*, constitutes a highlighted finding as the first report of this toxin in sea anemones.

Furthermore, others 53 non-redundant peptide sequences matched to 53 potential toxins, but with only one peptide identified for each protein (Table S2). Of all, 21 peptides sequences matched to 21 potential neurotoxins comprising presynaptic and postsynaptic toxins like ion channel blockers, mostly voltage-dependent potassium and calcium channels. Among them was found a Kunitz-type serine protease inhibitor, which can act as inhibitor of both serine proteases and voltage-gated potassium channels (Kv) [89]. Besides, three metalloproteases, two hyalunoridases, and a Beta-fibrinogenase were detected. On the other hand, seven PLA$_2$s and three PLA$_D$ occurred within potential toxins. Another potential toxin identified with PLA$_2$s activity, was the Helofensin-1 characterized from the genus *Heloderma* [96,97]. This toxin has no hemorrhagic nor hemolytic activities, instead directly inhibited the electrical stimulation of the isolated hemi-diaphragm of mice [96]. Finally, four hemolitic/cytolytic proteins and five additional proteins involved in the coagulation pathway (including two "snaclec") were found.

2.6. Putative Use of Toxins by B. verrucosa in Prey Catching and Feeding

Sea anemones are ancient active predators, belonging to what is considered "the oldest extant lineage of venomous animals" [98]. The *B. verrucosa* inhabits tidepools in rocks, crevices in shallow water [99], where occurs mussels, gastropods, small crabs, and goby fishes as potential preys. This sea anemone feeds on mussels and small gastropods at least, since we found specimens regurgitating one or more empty mussel shells, after removal from the substrate during sampling (Figure 6). Moreover, we found some specimens containing mussels' shells and gastropods into the gastrovascular cavity. It is noteworthy that in the sampling area mussels were abundant covering rocks, even in the pools where sea anemones grow (Figure 1). Therefore, these bivalves may constitute the main food source for *B. verrucosa*. This is not an isolated fact, since mussels seem to be the main food source of other intertidal sea anemones like *Anthopleura elegantissima* and *Anthopleura xanthogrammica* [100,101]. Moreover, mussels are suitable to be fed by sea anemones in home aquariums [102]. However, bivalves can close their valves for prolonged periods of time under adverse environmental condition [103,104]. In other words, how can sea anemones obtain nourishments from mussels, if these bivalves tightly close the valves when feel the predator attack?

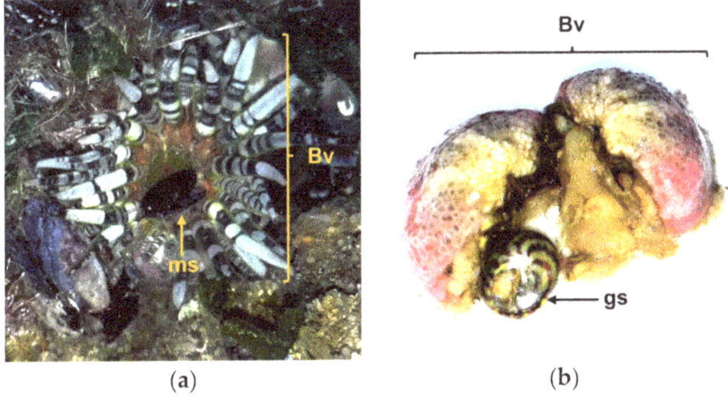

(a)   (b)

Figure 6. Evidence found relating the *Bunodactis verrucosa* (Bv) feeding on mollusks: (**a**) Specimen of *B. verrucosa* regurgitating an empty mussel's shell (ms) after body squeezing; (**b**) Gastropod (gs) found into the gastrovascular cavity of *B. verrucosa* after its body dissection.

Mussels are abundant in the intertidal community (Figure 1) and their movements are limited. In this scenario, sea anemones can capture a close mussel with its tentacles and introduce it into the gastrovascular cavity. Once the mussel is captured, it immediately closes its valves and stops filtering. Nonetheless, the sea anemones have cnidocytes in the gastrovascular cavity [8] capable of breaking mussels' protection. First, hydrolytic enzymes like zinc metalloproteinase/disintegrin, hyaluronidases and proteases found in *B. verrucosa* may be poured into the gastrovascular cavity. The combination of such enzymes could degrade the tissues that seals the shell, probably a dorsal elastic proteinaceous-ligament extending for the length of the hinge [105]; or through the ventral margin of the mussel. The tissues degradation by metalloproteinases can facilitate the diffusion of neurotoxins inside the prey. Then, neurotoxins could act on the adductor muscle, whose loss of function will lead to valve opening.

Specifically, SE-cephalotoxin can diffuse inside the valves, inhibiting the adductor muscles, thus producing flaccid paralysis increasing the valves gape aperture. The high solubility previous reported for SE-cephalotoxin seems to play an important role in the diffusion of this toxin in sea water. This property should be useful whether preys are nearby the sea anemone, because SE-cephalotoxin could disperse around or in the sea water remnant inside the shell after enclosed its valves. Besides, this feature can be used as an advantage to subdue prey prior to eating. Other neurotoxins detected, and the PLA$_2$ vurtoxin, are also able to block the adductor muscles. However, the diversity of toxins found is likely related to others potential preys as crabs and goby fishes (Gobiidae, Perciformes), polychaetes worms and starfish. Interestingly, other cephalotoxins have been previously purified from species of octopodiform cephalopods [106–109], which are likely used to neutralize crabs and bivalves.

Altogether, toxins found seemingly act synergistically to subdue mussels. Indeed, a similar mechanism in which hydrolytic enzymes like metalloproteinase facilitate the access of neurotoxic peptides to synaptic targets was previously proposed for the spider *T. loki* [83].

3. Materials and Methods

3.1. Protein Extraction

Specimens of *B. verrucosa* were sampled at Praia da Memória, Porto, Portugal (Lat/Long WGS84; 41°14′00.0″ N 8°43′27.0″ W). Then whole animal bodies (four specimens) were kept at −80 °C, freeze dried and subsequently homogenized in a blender until obtaining a dry powder. Lyophilized material of *B. verrucosa* (0.1 g) was mixed with 500 µL Tris-HCl (40 mM), $MgCl_2$ (5 mM), Dithiothreitol (DTT) (1 mM), protease inhibitors (87,785, Thermo Scientific, Waltham, MA, USA), at pH 8.0, (buffer 1) in vortex (2 × 30 s). The mixture was centrifuged at 16,000× g, during 20 min at 4 °C. The supernatant (soluble protein fraction, SF) was stored at −20 °C and the pellet was homogenized with 500 µL urea (7 M), thiourea (2 M), CHAPS (4%, *w/v*), dithiothreitol (65 mM) and ampholytes (0.8%, *v/v*), at pH 4–7 in vortex (2 × 30 s) and incubated overnight, at 4 °C. The homogenate was centrifuged at 16,000× g, during 20 min at 4 °C, and the supernatant (insoluble protein fraction, IF) collected and stored at −20 °C. Total protein concentration was estimated according to the Bradford method [110].

3.2. Two-Dimensional Gel Electrophoresis

Two-dimensional gel electrophoresis (2DE) was performed as described previously [49]. Duplicate IF and SF (~400 µg of protein) were diluted to 300 µL urea (7 M), thiourea (2 M), CHAPS (4%, *w/v*), dithiothreitol (65 mM) and ampholytes (0.8%, *v/v*), at pH 4–7 and loaded onto 17 cm, pH 4–7 immobiline dry strips (Bio-Rad, Hercules, CA, USA) with active hydration (50 Volt) for 12 h. Proteins were separated by isoelectric focusing (IEF) in a Protean IEF cell (Bio-Rad) with the following program: step 1, 15 min at 250 V; step 2, 3 h voltage gradient to 10,000 V (linear ramp); step 3, 10,000 V until achieving 60,000 V/h (linear ramp). Second-dimension sodium dodecyl sulfate-polyacrylamide gel electrophoresis (SDS-PAGE) was performed in a Hoefer SE 900 vertical slab electrophoresis system (Hoefer, Holliston, MA, USA), with 12% (*w/v*) acrylamide gels, at 480 mA and 20 °C. After electrophoresis run the gels were stained with colloidal Coomassie blue G-250 [111]. The 2DE protein profiles were analyzed by gel scanning with a GS-800 calibrated densitometer (Bio-Rad) and the D analysis software (Bio-Rad) as described previously [49]. Protein spots detected by this procedure were excised from the gels for subsequent identification.

3.3. MALDI-TOF MS Analysis

Matrix-assisted laser desorption/ionization time-of-flight (MALDI-TOF/TOF) mass spectrometry (MS) measurements were performed to identify protein spots from 2DE gels. Protein spots were washed, distained, reduced, alkylated, and digested with trypsin following the procedure described by Osório and Reis [112]. The solution containing the peptides was collected and stored at −20 °C until application to a MALDI plate. Peptides were acidified with trifluoroacetic acid (TFA) and concentrated using C18 micro-columns (C18 Tips, 10 µL, Thermo Scientific, 87782). Peptides were thereafter eluted from the micro-column directly onto the MALDI plate with 1.5 µL of α-CHCA matrix (8 mg/mL) prepared in acetonitrile (50%, *v/v*), TFA (0.1%, *v/v*) and 6 mM ammonium phosphate. MALDI mass spectra were externally calibrated following the manufacturer's instructions (TOF/TOF calibration mixture, AB SCIEX) and internal calibration was applied using trypsin autolysis peaks. Peptide mass spectra data was collected in positive ion reflector mode in the range of m/z 700–4000 (4800 Plus MALDI TOF/TOF Analyzer, AB SCIEX).

Proteins were identified by combining Peptide Mass Fingerprint and MS/MS information. Proteins were searched in a locally stored NCBI copy of protein sequences of the genomes of the sea anemones *Exaiptasia pallida* (26,042 protein count, GenBank accession: GCA_001417965.1) and

Nematostella vectensis (24,780 protein count, GenBank accession: GCA_000209225.1), using the Mascot search engine (Version 2.4). The search included peaks with a signal-to-noise ratio greater than 10 and allowed for up to two missed trypsin cleavage sites, mass tolerance of 50 ppm, cysteine carbamidomethylation (fixed modification), methionine oxidation (variable modification), and a charge state of +1. For a match to be considered significant, protein scores with a probability greater than 95% ($p < 0.05$), calculated by the Mascot software, were required [112]. The data generated from 2D-MALDI procedures were also searched against UniProtKB protein sequence database in the Metazoa section [113,114], using the same parameters mentioned before.

3.4. In Solution Protein Digestion and MS/MS Analysis

For LC-MS/MS analysis, SF and IF protein samples were processed by filter aided sample preparation (FASP) method [115] with the following modifications. Protein samples (40 μg) were alkylated and digested with trypsin (recombinant, proteomics grade, Roche, Basel, Switzerland), at enzyme to protein ratio of 1:100 (w/w), for 16 h at 37 °C, in centrifugal filter units with nominal molecular weight limit (NMWL) of 30 kDa (MRCF0R030, Millipore, Billerica, MA, USA). Peptides were subsequently recovered by centrifugal filtration, acidified with formic acid (FA) (10%, v/v), desalted and concentrated by reversed-phase extraction (C18 Tips, 100 μL, Thermo Scientific, 87784) using acetonitrile (ACN) (70%, v/v) and TFA (0.1%, v/v) for peptide elution. Before LC–MS/MS, the peptides were recovered in 0.1% (v/v) Formic acid (FA) to the concentration of 0.04–0.06 μg/μL.

FASP protein digests (duplicate samples) were analyzed by nano-LC coupled to a hybrid Ion-trap mass spectrometer (LTQ Orbitrap Velos Pro—ETD, Thermo Scientific) as described previously [68]. Peptides were separated by reverse-phase chromatography (20 mm × 100 μm C18 precolumn followed by a 100 mm × 75 μm C18 column with particle size 5 μm, NanoSeparations, Nieuwkoop, The Netherlands) using a linear ascending gradient of buffer B (ACN + FA, 0.1%, v/v), being buffer A TFA, 0.1%, v/v in water. The gradient started from 2% B to 30% B in 40 min and to 95% B (v/v) in 30 min, at a flow rate of 0.3 μL/min (total elution time 70 min). Peptides were analyzed by on-line nano-electrospray ionization (easy nano-ESI) in positive mode, with Xcalibur software (version 2.6, Thermo Scientific). Full scans were performed at a resolution of 30,000 with scan ranges of 380–2000 m/z. The top 20 most intense ions were isolated and fragmented with CID by applying normalized collision energy of 30% value, isolation width of 2.0, activation time of 10 milliseconds and Q-value of 0.25. In total 4 nano-LC-MS/MS runs were performed.

3.5. Peptide Identification

The resulting ion-trap raw data (LTQ Orbitrap) were searched against custom cnidarians protein databases using MaxQuant freeware software (version 1.5.5.1) with the Andromeda search engine. MS and MS/MS tolerances were set to 10 ppm and 0.6 Da, respectively. Trypsin was selected for protein cleavage allowing for one missed cleavage. Carbamidomethylation and oxidation were selected as static and dynamic modifications, respectively. Identifications were validated by performing a decoy database search for the estimation of False Discovery Rate (FDR) and peptide identifications were accepted if they could be established at a probability greater than 95.0%. Protein identifications were accepted if they could be established at a probability greater than 99.9% and contained at least two identified peptides (Razor + unique peptides) [116,117], based on Occam's razor principle). The protein database utilized was the locally stored NCBI copy of protein sequences of the genomes of the sea anemones *E. pallida* (26,042 protein count, GenBank accession: GCA_001417965.1), *N. vectensis* (24,780 protein count, GenBank accession: GCA_000209225.1), *Hydra vulgaris* (21,993 protein count, GenBank accession: GCF_000004095.1) and *Acropora digitifera* (33,878 protein count, GenBank accession: GCF_000222465.1). The identification of potential toxins was done against the manually reviewed venom proteins and toxins database, from the animal toxin annotation project of the UniProtKB/Swiss-Prot protein knowledgebase [118–120] (database size 1.20 MB, downloaded on 16 June 2016).

3.6. Protein Homology Search and GO Analysis

Protein sequences with unknown function were annotated with a blast search in the National Centre for Biotechnology Information database (NCBI, http://www.ncbi.nlm.nih.gov/) using blastp algorithm employing a threshold e-value of 1×10^{-10}. Total of proteins identified with Maxquant software, were also blasted and mapped using the Blast2Go software (version 2.4.4) [73]. Gene ontology (GO) terms were used to group proteins within the domains of BP, CC, andMF.

4. Conclusions

The present work revealed for the first time a draft of the whole proteome of the sea anemone *B. verrucosa*. The shotgun proteomics analysis yielded most of the protein identified in a total of 412, whereas gel-based analyses provided less data but useful as complementary information. Altogether, both gel-based and gel-free approaches of proteomics analyses and functional bioinformatics analyses revealed three major groups of proteins belonging to "metabolic process", "binding" and "cell parts" GO categories. Unlike throughput analyses, only eight proteins were identified from two-dimensional electrophoresis combined with MALDI-TOF/TOF. These eight proteins comprised enzymes mainly involved in the glycolytic pathway, antioxidants activities and RNA degradation. Notably, according to the results of KEGG analysis a significant number of enzymes corresponded to the Biosynthesis of antibiotics pathway indicating the importance of the biological antimicrobial chemical defense mechanisms. Moreover, some potential toxins such as metalloproteinases, and neurotoxin such as SE-cephalotoxin were identified. The combination of proteomic evidences and the ecology of the species, shed light about its strategy to subdue preys like mussels. In this sense, the toxins seemingly act synergically. Metalloproteinase may produce a degradation of the tissues, aiding the diffusion of the neurotoxins to the target, producing muscle paralysis. Hence, this work constitutes a reference proteome for future studies in sea anemones, also given insight about its potential toxin production and its putative mechanism of action in feeding.

Supplementary Materials: The following are available online at www.mdpi.com/1660-3397/16/2/42/s1, Figure S1: Two-dimensional gel electrophoresis of insoluble fraction (IF) from *Bunodactis verrucosa*, Figure S2: Combined Graph obtained for GO Distribution by Level (2); Table S1: Proteins identified against custom cnidarians databases title; Table S2: Proteins identified as potential toxins; Table S3: Details of GO annotation and protein accession number obtained with the Balst2Go software; Table S4: Details of the KEGG analyses obtained with the Balst2Go software.

Acknowledgments: Dany Domínguez Pérez was supported by a Ph.D. grant (SFRH/BD/80592/2011) from the Portuguese Foundation for Science and Technology (FCT—Fundação para a Ciência e a Tecnologia, Portugal). A. Campos work were supported respectively by Postdoc grants SFRH/BPD/92978/2013 and SFRH/BPD/103683/2014 from the FCT. Armando A Rodríguez was supported by an Alexander von Humboldt postdoctoral fellowship (3.2-KUB/1153731 STP and the Collaborative Research Centre 1279 funded by the German Research Foundation). We are grateful to Isabel Cunha and Daniela Almeida from CIIMAR, FCUP, University of Porto, for the help in the identification, sampling and transporting of *B. verrucosa* specimens. To Barbara Frazão, from Instituto Português do Mar e da Atmosfera, Lisbon, Portugal for suggestions and by providing information related the species studied. This study was funded in part by the Strategic Funding UID/Multi/04423/2013 through national funds provided by FCT and the European Regional Development Fund (ERDF) in the framework of the program PT2020, by the European Structural and Investment Funds (ESIF) through the Competitiveness and Internationalization Operational Program—COMPETE 2020 and by National Funds through the FCT under the project PTDC/AAG-GLO/6887/2014 (POCI-01-0124-FEDER-016845), and by the Structured Programs of R&D&I INNOVMAR—Innovation and Sustainability in the Management and Exploitation of Marine Resources (NORTE-01-0145-FEDER-000035, Research Line NOVELMAR) and CORAL NORTE (NORTE-01-0145-FEDER-000036), and funded by the Northern Regional Operational Program (NORTE2020) through the ERDF.

Author Contributions: Conceived and Designed the experiments: D.D.-P., V.V., A.C., A.A. Supervised and Contributed Reagents/Materials: V.V., A.A. Two-dimensional gel electrophoresis and MALDI-TOF/TOF analysis: T.R.; A.C., H.O. Shotgun proteomic analysis: M.V.T., D.D.-P., A.C. Wrote the paper: D.D.-P., A.C., A.A.R. Performed the figures and tables: D.D.-P., A.C. Analyzed, discussed and revised the manuscript: A.A.R., M.T., H.O., T.R., V.V., A.A.

Conflicts of Interest: The authors declare no conflict of interest.

References

1. Yan, L.; Herrington, J.; Goldberg, E.; Dulski, P.M.; Bugianesi, R.M.; Slaughter, R.S.; Banerjee, P.; Brochu, R.M.; Priest, B.T.; Kaczorowski, G.J.; et al. *Stichodactyla helianthus* peptide, a pharmacological tool for studying Kv3.2 channels. *Mol. Pharmacol.* **2005**, *67*, 1513–1521. [CrossRef] [PubMed]
2. Tejuca, M.; Diaz, I.; Figueredo, R.; Roque, L.; Pazos, F.; Martinez, D.; Iznaga-Escobar, N.; Perez, R.; Alvarez, C.; Lanio, M.E. Construction of an immunotoxin with the pore forming protein sti and ior C5, a monoclonal antibody against a colon cancer cell line. *Int. Immunopharmacol.* **2004**, *4*, 731–744. [CrossRef] [PubMed]
3. Beeton, C.; Pennington, M.W.; Wulff, H.; Singh, S.; Nugent, D.; Crossley, G.; Khaytin, I.; Calabresi, P.A.; Chen, C.Y.; Gutman, G.A.; et al. Targeting effector memory t cells with a selective peptide inhibitor of Kv1.3 channels for therapy of autoimmune diseases. *Mol. Pharmacol.* **2005**, *67*, 1369–1381. [CrossRef] [PubMed]
4. Chi, V.; Pennington, M.W.; Norton, R.S.; Tarcha, E.J.; Londono, L.M.; Sims-Fahey, B.; Upadhyay, S.K.; Lakey, J.T.; Iadonato, S.; Wulff, H.; et al. Development of a sea anemone toxin as an immunomodulator for therapy of autoimmune diseases. *Toxicon* **2012**, *59*, 529–546. [CrossRef] [PubMed]
5. Turk, T.; Kem, W.R. The phylum cnidaria and investigations of its toxins and venoms until 1990. *Toxicon* **2009**, *54*, 1031–1037. [CrossRef] [PubMed]
6. WoRMS. World Register of Marine Species. Available online: http://www.marinespecies.org/aphia.php?p= taxdetails&id=1267 (accessed on 24 October 2017).
7. Daly, M.; Brugler, M.R.; Cartwright, P.; Collins, A.G.; Dawson, M.N.; Fautin, D.G.; France, S.C.; McFadden, C.S.; Opresko, D.M.; Rodriguez, E. The phylum cnidaria: A review of phylogenetic patterns and diversity 300 years after Linnaeus. *Zootaxa* **2007**, *1668*, 127–182. [CrossRef]
8. Fautin, D.G. Structural diversity, systematics, and evolution of cnidae. *Toxicon* **2009**, *54*, 1054–1064. [CrossRef] [PubMed]
9. Frazão, B.; Vasconcelos, V.; Antunes, A. Sea anemone (Cnidaria, Anthozoa, Actiniaria) toxins: An overview. *Mar. Drugs* **2012**, *10*, 1812–1851. [CrossRef] [PubMed]
10. Beress, L. Biologically active compounds from coelenterates. *Pure Appl. Chem.* **1982**, *54*, 1981–1994. [CrossRef]
11. Aneiros, A.; Garateix, A. Bioactive peptides from marine sources: Pharmacological properties and isolation procedures. *J. Chromatogr. B Anal. Technol. Biomed. Life Sci.* **2004**, *803*, 41–53. [CrossRef] [PubMed]
12. Aneiros, A.; Garcia, I.; Martinez, J.R.; Harvey, A.L.; Anderson, A.J.; Marshall, D.L.; Engstrom, A.; Hellman, U.; Karlsson, E. A potassium channel toxin from the secretion of the sea anemone *Bunodosoma granulifera*. Isolation, amino acid sequence and biological activity. *Biochim. Biophys. Acta* **1993**, *7*, 86–92. [CrossRef]
13. Salinas, E.M.; Cebada, J.; Valdes, A.; Garateix, A.; Aneiros, A.; Alvarez, J.L. Effects of a toxin from the mucus of the Caribbean sea anemone (*Bunodosoma granulifera*) on the ionic currents of single ventricular mammalian cardiomyocytes. *Toxicon* **1997**, *35*, 1699–1709. [CrossRef]
14. Garateix, A.; Castellanos, M.; Hernandez, J.L.; Mas, R.; Menendez, R.; Romero, L.; Chavez, M. Effects of a high molecular weight toxin from the sea anemone *Condylactis gigantea* on cholinergic responses. *Comp. Biochem. Physiol. C* **1992**, *103*, 403–409. [CrossRef]
15. Garateix, A.; Flores, A.; Garcia-Andrade, J.M.; Palmero, A.; Aneiros, A.; Vega, R.; Soto, E. Antagonism of glutamate receptors by a chromatographic fraction from the exudate of the sea anemone *Phyllactis flosculifera*. *Toxicon* **1996**, *34*, 443–450. [CrossRef]
16. Garateix, A.; Salceda, E.; López, O.; Salazar, H.; Aneiros, A.; Zaharenko, A.J.; de Freitas, J.C. Pharmacological characterization of *Bunodosoma* toxins on mammalian voltage dependt sodium channels. *Pharmacologyonline* **2006**, *3*, 507–513.
17. Garateix, A.; Vega, R.; Salceda, E.; Cebada, J.; Aneiros, A.; Soto, E. Bgk anemone toxin inhibits outward K(+) currents in snail neurons. *Brain Res.* **2000**, *864*, 312–314. [CrossRef]
18. Castañeda, O.; Harvey, A.L. Discovery and characterization of cnidarian peptide toxins that affect neuronal potassium ion channels. *Toxicon* **2009**, *54*, 1119–1124. [CrossRef] [PubMed]
19. Castañeda, O.; Sotolongo, V.; Amor, A.M.; Stöcklin, R.; Anderson, A.J.; Harvey, A.L.; Engström, Å.; Wernstedt, C.; Karlsson, E. Characterization of a potassium channel toxin from the caribbean sea anemone *Stichodactyla helianthus*. *Toxicon* **1995**, *33*, 603–613. [CrossRef]

20. Cotton, J.; Crest, M.; Bouet, F.; Alessandri, N.; Gola, M.; Forest, E.; Karlsson, E.; Castaneda, O.; Harvey, A.L.; Vita, C.; et al. A potassium-channel toxin from the sea anemone *Bunodosoma granulifera*, an inhibitor for Kv1 channels. Revision of the amino acid sequence, disulfide-bridge assignment, chemical synthesis, and biological activity. *Eur. J. Biochem.* **1997**, *244*, 192–202. [CrossRef] [PubMed]

21. Standker, L.; Beress, L.; Garateix, A.; Christ, T.; Ravens, U.; Salceda, E.; Soto, E.; John, H.; Forssmann, W.G.; Aneiros, A. A new toxin from the sea anemone *Condylactis gigantea* with effect on sodium channel inactivation. *Toxicon* **2006**, *48*, 211–220. [CrossRef] [PubMed]

22. Rodríguez, A.A.; Cassoli, J.S.; Sa, F.; Dong, Z.Q.; de Freitas, J.C.; Pimenta, A.M.C.; de Lima, M.E.; Konno, K.; Lee, S.M.Y.; Garateix, A.; et al. Peptide fingerprinting of the neurotoxic fractions isolated from the secretions of sea anemones *Stichodactyla helianthus* and *Bunodosoma granulifera*. New members of the APETx-like family identified by a 454 pyrosequencing approach. *Peptides* **2012**, *34*, 26–38. [CrossRef] [PubMed]

23. Rodriguez, A.A.; Salceda, E.; Garateix, A.G.; Zaharenko, A.J.; Peigneur, S.; Lopez, O.; Pons, T.; Richardson, M.; Diaz, M.; Hernandez, Y.; et al. A novel sea anemone peptide that inhibits acid-sensing ion channels. *Peptides* **2014**, *53*, 3–12. [CrossRef] [PubMed]

24. Lanio, M.E.; Morera, V.; Alvarez, C.; Tejuca, M.; Gomez, T.; Pazos, F.; Besada, V.; Martinez, D.; Huerta, V.; Padron, G.; et al. Purification and characterization of two hemolysins from *Stichodactyla helianthus*. *Toxicon* **2001**, *39*, 187–194. [CrossRef]

25. Rodriguez, A.A.; Standker, L.; Zaharenko, A.J.; Garateix, A.G.; Forssmann, W.G.; Beress, L.; Valdes, O.; Hernandez, Y.; Laguna, A. Combining multidimensional liquid chromatography and MALDI-TOF-MS for the fingerprint analysis of secreted peptides from the unexplored sea anemone species *Phymanthus crucifer*. *J. Chromatogr. B Anal. Technol. Biomed. Life Sci.* **2012**, *903*, 30–39. [CrossRef] [PubMed]

26. Lanio, M.E.; Alvarez, C.; Ochoa, C.; Ros, U.; Pazos, F.; Martinez, D.; Tejuca, M.; Eugenio, L.M.; Casallanovo, F.; Dyszy, F.H.; et al. Sticholysins I and II interaction with cationic micelles promotes toxins' conformational changes and enhanced hemolytic activity. *Toxicon* **2007**, *50*, 731–739. [CrossRef] [PubMed]

27. Alvarez, C.; Casallanovo, F.; Shida, C.S.; Nogueira, L.V.; Martinez, D.; Tejuca, M.; Pazos, I.F.; Lanio, M.E.; Menestrina, G.; Lissi, E.; et al. Binding of sea anemone pore-forming toxins Sticholysins I and II to interfaces—Modulation of conformation and activity, and lipid-protein interaction. *Chem. Phys. Lipids* **2003**, *122*, 97–105. [CrossRef]

28. Álvarez, C.; Mancheño, J.M.; Martínez, D.; Tejuca, M.; Pazos, F.; Lanio, M.E. Sticholysins, two pore-forming toxins produced by the Caribbean sea anemone *Stichodactyla helianthus*: Their interaction with membranes. *Toxicon* **2009**, *54*, 1135–1147. [CrossRef] [PubMed]

29. Moran, Y.; Gordon, D.; Gurevitz, M. Sea anemone toxins affecting voltage-gated sodium channels—Molecular and evolutionary features. *Toxicon* **2009**, *54*, 1089–1101. [CrossRef] [PubMed]

30. Moran, Y.; Weinberger, H.; Sullivan, J.C.; Reitzel, A.M.; Finnerty, J.R.; Gurevitz, M. Concerted evolution of sea anemone neurotoxin genes is revealed through analysis of the *Nematostella vectensis* genome. *Mol. Biol. Evol.* **2008**, *25*, 737–747. [CrossRef] [PubMed]

31. Nesher, N.; Shapira, E.; Sher, D.; Moran, Y.; Tsveyer, L.; Turchetti-Maia, A.L.; Horowitz, M.; Hochner, B.; Zlotkin, E. Ade-1, a new inotropic Na(+) channel toxin from *Aiptasia diaphana*, is similar to, yet distinct from, known anemone Na(+) channel toxins. *Biochem. J.* **2013**, *451*, 81–90. [CrossRef] [PubMed]

32. Orts, D.J.; Moran, Y.; Cologna, C.T.; Peigneur, S.; Madio, B.; Praher, D.; Quinton, L.; De Pauw, E.; Bicudo, J.E.; Tytgat, J.; et al. Bcstx3 is a founder of a novel sea anemone toxin family of potassium channel blocker. *FEBS J.* **2013**, *280*, 4839–4852. [CrossRef] [PubMed]

33. Hasegawa, Y.; Honma, T.; Nagai, H.; Ishida, M.; Nagashima, Y.; Shiomi, K. Isolation and cDNA cloning of a potassium channel peptide toxin from the sea anemone *Anemonia erythraea*. *Toxicon* **2006**, *48*, 536–542. [CrossRef] [PubMed]

34. Honma, T.; Kawahata, S.; Ishida, M.; Nagai, H.; Nagashima, Y.; Shiomi, K. Novel peptide toxins from sea anemone *Stichodactyla haddoni*. *Peptides* **2008**, *29*, 536–544. [CrossRef] [PubMed]

35. Honma, T.; Shiomi, K. Peptide toxins in sea anemones: Structural and functional aspects. *Mar. Biotechnol.* **2006**, *8*, 1–10. [CrossRef] [PubMed]

36. Minagawa, S.; Ishida, M.; Nagashima, Y.; Shiomi, K. Primary structure of a potassium channel toxin from the sea anemone *Actinia equina*. *FEBS Lett.* **1998**, *427*, 149–151. [CrossRef]

37. Oliveira, J.S.; Fuentes-Silva, D.; King, G.F. Development of a rational nomenclature for naming peptide and protein toxins from sea anemones. *Toxicon* **2012**, *60*, 539–550. [CrossRef] [PubMed]

38. Urbarova, I.; Karlsen, B.O.; Okkenhaug, S.; Seternes, O.M.; Johansen, S.D.; Emblem, Å. Digital marine bioprospecting: Mining new neurotoxin drug candidates from the transcriptomes of cold-water sea anemones. *Mar. Drugs* **2012**, *10*, 2265–2279. [CrossRef] [PubMed]

39. Macrander, J.; Brugler, M.R.; Daly, M. A RNA-seq approach to identify putative toxins from acrorhagi in aggressive and non-aggressive *Anthopleura elegantissima* polyps. *BMC Genom.* **2015**, *16*, 221. [CrossRef] [PubMed]

40. Oliveira, J.S.; Fuentes-Silva, D.; Zaharenko, A.J. Sea anemone peptides. Biological activities, structure-function relationships and phylogenetic aspects. In *Animal Toxins: State of the Art. Perspective in Health and Biotechnology*, 1st ed.; de Lima, M.E., Pimenta, A.M., Martin-Eauclaire, M.F., Zingali, R.B., Rochat, H., Eds.; Editora UFMG: Belo Horizonte, Brazil, 2009; 800p, ISBN 978-85-7041-735-0.

41. Norton, R.S. Structure and structure-function relationships of sea anemone proteins that interact with the sodium channel. *Toxicon* **1991**, *29*, 1051–1084. [CrossRef]

42. Norton, R.S. Structures of sea anemone toxins. *Toxicon* **2009**, *54*, 1075–1088. [CrossRef] [PubMed]

43. Fautin, D.G.; Malarky, L.; Soberón, J. Latitudinal diversity of sea anemones (Cnidaria: Actiniaria). *Biol. Bull.* **2013**, *224*, 89–98. [CrossRef] [PubMed]

44. Frazão, B. A Genomic and Proteomic Study of Sea Anemones and Jellyfish from Portugal. Ph.D. Thesis, University of Porto, Porto, Portugal, 2017.

45. Fautin, D. Bunodactis verrucosa. *In: Hexacorallians (Actiniaria) of the World. Accessed through: World Register of Marine Species.* Available online: http://www.marinespecies.org/aphia.php?p=taxdetails&id=100819 (accessed on 5 May 2017).

46. Frazão, B.; Antunes, A. Jellyfish bioactive compounds: Methods for wet-lab work. *Mar. Drugs* **2016**, *14*, 75. [CrossRef] [PubMed]

47. Putnam, N.H.; Srivastava, M.; Hellsten, U.; Dirks, B.; Chapman, J.; Salamov, A.; Terry, A.; Shapiro, H.; Lindquist, E.; Kapitonov, V.V. Sea anemone genome reveals ancestral eumetazoan gene repertoire and genomic organization. *Science* **2007**, *317*, 86–94. [CrossRef] [PubMed]

48. Zhuang, Y.; Hou, H.; Zhao, X.; Zhang, Z.; Li, B. Effects of collagen and collagen hydrolysate from jellyfish (*Rhopilema esculentum*) on mice skin photoaging induced by uv irradiation. *J. Food Sci.* **2009**, *74*, H183–H188. [CrossRef] [PubMed]

49. Campos, A.; Puerto, M.; Prieto, A.; Cameán, A.; Almeida, A.M.; Coelho, A.V.; Vasconcelos, V. Protein extraction and two-dimensional gel electrophoresis of proteins in the marine mussel *Mytilus galloprovincialis*: An important tool for protein expression studies, food quality and safety assessment. *J. Sci. Food Agric.* **2013**, *93*, 1779–1787. [CrossRef] [PubMed]

50. Puerto, M.; Campos, A.; Prieto, A.; Cameán, A.; de Almeida, A.M.; Coelho, A.V.; Vasconcelos, V. Differential protein expression in two bivalve species, *Mytilus galloprovincialis* and *Corbicula fluminea*; exposed to *Cylindrospermopsis raciborskii* cells. *Aquat. Toxicol.* **2011**, *101*, 109–116. [CrossRef] [PubMed]

51. Frazão, B.; Campos, A.; Osório, H.; Thomas, B.; Leandro, S.; Teixeira, A.; Vasconcelos, V.; Antunes, A. Analysis of *Pelagia noctiluca* proteome reveals a red fluorescent protein, a zinc metalloproteinase and a peroxiredoxin. *Protein J.* **2017**, *36*, 77–97. [CrossRef] [PubMed]

52. Zhang, X. Less is more: Membrane protein digestion beyond urea–trypsin solution for next-level proteomics. *Mol. Cell. Proteom.* **2015**, *14*, 2441–2453. [CrossRef] [PubMed]

53. Wu, C.C.; MacCoss, M.J.; Howell, K.E.; Yates, J.R. A method for the comprehensive proteomic analysis of membrane proteins. *Nat. Biotechnol.* **2003**, *21*, 532–538. [CrossRef] [PubMed]

54. Ji, H.; Wang, J.; Guo, J.; Li, Y.; Lian, S.; Guo, W.; Yang, H.; Kong, F.; Zhen, L.; Guo, L. Progress in the biological function of alpha-enolase. *Anim. Nutr.* **2016**, *2*, 12–17. [CrossRef]

55. Iida, H.; Yahara, I. Yeast heat-shock protein of Mr 48,000 is an isoprotein of enolase. *Nature* **1985**, *315*, 688–690. [CrossRef]

56. Butterworth, P. Book review: Ribonucleases: Structure and function. G. D'Alessio and J. F. Riordan (Eds.). Academic press, new york and london. 670 + xix pages (1997). *Cell Biochem. Funct.* **1998**, *16*, 225. [CrossRef]

57. Davenport, R.C.; Bash, P.A.; Seaton, B.A.; Karplus, M.; Petsko, G.A.; Ringe, D. Structure of the triosephosphate isomerase-phosphoglycolohydroxamate complex: An analog of the intermediate on the reaction pathway. *Biochemistry* **1991**, *30*, 5821–5826. [CrossRef] [PubMed]

58. Cooper, S.J.; Leonard, G.A.; McSweeney, S.M.; Thompson, A.W.; Naismith, J.H.; Qamar, S.; Plater, A.; Berry, A.; Hunter, W.N. The crystal structure of a class II fructose-1, 6-bisphosphate aldolase shows a novel binuclear metal-binding active site embedded in a familiar fold. *Structure* **1996**, *4*, 1303–1315. [CrossRef]

59. Horecker, B.; Tsolas, O.; Lai, C. 6 aldolases. *Enzymes* **1972**, *7*, 213–258. [CrossRef]
60. Harrison, P.M.; Arosio, P. The ferritins: Molecular properties, iron storage function and cellular regulation. *Biochim. Biophys. Acta Bioenerg.* **1996**, *1275*, 161–203. [CrossRef]
61. Arosio, P.; Levi, S. Ferritin, iron homeostasis, and oxidative damage. *Free Radic. Biol. Med.* **2002**, *33*, 457–463. [CrossRef]
62. Kim, K.; Kim, I.; Lee, K.-Y.; Rhee, S.; Stadtman, E. The isolation and purification of a specific "protector" protein which inhibits enzyme inactivation by a thiol/Fe (III)/O2 mixed-function oxidation system. *J. Biol. Chem.* **1988**, *263*, 4704–4711. [PubMed]
63. Rhee, S.G.; Woo, H.A.; Kil, I.S.; Bae, S.H. Peroxiredoxin functions as a peroxidase and a regulator and sensor of local peroxides. *J. Biol. Chem.* **2012**, *287*, 4403–4410. [CrossRef] [PubMed]
64. Cui, H.; Wang, Y.; Wang, Y.; Qin, S. Genome-wide analysis of putative peroxiredoxin in unicellular and filamentous cyanobacteria. *BMC Evol. Biol.* **2012**, *12*, 220. [CrossRef] [PubMed]
65. Maksimenko, A.V.; Vavaev, A.V. Antioxidant enzymes as potential targets in cardioprotection and treatment of cardiovascular diseases. Enzyme antioxidants: The next stage of pharmacological counterwork to the oxidative stress. *Heart Int.* **2012**, *7*, 14–19. [CrossRef] [PubMed]
66. Ruan, Z.; Liu, G.; Wang, B.; Zhou, Y.; Lu, J.; Wang, Q.; Zhao, J.; Zhang, L. First report of a peroxiredoxin homologue in jellyfish: Molecular cloning, expression and functional characterization of CcPrx4 from *Cyanea capillata*. *Mar. Drugs* **2014**, *12*, 214–231. [CrossRef] [PubMed]
67. Fukai, T.; Ushio-Fukai, M. Superoxide dismutases: Role in redox signaling, vascular function, and diseases. *Antioxid. Redox Signal.* **2011**, *15*, 1583–1606. [CrossRef] [PubMed]
68. Campos, A.; Apraiz, I.; da Fonseca, R.R.; Cristobal, S. Shotgun analysis of the marine mussel *Mytilus edulis* hemolymph proteome and mapping the innate immunity elements. *Proteomics* **2015**, *15*, 4021–4029. [CrossRef] [PubMed]
69. Campos, A.; Danielsson, G.; Farinha, A.P.; Kuruvilla, J.; Warholm, P.; Cristobal, S. Shotgun proteomics to unravel marine mussel (*Mytilus edulis*) response to long-term exposure to low salinity and propranolol in a baltic sea microcosm. *J. Proteom.* **2016**, *137*, 97–106. [CrossRef] [PubMed]
70. Culma, M.F.; Pereanez, J.A.; Rangel, V.N.; Lomonte, B. Snake venomics of *Bothrops punctatus*, a semiarboreal pitviper species from Antioquia, Colombia. *PeerJ* **2014**, *2*, e246. [CrossRef] [PubMed]
71. Zhang, Y.; Fonslow, B.R.; Shan, B.; Baek, M.-C.; Yates III, J.R. Protein analysis by shotgun/bottom-up proteomics. *Chem. Rev.* **2013**, *113*, 2343–2394. [CrossRef] [PubMed]
72. Keller, B.O.; Sui, J.; Young, A.B.; Whittal, R.M. Interferences and contaminants encountered in modern mass spectrometry. *Anal. Chim. Acta* **2008**, *627*, 71–81. [CrossRef] [PubMed]
73. Conesa, A.; Götz, S.; García-Gomez, J.; Terol, J.; Talon, M.; Robles, M. Blast2go: A universal tool for annotation, visualization and analysis in functional genomics research. *Bioinformatics* **2005**, *21*, 3674–3676. [CrossRef] [PubMed]
74. Daly, M.; Gusmão, L.C.; Reft, A.J.; Rodríguez, E. Phylogenetic signal in mitochondrial and nuclear markers in sea anemones (Cnidaria, Actiniaria). *Integr. Comp. Biol.* **2010**, *50*, 371–388. [CrossRef] [PubMed]
75. Moffatt, B.A.; Ashihara, H. Purine and pyrimidine nucleotide synthesis and metabolism. *Arabidopsis Book* **2002**, *1*, e0018. [CrossRef] [PubMed]
76. Lonsdale, D. A review of the biochemistry, metabolism and clinical benefits of thiamin (e) and its derivatives. *Evid. Based Complement. Altern. Med.* **2006**, *3*, 49–59. [CrossRef] [PubMed]
77. Borbón, H.; Váldes, S.; Alvarado-Mesén, J.; Soto, R.; Vega, I. Antimicrobial properties of sea anemone anthopleura nigrescens from pacific coast of costa rica. *Asian Pac. J. Trop. Biomed.* **2016**, *6*, 418–421. [CrossRef]
78. Rocha, J.; Peixe, L.; Gomes, N.; Calado, R. Cnidarians as a source of new marine bioactive compounds—An overview of the last decade and future steps for bioprospecting. *Mar. Drugs* **2011**, *9*, 1860–1886. [CrossRef] [PubMed]
79. Mariottini, G.L.; Grice, I.D. Antimicrobials from cnidarians. A new perspective for anti-infective therapy? *Mar. Drugs* **2016**, *14*, 48. [CrossRef] [PubMed]
80. Davy, S.K.; Allemand, D.; Weis, V.M. Cell biology of cnidarian-dinoflagellate symbiosis. *Microb. Mol. Biol. Rev.* **2012**, *76*, 229–261. [CrossRef] [PubMed]
81. Bellis, E.S.; Howe, D.K.; Denver, D.R. Genome-wide polymorphism and signatures of selection in the symbiotic sea anemone *Aiptasia*. *BMC Genom.* **2016**, *17*, 160. [CrossRef] [PubMed]

82. Tsai, I.H.; Wang, Y.M.; Chiang, T.Y.; Chen, Y.L.; Huang, R.J. Purification, cloning and sequence analyses for pro-metalloprotease-disintegrin variants from *Deinagkistrodon acutus* venom and subclassification of the small venom metalloproteases. *Eur. J. Biochem.* **2000**, *267*, 1359–1367. [CrossRef] [PubMed]

83. Undheim, E.A.; Sunagar, K.; Herzig, V.; Kely, L.; Low, D.H.; Jackson, T.N.; Jones, A.; Kurniawan, N.; King, G.F.; Ali, S.A. A proteomics and transcriptomics investigation of the venom from the barychelid spider *Trittame loki* (brush-foot trapdoor). *Toxins* **2013**, *5*, 2488–2503. [CrossRef] [PubMed]

84. Sarras, M.P.; Li, Y.; Leontovich, A.; Zhang, J.S. Structure, expression, and developmental function of early divergent forms of metalloproteinases in hydra. *Cell Res.* **2002**, *12*, 163–176. [CrossRef] [PubMed]

85. Leontovich, A.A.; Zhang, J.; Shimokawa, K.-I.; Nagase, H.; Sarras, M. A novel hydra matrix metalloproteinase (hmmp) functions in extracellular matrix degradation, morphogenesis and the maintenance of differentiated cells in the foot process. *Development* **2000**, *127*, 907–920. [PubMed]

86. Yan, L.; Fei, K.; Zhang, J.; Dexter, S.; Sarras, M. Identification and characterization of hydra metalloproteinase 2 (HMP2): A meprin-like astacin metalloproteinase that functions in foot morphogenesis. *Development* **2000**, *127*, 129–141. [PubMed]

87. Pan, T.-L.; Gröger, H.; Schmid, V.; Spring, J. A toxin homology domain in an astacin-like metalloproteinase of the jellyfish podocoryne carnea with a dual role in digestion and development. *Dev. Genes Evol.* **1998**, *208*, 259–266. [CrossRef] [PubMed]

88. Lee, H.; Jung, E.-S.; Kang, C.; Yoon, W.D.; Kim, J.-S.; Kim, E. Scyphozoan jellyfish venom metalloproteinases and their role in the cytotoxicity. *Toxicon* **2011**, *58*, 277–284. [CrossRef] [PubMed]

89. Moran, Y.; Praher, D.; Schlesinger, A.; Ayalon, A.; Tal, Y.; Technau, U. Analysis of soluble protein contents from the nematocysts of a model sea anemone sheds light on venom evolution. *Mar. Biotechnol.* **2013**, *15*, 329–339. [CrossRef] [PubMed]

90. Tsai, I.-H.; Wang, Y.-M.; Cheng, A.C.; Starkov, V.; Osipov, A.; Nikitin, I.; Makarova, Y.; Ziganshin, R.; Utkin, Y. cDNA cloning, structural, and functional analyses of venom phospholipases a 2 and a kunitz-type protease inhibitor from steppe viper *Vipera usinii renardi*. *Toxicon* **2011**, *57*, 332–341. [CrossRef] [PubMed]

91. Fry, B.G.; Roelants, K.; Champagne, D.E.; Scheib, H.; Tyndall, J.D.; King, G.F.; Nevalainen, T.J.; Norman, J.A.; Lewis, R.J.; Norton, R.S. The toxicogenomic multiverse: Convergent recruitment of proteins into animal venoms. *Annu. Rev. Genom. Hum. Genet.* **2009**, *10*, 483–511. [CrossRef] [PubMed]

92. Valentin, E.; Ghomashchi, F.; Gelb, M.H.; Lazdunski, M.; Lambeau, G. Novel human secreted phospholipase A2 with homology to the group III bee venom enzyme. *J. Biol. Chem.* **2000**, *275*, 7492–7496. [CrossRef] [PubMed]

93. Fry, B. *Venomous Reptiles and Their Toxins: Evolution, Pathophysiology and Biodiscovery*; Oxford University Press: New York, NY, USA, 2015; p. 576, ISBN-13 978-0199309399.

94. Nevalainen, T.J. Phospholipases a 2 in the genome of the sea anemone *Nematostella vectensis*. *Comp. Biochem. Physiol. Part D Genom. Proteom.* **2008**, *3*, 226–233. [CrossRef] [PubMed]

95. Fry, B.G.; Wüster, W. Assembling an arsenal: Origin and evolution of the snake venom proteome inferred from phylogenetic analysis of toxin sequences. *Mol. Biol. Evol.* **2004**, *21*, 870–883. [CrossRef] [PubMed]

96. Komori, Y.; Nikai, T.; Sugihara, H. Purification and characterization of a lethal toxin from the venom of *Heloderma horridum horridum*. *Biochem. Biophys. Res. Commun.* **1988**, *154*, 613–619. [CrossRef]

97. Fry, B.G.; Roelants, K.; Winter, K.; Hodgson, W.C.; Griesman, L.; Kwok, H.F.; Scanlon, D.; Karas, J.; Shaw, C.; Wong, L. Novel venom proteins produced by differential domain-expression strategies in beaded lizards and Gila monsters (genus *Heloderma*). *Mol. Biol. Evol.* **2009**, *27*, 395–407. [CrossRef] [PubMed]

98. Jouiaei, M.; Yanagihara, A.A.; Madio, B.; Nevalainen, T.J.; Alewood, P.F.; Fry, B.G. Ancient venom systems: A review on cnidaria toxins. *Toxins* **2015**, *7*, 2251–2271. [CrossRef] [PubMed]

99. Skewes, M. *Aulactinia verrucosa Gem Anemone*; Tyler-Walters, H., Hiscock, K., Eds.; Marine Life Information Network: Biology and Sensitivity Key Information Reviews, [on-line]; Marine Biological Association of the United Kingdom: Plymouth, UK, 2007; Available online: http://www.marlin.ac.uk/species/detail/1601 (accessed on 12 May 2017).

100. White, B. *Anthopleura xanthogrammica*; Walla Walla University: College Place, WA, USA, 2004; Available online: http://www.wallawalla.edu/academics/departments/biology/rosario/inverts/Cnidaria/Class-Anthozoa/Subclass_Zoantharia/Order_Actiniaria/Anthopleura_xanthogrammica.html (accessed on 12 May 2017).

101. Sebens, K.P. The allometry of feeding, energetics, and body size in three sea anemone species. *Biol. Bull.* **1981**, *161*, 152–171. [CrossRef]

102. Hand, C.; Uhlinger, K.R. The culture, sexual and asexual reproduction, and growth of the sea anemone *Nematostella vectensis*. *Biol. Bull.* **1992**, *182*, 169–176. [CrossRef] [PubMed]

103. Cope, W.G.; Bringolf, R.B.; Buchwalter, D.B.; Newton, T.J.; Ingersoll, C.G.; Wang, N.; Augspurger, T.; Dwyer, F.J.; Barnhart, M.C.; Neves, R.J.; et al. Differential exposure, duration, and sensitivity of unionoidean bivalve life stages to environmental contaminants. *J. N. Am. Benthol. Soc.* **2008**, *27*, 451–462. [CrossRef]

104. Gilroy, È.A.; Gillis, P.L.; King, L.E.; Bendo, N.A.; Salerno, J.; Giacomin, M.; de Solla, S.R. The effects of pharmaceuticals on a unionid mussel (*Lampsilis siliquoidea*): An examination of acute and chronic endpoints of toxicity across life stages. *Environ. Toxicol. Chem.* **2017**, *36*, 1572–1583. [CrossRef] [PubMed]

105. Bleam, D.E.; Couch, K.J.; Distler, D.A. Key to the unionid mussels of kansas. *Trans. Kans. Acad. Sci. (1903)* **1999**, *102*, 83–91. [CrossRef]

106. Ghiretti, F. Cephalotoxin: The crab-paralysing agent of the posterior salivary glands of cephalopods. *Nature* **1959**, *183*, 1192–1193. [CrossRef]

107. McDonald, N.; Cottrell, G. Purification and mode of action of toxin from *Eledone cirrosa*. *Comp. Gen. Pharmacol.* **1972**, *3*, 243–248. [CrossRef]

108. Songdahl, J.; Shapiro, B. Purification and composition of a toxin from the posterior salivary gland of *Octopus dofleini*. *Toxicon* **1974**, *12*, 109–112. [CrossRef]

109. Cariello, L.; Zanetti, L. α- and β-cephalotoxin: Two paralysing proteins from posterior salivary glands of *Octopus vulgaris*. *Comp. Biochem. Physiol. C Comp. Pharmacol.* **1977**, *57*, 169–173. [CrossRef]

110. Bradford, M.M. A rapid and sensitive method for the quantitation of microgram quantities of protein utilizing the principle of protein-dye binding. *Anal. Biochem.* **1976**, *72*, 248–254. [CrossRef]

111. Neuhoff, V.; Arold, N.; Taube, D.; Ehrhardt, W. Improved staining of proteins in polyacrylamide gels including isoelectric focusing gels with clear background at nanogram sensitivity using Coomassie Brilliant Blue G-250 and R-250. *Electrophoresis* **1988**, *9*, 255–262. [CrossRef] [PubMed]

112. Osório, H.; Reis, C.A. Mass Spectrometry Methods for Studying Glycosylation in Cancer. In *Mass Spectrometry Data Analysis in Proteomics, Methods in Molecular Biology (Methods and Protocols)*; Matthiesen, R., Ed.; Humana Press: Totowa, NJ, USA, 2013; Volume 1007, pp. 301–316. ISBN 978-1-62703-391-6.

113. Apweiler, R.; Bairoch, A.; Wu, C.H.; Barker, W.C.; Boeckmann, B.; Ferro, S.; Gasteiger, E.; Huang, H.; Lopez, R.; Magrane, M. Uniprot: The universal protein knowledgebase. *Nucleic Acids Res.* **2004**, *32*, D115–D119. [CrossRef] [PubMed]

114. The UniProt Consortium. Uniprot: The universal protein knowledgebase. *Nucleic Acids Res.* **2017**, *45*, D158–D169. [CrossRef]

115. Wisniewski, J.R.; Zougman, A.; Nagaraj, N.; Mann, M. Universal sample preparation method for proteome analysis. *Nat. Methods* **2009**, *6*, 359. [CrossRef] [PubMed]

116. Cox, J.; Mann, M. Maxquant enables high peptide identification rates, individualized ppb-range mass accuracies and proteome-wide protein quantification. *Nat. Biotechnol.* **2008**, *26*, 1367–1372. [CrossRef] [PubMed]

117. Cox, J.; Matic, I.; Hilger, M.; Nagaraj, N.; Selbach, M.; Olsen, J.V.; Mann, M. A practical guide to the maxquant computational platform for silac-based quantitative proteomics. *Nat. Protoc.* **2009**, *4*, 698–705. [CrossRef] [PubMed]

118. Jungo, F.; Estreicher, A.; Bairoch, A.; Bougueleret, L.; Xenarios, I. Animal toxins: How is complexity represented in databases? *Toxins* **2010**, *2*, 262–282. [CrossRef] [PubMed]

119. Jungo, F.; Bairoch, A. Tox-Prot, the toxin protein annotation program of the Swiss-Prot protein knowledgebase. *Toxicon* **2005**, *45*, 293–301. [CrossRef] [PubMed]

120. Jungo, F.; Bougueleret, L.; Xenarios, I.; Poux, S. The Uniprotkb/Swiss-Prot Tox-Prot program: A central hub of integrated venom protein data. *Toxicon* **2012**, *60*, 551–557. [CrossRef] [PubMed]

Article

Barrel Jellyfish (*Rhizostoma pulmo*) as Source of Antioxidant Peptides

Stefania De Domenico [1], Gianluca De Rinaldis [1,2], Mélanie Paulmery [3], Stefano Piraino [4,5] and Antonella Leone [1,5,*]

[1] Istituto di Scienze delle Produzioni Alimentari, Consiglio Nazionale delle Ricerche (CNR-ISPA) Unit of Lecce, Via Monteroni, 73100 Lecce, Italy; stefania.dedomenico@ispa.cnr.it (S.D.D.); gianluca.derinaldis@ispa.cnr.it (G.D.R.)
[2] Dipartimento di Biotecnologia, Chimica e Farmacia (DBCF), Università Degli Studi Di Siena, Via A. Moro, 2, 53100 Siena, Italy
[3] Département des Sciences et Technologies, Université de Lille, Cité Scientifique, F-59655 Villeneuve d'Ascq, France; melanie.paulmery@gmail.com
[4] Dipartimento di Scienze e Tecnologie Biologiche ed Ambientali (DiSTeBA), University of Salento, 73100 Lecce, Italy; stefano.piraino@unisalento.it
[5] Consorzio Nazionale Interuniversitario per le Scienze del Mare (CoNISMa), Local Unit of Lecce, Via Monteroni, 73100 Lecce, Italy
* Correspondence: antonella.leone@ispa.cnr.it; Tel.: +39-0832-422615

Received: 31 January 2019; Accepted: 19 February 2019; Published: 23 February 2019

Abstract: The jellyfish *Rhizostoma pulmo*, Macrì 1778 (Cnidaria, Rhizostomae) undergoes recurrent outbreaks in the Mediterranean coastal waters, with large biomass populations representing a nuisance or damage for marine and maritime activities. A preliminary overview of the antioxidant activity (AA) of *R. pulmo* proteinaceous compounds is provided here based on the extraction and characterization of both soluble and insoluble membrane-fractioned proteins, the latter digested by sequential enzymatic hydrolyses with pepsin and collagenases. All jellyfish proteins showed significant AA, with low molecular weight (MW) proteins correlated with greater antioxidant activity. In particular, collagenase-hydrolysed collagen resulted in peptides with MW lower than 3 kDa, ranging 3–10 kDa or 10–30 kDa, with AA inversely proportional to MW. No cytotoxic effect was detected on cultured human keratinocytes (HEKa) in a range of protein concentration 0.05–20 µg/mL for all tested protein fractions except for soluble proteins higher than 30 kDa, likely containing the jellyfish venom compounds. Furthermore, hydrolyzed jellyfish collagen peptides showed a significantly higher AA and provided a greater protective effect against oxidative stress in HEKa than the hydrolyzed collagen peptides from vertebrates. Due to a high reproductive potential, jellyfish may represent a potential socioeconomic opportunity as a source of natural bioactive compounds, with far-reaching beneficial implications. Eventually, improvements in processing technology will promote the use of untapped marine biomasses in nutraceutical, cosmeceutical, and pharmaceutical fields, turning marine management problems into a more positive perspective.

Keywords: invertebrate proteins; biological activity; antioxidants; collagen; pepsin hydrolysis; collagenase hydrolysis; oxidative stress; keratinocytes; cytotoxicity

---

1. Introduction

In European populations, jellyfish evoke unpleasant or disgusting feelings, meanwhile in Asia, they are recognized as an important source of bioactive compounds used in traditional food and medicine [1].

Variations in water mass, high salinity, and warm temperature associated with the current global climatic change, in combination with multiple anthropogenic impacts such as overfishing and coastal sprawl, led to increases in jellyfish peak abundances (blooms) and frequencies in the world's oceans [2,3]. Jellyfish blooms usually negatively impact human health and activities in coastal waters [4–6]. Instead, marine gelatinous organisms should be regarded through a more positive perspective as new important bio-resource [7,8].

Known for its nutritional and medical value in the Chinese pharmacopeia, increasing attention has been pointed to medusozoan jellyfish as an untapped source of essential nutrients [9–11], novel bioactive metabolites, and lead compounds, so to have been recently appointed as novel food in western Countries [12–15].

In the last decade, different kinds of extracts obtained from several specimens of jellyfish were analyzed and many pharmacological activities were found: for instance, several studies have been focused on box jellyfish venoms, which contain a great variety of bioactive proteins and are shown to have hemolytic, cytotoxic, cardiovascular [16–20], neurotoxic [17,21–23], and anti-tumoral [24] activities, both *in vivo* and *in vitro*. Furthermore, similar analysis has been performed using extracts carried out on the whole jellyfish biomass. Jellyfish tissue components showed others biological activities linked to proteins components on both cell cultures and *in vivo*, such as anti-fatigue activity [25] cytotoxicity on cancer cells [10,26], apoptosis and anti-cancer effects [27], and antioxidant properties [11,28–31], as well as anti-microbial activity [32].

Studies on the biochemical composition of wild jellyfish biomass, however, have been available only in recent decades. Despite the large biomasses, the dry weight of most rhizostomeae jellyfish (Cnidaria, Scyphozoa) ranges from 2–5%, mainly composed of proteins, while carbohydrates and lipids represent minor components [11,33,34].

New biological functions are increasingly attributed to protein hydrolysates and derived peptides, obtained from vegetable and animal sources [35,36], including antihypertensive, antitumoral, antiproliferative, hypocholesterolemic, anti-inflammatory and antioxidant activities [37,38]. Bioactive peptides are released during food processing or as result of enzymatic or chemical hydrolyses. Their functions are largely influenced by the nature of proteins, hydrolytic enzymes, enzyme-substrate ratio, temperature and time of reaction. All these conditions affect the molecular weight and the amino acid composition of peptides and, as a consequence, their activities [39].

Scientific evidence has clearly indicated a link between oxidative stress and various chronic diseases and aging, involving both intrinsic and extrinsic sources of Reactive Oxygen Species (ROS) as key mechanisms of these processes [40–42].

Recently, attention has been paid to antioxidant activity associated to a single or mixtures of molecules deriving from natural sources, according to their recognized role in the prevention of oxidative stresses mechanisms associated with numerous degenerative diseases, such as diabetes, cardiovascular and neurodegenerative disorders, and cancer [43–46]. Natural antioxidants may exhibit a reduced potential health hazard compared with synthetic compounds and they are already used in food industry as dietary supplements and in pharmaceuticals or cosmeceutical products, as replacement for synthetic antioxidants. Indeed, protein hydrolysates from plant and animal sources have been found to possess strong antioxidant activities [47–50].

Enriched by an enormous but still poorly explored biodiversity, the oceans represent an immense reservoir of bioactive peptides [51], extracted from a variety of diverse marine organisms, from invertebrates such as sponges, tunicates, bivalves, cephalopods [51–55] to vertebrates, such as the hairtail fish *Trichiurus lepturus* [56]. Protein hydrolysates of seafood and their by-products are known to have different functional properties and great potential for nutraceutical and pharmaceutical applications [57] including oxidative stress protection. Moreover, a considerable number of these marine peptides have been identified and characterized: they are generally short with low molecular weight [58]; they seem resistant to gastrointestinal hydrolysis, enhancing their absorption in intact form [56,58–60].

Among marine invertebrates, jellyfish could represent an abundant source of new bioactive peptides, due to the high protein content, especially in collagen that accounts for up to 40–60% of dry weight [34,61]. Collagen is a group of fibrous proteins and it is the main component of extracellular matrix with a structure highly conserved and characterized by triple helical structure with repeating sequence of Gly-X-Y, where generally X is proline and Y is hydroxyproline [62,63]. Jellyfish collagen shares several features with its vertebrate counterpart that makes it highly biocompatible [64]. A few recent research studies are focused on jellyfish collagen and hydrolyzed collagen, which was shown to have biological activities as angiotensin-converting enzyme inhibitory action [65], immune-stimulation effects [66], anti-fatigue [25] and antioxidant proprieties [25,65]. Furthermore, collagen molecules extracted from the Mediterranean Sea barrel jellyfish, *Rhizostoma pulmo* [67], seem to have an effect on human cell comparable to the mammalian type I collagen [61].

In this study, novel information is provided about proteins extraction and hydrolyzed peptides isolation from *Rhizostoma pulmo*, one of the most abundant jellyfish species along the Mediterranean coasts. This jellyfish is characterized by a rather harmless envenomation potential for humans, and by typically possessing a greater body size and body texture than other scyphozoan jellyfish (e.g., compared with the highly watery moon jellyfish, *Aurelia*). Several molecular weight proteins fractions, including hydrolyzed collagen peptides, were here analyzed for their antioxidant activity *in vitro,* including on human keratinocytes cultures under oxidative stress conditions. Our results strongly suggest that the Mediterranean Sea barrel jellyfish, due to its metagenetic life cycle and high proliferative potential, may well represent a sustainable source for natural antioxidant bioprospecting and, more generally, for the isolation of bioactive compounds.

2. Results and Discussion

2.1. Proteins Content and SDS-PAGE Separation

The lyophilized *Rhizostoma pulmo* whole jellyfish samples (umbrella and oral arms) were subjected to aqueous protein extraction by phosphate-buffered saline (PBS) to separate the hydro-soluble fractions from the insoluble ones, which were eventually exposed to a two-step sequential enzymatic digestion.

Soluble proteins and hydrolysed peptides were molecular weight (MW)-fractionated by membrane filtration and each fraction was analysed for antioxidant activity and for their effect on cultures of human keratinocyte adult (HEKa) cells. Membrane ultrafiltration was here used as the first step in the *R. pulmo* peptide purification, as reported for purification of bioactive peptides from the edible jellyfish *Rhopilema esculentum* [68] and other invertebrates [58].

The average concentration of *R. pulmo* proteins was 34.1±2 mg/g of dry weight (DW) (Table 1), with a small majority (56%) composed by of PBS-soluble peptides and the remaining fraction (44%) by insoluble proteins. In order to roughly characterize the protein fractions, hydrosoluble proteins (SP) were separated in four sub-fractions at different molecular weight (MW) ranges: higher than 30 kDa (SP > 30), between 30 and 10 kDa (SP 10–30), between 10 and 3 kDa (SP 3–10), and lower than 3 kDa (SP < 3). Near 93% of soluble proteins had MW higher than 30 kDa, whereas only 0.5% of the total SP showed MW between 3 and 10 kDa; approximately 4% of the total SP had MW between 10 or 30 kDa and less than 3% with MW < 3 kDa. The latter sub-fraction was no further considered because of the high salt content. In our samples, the largest sub-fraction—i.e., containing peptides with high MW—most likely includes proteinaceous components of the jellyfish mucus, which was found to have MW higher than 40 kDa in various jellyfish species and other coelenterates [69–71]. In agreement with this hypothesis, the mucus fraction of the jellyfish *Aurelia coerulea* contained proteins falling within three MW ranges, i.e., 100–250 kDa, 50–100 kDa and 37–50 kDa, while the tissue proteins were dispersed in a wider range [71]. Interestingly, hydroalcoholic extracts of the zooxanthellate jellyfish *Cotylorhiza tuberculata* showed only low MW proteins ranging 10–14 kDa [10].

Table 1. Protein contents in soluble and insoluble fractions from lyophilized *Rhizostoma pulmo* whole jellyfish samples. Soluble proteins were separated in four sub-fractions at different molecular weight (MW) ranges by membrane filtration. Data expressed as mean ± standard deviation (SD) of six experiments.

Fractions	Protein Concentration	
	mg/g DW ± SD	% Total Proteins
Soluble Proteins (SP)	19.2 ± 1.9	56.3
SP > 30 (MW > 30 kDa)		(52.1)
SP 10–30 (10 kDa < MW < 30 kDa)		(2.3)
SP 3–10 (3 kDa < MW < 10 kDa)		(0.3)
SP < 3 (MW < 3 kDa)		(1.6)
Insoluble Proteins (IP)	14.9 ± 0.9	43.7
Total	34.1 ± 2	100

The total jellyfish proteins (JFP), total soluble proteins (SP), SP sub-fractions with MW > 30 kDa (SP > 30), between 10 and 30 kDa (SP 10–30), between 3 and 10 kDa (SP 3–10) and <3 kDa (SP < 3) were analyzed by SDS-PAGE (Figure 1). A commercial purified vertebrate collagen from calf skin (vertebrate collagen, VC) was also analyzed as standard collagen (line 7). To identify the protein fractions with high sensitivity, *R. pulmo* soluble protein fractions were separated by SDS-PAGE and visualized by stain-free system (Figure 1A) and by the standard Coomassie Brillian Blue (CBB) staining (Figure 1B). The 2,2,2-Trichloroethanol (TCE) gel and Stain-Free system [72,73] allowed a higher sensitivity as compared to standard CBB staining, except for proteins with low tryptophan content, such as collagen, since the staining is based on the reaction among the tryptophan residues and trihalo compounds in the gel (see Materials and Method). The calf skin collagen (type I collagen) was considered as control also for its similarity with the collagen isolated from the jellyfish *R. esculentum* [74]. Bovine collagen (VC) is clearly detected in Figure 1B by CBB staining method due to the specific mechanism of staining. The lyophilized whole jellyfish sample (JFP, lane 1 Figure 1A) shows a large number of polypeptides in a wide MW range, and two main bands: at apparent MW about 39 kDa and a large, unstained band of about 150–250 kDa. The latter could be related to largely insoluble collagen proteins. Both these proteins seem to be insoluble in aqueous solution as they are not present in the total SP (lane 3) and in SP > 30 (lane 2) fractions. Therefore, the total SP sub-fraction (lane 3) is composed by proteins in a wide range of apparent MW with five main bands at about 60 kDa, 40 kDa, 35 kDa, 30 kDa and 12 kDa. Except for the latter band (12 kDa), all bands are also present in the SP > 30 fractions. Similar protein diversity at wide range of MW was found in aqueous extracts from deep-sea jellyfish [26]. Although faintly visible, the electrophoretic separation of proteins with apparent MW ranging 10–30 kDa (SP 10–30), 3–10 kDa (SP 3–10) and <3 kDa (SP < 3) is shown by lanes 4, 5 and 6, respectively. In Figure 1B the total insoluble jellyfish proteins are also separated by SDS-PAGE showing proteins in a wide apparent MW range, mainly in MW > 50 kDa.

After aqueous extraction, the insoluble proteins (IP) were submitted to sequential digestions with pepsin followed by collagenase, as in Leone et al. [11]. Hydrolysed peptides were sub-fractionated by membrane filtration to obtain peptides at different MW ranging 10–30 kDa, 3–10 kDa and <3 kDa.

In order to obtain reproducible and relatively easy to purify compounds, in each step a single commercial protease was used for the enzymatic hydrolysis. Single-protease hydrolysis is simpler when compared with a combination of several proteases and likely allows a better control of the physico-chemical conditions of the digestion, providing a relatively controlled composition of the resulting peptides [75]. These were then separated by membrane filtration in sub-fractions at different molecular weight (10–30, 3–10 and <3 kDa). All sub-fractions, soluble proteins and not-hydrolysed components were assayed *in vitro* for their antioxidant activity (AA) as a term of reference of the possible content of active substances.

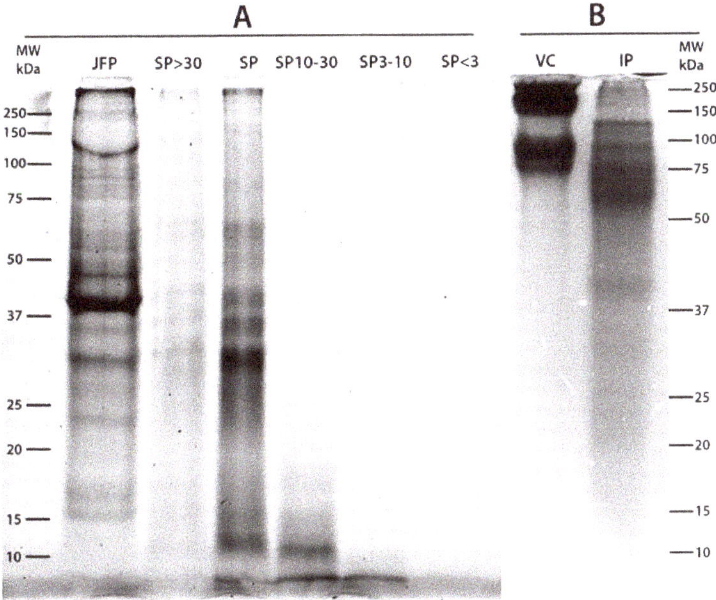

Figure 1. SDS-PAGE analysis of *Rhizostoma pulmo* jellyfish soluble proteins (20 µg) imaged with ChemiDoc MP Imaging System (**A**) and stained with Comassie Brillant Blu (**B**). (**A**) JFP, Total Jellyfish Proteins; SP > 30, Soluble Protein fraction with MW > 30 kDa; SP, Total SP; SP 10–30, Soluble Protein fraction 10 < MW < 30 kDa; SP 3–10, Soluble Protein fraction with 3 < MW < 10 kDa; SP < 3, Soluble Protein fraction with MW < 3 kDa; (**B**) VC, Calf skin Collagen (VC); IP, Insoluble Proteins; MW = Molecular-weight size marker.

2.2. In Vitro Antioxidant Activity of Soluble and Hydrolysed Protein Fractions

Increasing scientific evidence demonstrates that peptides with antioxidant properties can be obtained from marine vertebrate and invertebrate proteins, hydrolysed proteins, seafood by-products [11,56,58,60,76] as well as from terrestrial animal by-products [77]. In the present study, a preliminary overview of the antioxidant capacity of the Mediterranean Sea barrel jellyfish *Rhizostoma pulmo* proteins is provided. The soluble jellyfish proteins, fractionated according to their MW, as well as the insoluble proteins, both native and enzymatically hydrolysed and subsequently fractionated by MW, were analysed for their antioxidant activity (AA). The radical scavenging activity was evaluated by the ABTS assay, widely used as a screening assay for natural antioxidant compounds [78,79].

The AA of the considered protein fractions (Figure 2), expressed as nmol of Trolox equivalents (TE) per milligram of proteins, is remarkably higher in protein fractions with low MW than in fractions at high MW. The same AA pattern was observed both for PBS-soluble proteins and for the enzymatically hydrolysed peptides derived from insoluble proteins.

The AA evaluated in PBS soluble total protein (SP) was 92.9 nmol TE/mg of proteins, not significantly different from the AA measured in the fraction containing only high MW proteins (SP > 30) (145.0 nmol TE/mg of proteins). The peptides present in the fractions SP 10–30 and SP 3–10 showed AA values of 792.3 and 2459.0 nmol TE/mg of proteins, respectively. Both were significantly higher ($p < 0.05$ and $p < 0.01$, respectively) than AA values from SP and SP > 30 sub-fractions, and inversely proportional to the MW of peptides, demonstrating an enrichment in antioxidant compounds in low MW soluble protein sub-fractions.

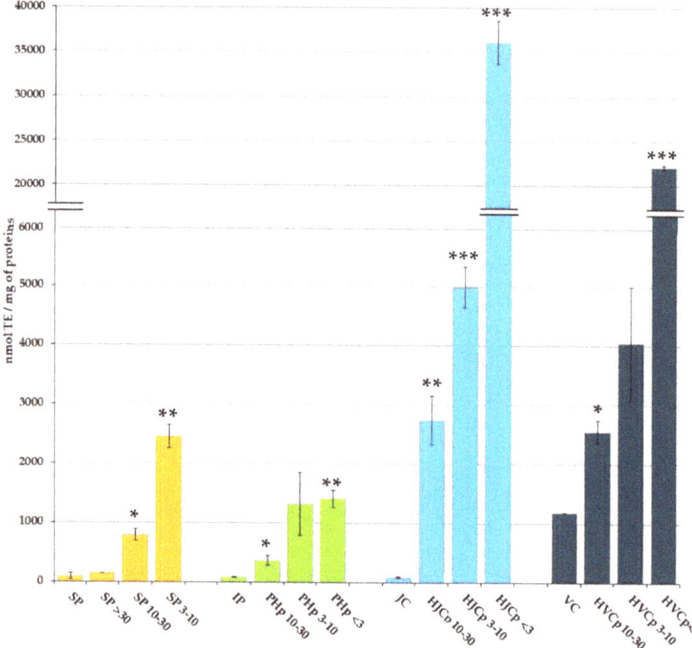

Figure 2. Antioxidant activity of *Rhizostoma pulmo* proteins. SP, Aqueous soluble proteins; IP, Insoluble proteins; JC, jellyfish collagen; VC, vertebrate (bovine) collagen. SP > 30, Soluble proteins with MW > 30 kDa; SP 10–30, Soluble proteins with MW 10–30 kDa; SP 3–10, Soluble proteins with MW 3–10 kDa; PHp 10–30, pepsin hydrolysed proteins with MW 10–30 kDa; PHp 3–10, pepsin hydrolysed proteins with MW 3–10 kDa; PHp < 3, pepsin hydrolysed proteins with MW < 3 kDa; HJCp 10–30, hydrolysed jellyfish collagen with MW 10–30 kDa; HJCp 3–10, hydrolysed jellyfish collagen with MW 3–10 kDa; HJCp < 3, hydrolysed jellyfish collagen with MW < 3 kDa; HVCp 10–30, hydrolysed vertebrate collagen with MW 10–30 kDa; HVCp 3–10, hydrolysed vertebrate collagen with MW between 3 and 10 kDa; HVCp < 3, hydrolysed vertebrate collagen with MW < 3 kDa. Data are the mean values of six independent experiments performed in three technical replicates. The antioxidant activity is expressed as nmol of TE per mg of protein ± standard deviation. Student's t-test * $p < 0.05$, ** $p < 0.01$ and *** $p < 0.001$.

Proteins insoluble in PBS (IP) showed an AA of 83.2 nmol TE/mg of proteins that was not significantly different from the extractable proteins; however, pepsin digestion produced a variable content of antioxidant peptides.

Depending on the source of protein and degree of hydrolysis, small MW sub-fractions of hydrolysed proteins were already demonstrated to possess the strongest AA activity [52–55,68,79,80], i.e., higher than high MW protein sub-fractions. Pepsin is able to hydrolyse a wide range of proteinaceous components including non-triple helical domains of collagen [81,82]. Therefore, pepsin digestion has been used here to aid the collagen solubilisation process [83–85]: indeed, the resulting extract consists mainly of non-collagen proteins and atelocollagen. In addition, pepsin cleaves peptides specifically in telopeptide region of collagen, which are non-helical ends; thus, by hydrolyzing some non-collagenous proteins, the pepsin treatment increases the purity of collagen and likely reduces its antigenicity [86]. In addition, the proteolytic effect of 2% pepsin (i.e., the concentration used here) cleaved cross-linked molecules without damaging the triple helix [87].

In our experiments, pepsin hydrolysed peptides (PHp) were separated by membrane filtration in sub-fractions containing peptides with different MW ranges: MW > 30 kDa, MW 10–30 kDa,

MW 3–10 kDa and MW < 3 kDa. The fraction with peptides at highest MW (PHp > 30 kDa) was not further considered because it contained the pepsin enzyme (theoretical MW 34.5 kDa).

The AA values measured in the fractions PHp 10–30, PHp 3–10 and PHp < 3 were 364.9 nmol TE/mg, 1321.7 nmol TE/mg, and 1403,8 nmol TE/mg of proteins, respectively (Figure 2). Comparing the hydrolysed peptides with the parent proteins, AA values in PHp 10–30, PHp 3–10 or PHp < 3 resulted respectively four, fifteen, or sixteen times higher than the AA measured in the IP fraction.

Undigested insoluble proteins, after pepsin hydrolysis, likely consist of helical domains of collagen presumably inaccessible to pepsin action. As demonstrated in Leone et al. [11], bacterial collagenase was able to digest the insoluble protein fraction remained after pepsin hydrolysis, confirming its nature. Peptides derived from the *R. pulmo* proteins after collagenase digestion ranged MW 20–70 kDa, with two main bands at about 35 kDa and 50 kDa [11] In the present work, a fractionation of the collagenase-digested peptides was performed by membrane filtration, and four fractions were obtained at different MW range: MW > 30 kDa (HJCp > 30), MW 10–30 kDa (HJCp 10–30), MW 3–10 kDa (HJCp 3–10) and MW < 3 kDa (HJCp < 3). The commercial purified bovine collagen from calf skin (vertebrate collagen, VC) was also subjected to the same sequential pepsin-collagenase hydrolysis and MW-fractionation (HVCp > 30, HVCp 10–30, HVCp 3–10 and HVCp < 3) and was used as standard comparison with vertebrate collagen. The fractions HJCp > 30 and HVCp > 30 were not further considered because they contained the collagenase enzymes ranging in MW 68–130 kDa (as stated by the manufacturer). Proteins not hydrolysed by pepsin, mainly consisting of jellyfish collagen (JC) showed a very low AA (88 nmol TE/mg of proteins), which value was similar to that of undigested total protein (IP). The AA of not-hydrolysed bovine collagen (VC) was 1182 nmol TE/mg, but we could not determine whether it is an intrinsic characteristic or if the high AA value was due to the presence of other peptides.

The AA measured in the sub-fractions containing collagenase-hydrolysed jellyfish peptides resulted significantly higher as compared to proteins before collagenase digestion, namely 2741, 4980 and 36129 nmol TE/mg for proteins in the fractions HJCp 10–30, HJCp 3–10 and HJCp < 3, respectively. The AA of the fractions of peptides derived from vertebrate collagen were significantly lower as compared to jellyfish protein fractions, namely 2543, 4045, and 22,092 nmol TE/mg for peptide fractions HVCp 10–30, HVCp 3–10 and HVCp < 3, respectively. The main difference was between the fractions containing the smallest peptides, i.e., HJCp < 3 and HVCp < 3: AA appeared almost 2 times higher in low MW jellyfish-derived peptides than in low MW vertebrate-derived peptides. It is remarkable that, again, the hydrolysed peptides with lower MW showed higher AA in the order HJCp < 3 >>> HJCp 3–10 >> HJCp 10–30 >> JC. It is also notable that the value of AA of the smallest jellyfish peptides (HJCp < 3) was more than four hundred times the AA of collagenase-undigested peptides, the HJCp 3–10 more than twelve times higher and the HJCp 10–30 more than seven times higher. Similar differences, but less sharp, were evident in in vertebrate collagen fractions. These data confirmed and strengthened our previous findings about a higher AA of jellyfish collagens (from three species: *Aurelia coerulea*, *Cotylorhiza tuberculata* and *Rhizostoma pulmo*) as compared to chicken sternal cartilage collagen (Type II collagen) [11]. All jellyfish peptides derived from collagenase digestion showed considerably higher AA as compared to both peptides from pepsin digestion and PBS extracted (not hydrolysed) peptides [11].

The enzymatic hydrolysis of jellyfish proteins provided antioxidant peptides that can be further tested for their activity in cell culture systems. The enzymatic hydrolysis seems to be the most efficient method to produce homogeneous bioactive peptides, useful for further purification steps. As in this case, using specific enzymes and controlled reactions would help to release more homogeneous bioactive fragments than the chemical hydrolysis [88].

Although aqueous extraction can allow the solubilisation of antioxidant molecules other than proteins [11], the AA measured in the PBS extract and its fractions was lower as compared to hydrolysed fractions of the aqueous insoluble proteins. Therefore, the non-structural hydro-soluble proteins, the protein moiety of the mucus and maybe the proteinaceous toxins related to the abundant presence of nematocysts in the ectoderm and in the mucus of *R. pulmo* altogether play a marginal role in providing antioxidant activity compared to insoluble proteins. Although in this work the lyophilized whole jellyfish was considered, this finding seems to be in agreement with our recent finding that compounds from *R. pulmo* whole fresh jellyfish freely soluble in the PBS medium have low or no antioxidant activity [15].

Among *R. pulmo* hydrolysed proteins, the lower AA of pepsin-hydrolysed peptides, as compared to collagenase-digested proteins, could be due to the amino acid composition of the non-helical collagen as well as to other non-structural proteins of the jellyfish tissues. In addition, the pepsin-hydrolysed fractions could have a less homogeneous composition. Indeed, the activities of protein hydrolysates can be influenced by the amino acid composition, degree of hydrolysis, peptide size, peptide sequence, and type of used enzymes [65]. It is reasonable to assume that the differences in antioxidant activity could be related to differences in other biological activities.

2.3. Effect Jellyfish Proteins on HEKa Cell Cultures

2.3.1. Effect of Soluble Protein Fractions on HEKa Cell Cultures

To determine the potential biological effect of aqueous soluble proteins of *R. pulmo*, dose-response experiments were carried out using total soluble extract and their derived fractions on cultured Human Epidermal Keratinocytes isolated from adult skin (HEKa). To verify and compare the possible cytotoxic effect HEKa cultures were treated with different jellyfish protein concentrations ranging from 0.25 to 20 µg/mL for 24 h, and the effect on cell viability was measured by MTS assay. The cell viability (Figure 3) was measured in HEKa cell cultures after treatment with the PBS whole extract (SP), and its fractions SP > 30, SP 10–30, SP 3–10. As already mentioned, the fraction containing compounds with MW lower than 3 kDa was not considered because of its high salt content, which could be cytotoxic *per se*. Figure 3 shows that only the whole jellyfish aqueous extract (SP) and its derived fraction containing proteins with MW > 30 kDa (SP > 30) were cytotoxic at the assayed concentrations. SP was able to reduce the cell viability to about 60% of the control at concentrations higher than 2.5 µg/mL, while SP > 30 exerted its cytotoxic effect from the concentration of 1 µg/mL. The fractions of jellyfish soluble proteins with MW ranging 10–30 kDa and 3–10 kDa were both non-cytotoxic even at the highest tested concentration (Figure 3C,D). The half maximal inhibitory concentration (IC_{50}) for SP and SP > 30 fractions was determined at 2.7 + 1.5 µg/mL and 1.01 ± 0.06 µg/mL, respectively. Therefore, cytotoxic soluble compounds seem to be proteins at MW > 30 kDa, as the fraction SP > 30 resulted enriched of toxic compounds.

Different fractions of a hydro alcoholic extract from *Cotylorhiza tuberculata* were found non- toxic for HEKa cells until a concentration of 80 µg/mL while they were cytotoxic for breast cancer cells MCF7 [10]; however, the nature of the extracted compounds could be very different due to the solvent and fractionation method used.

Various molecules can be responsible of the cytotoxicity detected here on HEKa cell cultures; indeed, the aqueous extract of *R. pulmo* contains soluble proteins including nematocyst venom. Nematocysts are subcellular organelles produced by highly specialized mechano-sensory nerve cells, the nematocytes [89]. Functional nematocytes are distributed in the ectodermal layer of cnidarian tissues, at high concentrations particularly over tentacles and oral structures. Upon mechanic or chemical signals, each nematocyte can fire a syringe-like filament injecting a mixture of proteinaceous and non-proteinaceous compounds produced and stored in the nematocysts [5]. Generally, the toxicological properties of jellyfish venoms are species-specific: *R. pulmo* it is commonly considered mild stinger to humans since its effect is no more than a burning sensation. However, the severity of envenomation depends on the number of discharged nematocytes and the affected body part. Interestingly, nematocysts are also extremely abundant in the mucus of *R. pulmo* (Leone, personal observation), possibly released as defensive mechanism following interspecific contacts.

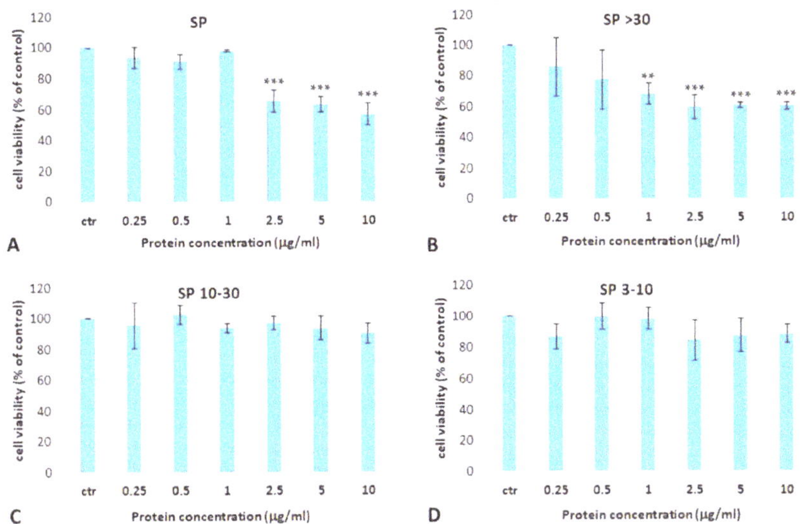

Figure 3. Cell viability of human epidermal keratinocytes (HEKa) treated for 24 h with different concentrations of (**A**) total PBS jellyfish extract (SP); (**B**) extract fraction containing molecules with a MW > 30 kDa (SP > 30), or (**C**) with MW 10–30 kDa (SP 10–30) or (**D**) MW 3–10 kDa (SP 3–10). Data are mean values of six independent experiments performed in five technical replicates, ±standard deviation. Ctr, control. ANOVA statistic test followed by Dunnett's post-test was used to compare each treatment with the control, * $p < 0.05$, ** $p < 0.01$ and *** $p < 0.001$.

Rhizolysin, a high molecular weight cytolysin of 260 kDa was isolated from the nematocysts of *R. pulmo* [90] and a 95 kDa metalloproteinase named rhizoprotease was identified in a tentacle extract fraction [91]. In *R. pulmo* a cytotoxic and hemolytic activity was also detected in tissue isolates free of nematocysts [92]. A number of venom proteins were characterized for their MW by SDS-PAGE separation: proteins with anticoagulant activity extracted from tentacles of *Aurelia* sp. (as *A. aurita*) showed MW ranging 50–160 kDa [93]; toxins with haemolytic activity with apparent MW of 12 kDa [94], 43 kDa and 45 kDa [95] were isolated from *Alatina* (*Carybdea*) *alata*; several nematocyst venom proteins with approximate MW of 35, 50, 55 and 100 kDa were found in *Chrysaora achlyos* [96]. FPLC gel filtration chromatography allowed the separation of venom proteins with molecular mass of about 102–107 kDa from the nematocysts of *Carybdea marsupialis* [97] and proteins peaks at 85 and 40 kDa obtained from the crude toxin of *Rhopilema nomadica* [98].

By functional assay (using fibrinolytic activity in zymography assay), Bae et al. [99] reported venom proteins of *Nemopilema nomurai* as characterized by MW of approximately 70, 35, 30, and 28 kDa. These authors compared *N. nomurai* with *Aurelia aurita* venoms, with similar banding patterns, distributed in 60–80 kDa and 25–37 kDa size bands, and with the siphonophoran *Physalia physalis* venom, with MW > 25 kDa.

In the present work, in order to verify the presence of proteinaceous venom in our extracts, the two cytotoxic fractions (SP and SP > 30) were heat-denatured by exposure to 100 °C for 10 min before their administration on cells. A loss of cytotoxicity on HEKa was observed after heat treatment of those fractions (Figure 4). This activity reduction was mitigated at the highest tested protein concentrations of 10 µg/mL in both, SP and SP > 30, with *p*-values of $p < 0.05$ and $p < 0.01$, respectively. The difference was likely due to the enrichment of toxins in the SP > 30 fraction, after membrane filtration, as compared to the whole extract (SP). In addition, as the cytotoxicity has not completely declined, at least at the highest concentration, some heat-resistant proteins should be still present. This confirms previous observations on the fibrinogenolytic activity of *R. pulmo* tentacle extract, which was significantly reduced but not abolished by heat treatment for 1 min at 100 °C [91], suggesting that some active components of the *R. pulmo* extracts are not completely thermolabile.

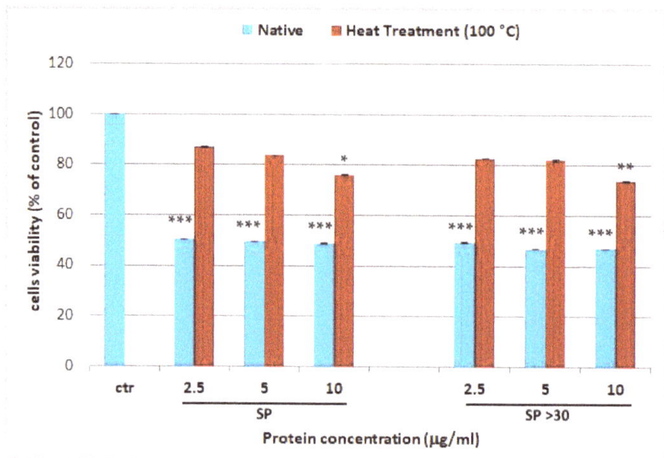

Figure 4. Cell viability of human epidermal keratinocytes (HEKa) treated with different concentrations of soluble proteins (SP) and soluble proteins with MW > 30 kDa (SP > 30), not treated (Native) and heat-denatured for 10 min at 100 °C (Heat Treatment 100 °C). Data are mean values of six independent experiment ± SD. Statistical analysis was carried out by ANOVA test followed by Dunnett's post-test, * $p < 0.05$, ** $p < 0.01$ and *** $p < 0.001$.

Our findings, together with literature data, suggest that a simple washing in aqueous solutions and the separation of high molecular weight proteins from the extract, e.g., by membrane filtration, could represent a crucial strategy for both removing possible toxic compounds from jellyfish extracts and to concentrate potentially bioactive soluble compounds. This basically simple procedure could be an easy starting point for the isolation of *R. pulmo* venom and for the development of a processing method of jellyfish biomasses suitable for isolation and characterization of potentially active soluble components, useful as nutraceutical and cosmeceutical ingredients.

2.3.2. Effect of Hydrolysed Proteins on HEKa Cell Cultures

The fractions containing hydrolyzed jellyfish peptides were tested for their activity on HEKa cultures. The fractions containing jellyfish peptides with MW > 30 kDa, PHp > 30 and HJCp > 30 were not assayed because of the occurrence of the enzymes pepsin and collagenase, respectively.

The cell viability was assayed after 24 h of treatment with the pepsin-hydrolyzed jellyfish peptides PHp 10–30, PHp 3–10 and PHp < 3 and with the undigested fraction IP, at concentrations ranging from 0.05 to 5 µg/mL (Figure 5). No cytotoxic effect and no significant changes of cell viability, as compared to the controls, was evident in keratinocytes treated with all the hydrolyzed protein fractions at all tested concentrations.

The cell viability of HEKa cells was also assayed after a 24 h-treatment with different concentrations of collagenase-hydrolyzed jellyfish peptides: HJCp 10–30, HJCp 3–10 and HJCp < 3 (Figure 6). The effects on HEKa of the bovine collagen fractions (HVCp 10–30, HVCp 3–10 and HVCp < 3), subjected to the identical fractionation procedure, HVCp 10–30, HVCp 3–10 and HVCp < 3, tested at the same concentrations, are also shown. Pepsin-digested fractions before collagenase digestion, JC and VC, were also tested.

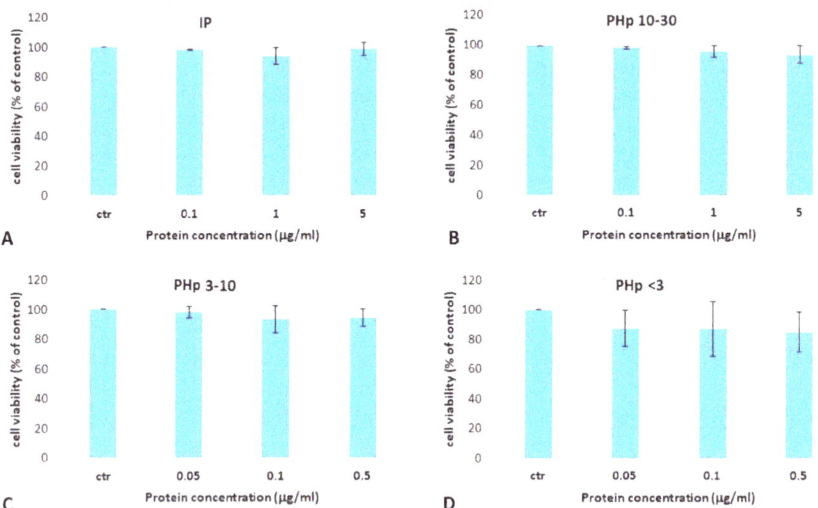

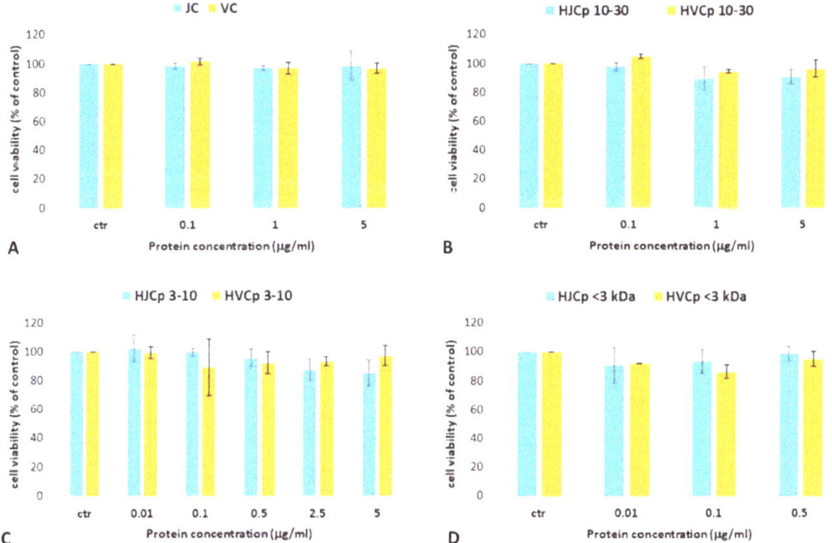

Figure 5. Cell viability of human epidermal keratinocytes (HEKa) treated with different concentrations of (**A**) insoluble jellyfish proteins (IP); (**B**) pepsin-hydrolysed fraction containing peptides with MW 10–30 kDa (PHp 10–30), (**C**) MW 3–10 kDa (PHp 3–10) and (**D**) MW < 3 kDa (PHp < 3). Data are mean values of six independent experiment ± SD. A statistical analysis was performed using ANOVA followed Dunnett's post-test ($p < 0.05$).

Figure 6. Cell viability of human epidermal keratinocytes (HEKa) treated with different concentrations of (**A**) jellyfish and calf skin collagen (JC and VC); (**B**) collagenase-hydrolysed fraction containing peptides with MW ranging 10–30 kDa (HJCp 10–30 and HVCp 10–30), (**C**) MW 3–10 kDa (HJCp 3–10 and HVCp 3–10) and (**D**) MW < 3 kDa (HJCp < 3 and HVCp < 3). Data are mean values ± SD of six independent experiments. Statistical analysis performed with ANOVA and Dunnett's test ($p < 0.05$).

Again, no cytotoxic effect and no significant changes of cell viability was observed in keratinocytes treated with all the hydrolyzed protein fractions at all the tested concentrations. In addition, there were no differences between the fractions of jellyfish collagen and the commercial bovine collagen (Figure 6). Over the 24-h course of the treatment, no significant increase in cell proliferation was found in our experiments. The aim of this preliminary work was to establish the maximum non-toxic dose or maximum tolerated dose for jellyfish derived proteins in human keratinocytes, in order to carry out a preliminary screening for bioactive compounds derived from *R. pulmo*.

It is known that bovine collagen increases cell adhesion and proliferation in murine primary keratinocytes [100] and pepsin-solubilized collagen from red sea cucumber (*Stichopus japonicus*) increased the cell migration in wound-healing test, fibronectin synthesis and cell proliferation in human keratinocyte (HaCaT) better that mammalian collagens [101]. A concentration of hydrolyzed fish collagen ranging from 0.76–1.53 µg/mL was found to increase keratinocytes proliferation [102]. To the best of our knowledge, there are no studies about jellyfish collagen and keratinocytes *in vitro*.

In mice, dietary supplementation with *Rhopilema asamushi* jellyfish collagen (JC) and jellyfish collagen hydrolysate showed *in vivo* protective effects on skin photoaging, alleviating the UV-induced changes of antioxidative enzy *in vivo* and *in vitro* mes and the content of glutathione, also protecting skin lipids and hydroxyproline content from the UV radiation damages [37]. Furthermore, collagens enhanced skin immunity, reduced water loss, restored cutaneous collagen and elastin levels and structure, and maintained type III to I collagen ratio in the model of chronic UVA + UVB irradiation of mice [38].

2.3.3. Effect of Fractions of Hydrolysed Jellyfish Collagen on HEKa Cell Cultures Subjected to Oxidative Stress

The effect of jellyfish collagen-derived peptides on keratinocytes was also evaluated in co-occurrence of a chemically induced oxidative stress, in order to verify the antioxidant capacity of jellyfish derived peptides in cells. Hydrogen peroxide solution (H_2O_2 0.1 mM) was used to induce reactive oxygen species (ROS) formation in cells. Keratinocytes were pre-treated with the highest concentration of jellyfish peptides for 24 h, and in the last hour of the experiment the H_2O_2 solution was added (Figure 7).

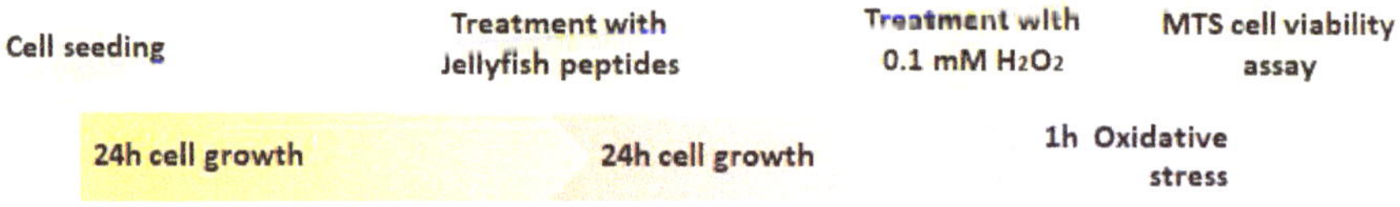

Figure 7. Experimental scheme of the cells treatments with collagen peptides and Hydrogen peroxide (0.1 mM). Controls without treatment and stress induction (1 h with H_2O_2) were run for each experiment ($n = 6$). Cells vitality was analyzed using MTS assay, after 1 h of the oxidative stress.

Specifically, the fractions HJCp 10–30 and HJCp 3–10 were tested at the protein concentration of 5 µg/mL and the fraction HJCp < 3 was tasted at 0.5 µg/mL (Figure 8). Same concentrations of the peptide fractions derived from bovine collagen (HVCp) were administered in parallel experiments. The main difference between jellyfish and vertebrate collagen peptides was observed when cells were pre-treated with peptides having a MW ranging 10–30 kDa, showing a higher protective effect against oxidative stress due to HJCp 10–30 fraction. Maybe the peptides HJCp 10–30 derived from jellyfish collagen could quench the ROSs molecules, preventing oxidative cell damaging, better than bovine collagen peptides.

Further analysis such as determination of amino acidic composition, biochemical features and health effects will pave the way to better characterize the peptide fractions. Available evidence suggests that protein hydrolysates, including hydrolyzed collagen, from different sources with antioxidant and other functional properties and biological activities, will increasingly receive attention from the large community of researchers working on health food, nutraceuticals and cosmeceuticals industries as well as on processing/preservation technologies.

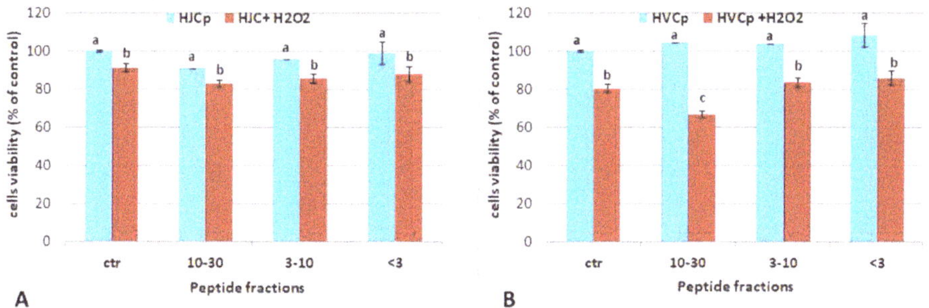

Figure 8. Effect of oxidative stress on HEKa cells of different concentration of jellyfish (**A**) and calf skin (**B**) collagen hydrolyzed peptides MW ranging 30–10 kDa (5 μg/mL), 3–10 kDa (5 μg/mL) and <3 kDa (0.5 μg/mL). Cells viability was valuated 1 h after H_2O_2 treatment. Data are mean values ± SD of six independent experiments, analysed with ANOVA and Bonferroni post-test ($p < 0.05$).

In conclusion, this work introduces the idea and the methodology for a safe use of proteinaceous compounds from *Rhizostoma pulmo* biomass, by reducing the potential cytotoxic fractions by aqueous extractions and producing antioxidant peptide fractions by protein hydrolyses of the insoluble fraction. A growing scientific evidence base demonstrates that jellyfish can be considered as a valuable source of new bioactive metabolites and the improvements in jellyfish processing technologies will grant the use of abundant jellyfish biomass as a sustainable resource for marine biotechnology applications. We predict that the large natural biomass of *R. pulmo* in the Mediterranean basin will sustain the development of new research and new applications of jellyfish-derived compounds in the cosmeceutical, nutraceutical and pharmaceutical fields.

3. Materials and Methods

3.1. Chemicals, Materials and Equipment

Amicon® Ultra-15 Centrifugal Filter Devices and acetic acid were purchased from Merck (Darmstadt, Germany). Bovine serum albumin (BSA), phosphate buffered saline (PBS), Pepsin from porcine gastric mucosa ($\geq$2500 U/mg), Collagenase from *Clostridium histolyticum* [0.5–5.0 furylacryloyl-Leu-Gly-Pro-Ala (FALGPA) units/mg solid, $\geq$125 collagen digestion unit (CDU)/mg solid], Collagen from calf skin (Sigma-Aldrich, Saint Louis, MS, USA), ABTS [2,20-Azinobis (3-ethylben-zothiazoline-6-sulfonic acid) diammonium salt], Cell Freezing Medium with DMSO serum-free, TES(2-{[1,3-Dihydroxy-2-(hydroxymethyl)-2-propanyl]amino}ethanesulfonic acid buffer), Hydrogen peroxide solution and Trypsin-EDTA solution were purchased from Sigma-Aldrich (Milan, Italy). Protein Assay Dye Reagent concentrates, TGX™ FastCast™ Acrylamide Solutions, Protein Standard for Electrophoresis and ChemiDoc™ MP Imaging System, Bio-Rad Protein Assay were purchased from Bio-Rad Laboratories (Munich, Germany). Potassium persulfate (dipotassium peroxdisulfate), 6-hydroxy-2,5,7,8-tetramethylchroman-2-carboxylic acid (Trolox) were purchased from Hoffman-La Roche (Basel, Switzerland). Dulbecco's phosphate buffered saline (DPBS); Cascade Biologics™ Epilife® with 60 μM calcium; HKGS, were purchased from Life Technologies (Carlsbad, CA, USA). Human Epidermal Keratinocytes adult (HEKa), Trypan blue solution 0.4%, Countess™ automated cell counter and Countess™ cell counting chamber slides were purchased from Invitrogen™ (Carlsbad, CA, USA). MTS CellTiter 96® AQueous Non-Radioactive Cell Proliferation Assay was purchased from Promega (Madison, WI, USA). Infinite M200, quad4 monochromator™ detection system was purchased from

Tecan group (Männedorf, Switzerland). Flat and round-bottom 96-well microplates were purchased from Corning (Corning, NY, USA).

3.2. Jellyfish Samples

Rhizostoma pulmo, Macrì 1778 [67] specimens were collected at Marina di Ginosa (Taranto, Italy) (Supplementary Materials, Video S1), in summers 2017–2018, by means of a nylon landing net with 3.5 cm mesh size, from an open type motorboat, and stored in refrigerated seawater in 100 L barrels for a maximum of 2 h. Specimens were adult both male and female jellyfish with a diameter ranging from 17 to 25 cm. Whole jellyfish were individually frozen in liquid nitrogen and stored at −80 °C. Frozen samples were then lyophilized in a freeze dryer (Freezone 4.5 L Dry System, Labconco Co. Thermo Scientific, Milan, Italy), at −55 °C for 4 days using a chamber pressure of 0.110 mbar and then stored at −20 °C until use. Each lyophilized jellyfish has been made homogeneous (oral arms and umbrella) by mixing its powder, and then the lyophilized powder from 5 individuals was pooled. Six different pools were considered as representative samples and used for independent experiments.

3.3. Protein Extraction and Sequential Hydrolysis

Lyophilized tissues were ground into a fine powder with liquid nitrogen and 1 g was used as described below (Figure 9). Soluble proteins (SP) were extracted by insoluble material (IP) by gentle stirring of the sample with 16 volumes (w/v) of PBS, (phosphate buffer saline) pH 7.4, at 4 °C for 2 h and then centrifuged at 9000× g for 30 min at 4 °C. Supernatant was separated from the insoluble material (IP) and subjected membrane fractionation as described below. Pellet was subjected to sequential enzymatic hydrolyses by pepsin (1 mg/mL) in 0.5 M acetic acid, using an enzyme/substrate ratio of 1:50 (w/w) and stirred for 48 h at 4 °C. The digested sample was centrifuged at 9000× g for 30 min and the pepsin-hydrolyzed peptides (PHp) were stored for further separations. The pellet was washed two times with bi-distilled water, and subjected to a second digestion with collagenase (6 mg/mL in TES buffer 50 mM, pH 7.4 and 0.36 mM of $CaCl_2$) using an enzyme/substrate ratio of 1:50 (w/w), by stirring 5 h at 37 °C. Collagenase cuts the peptide sequences as –R-Pro-X-Gly-Pro-R where X is generally a neutral amino acid. After hydrolysis, the sample was centrifuged at 9000× g for 30 min, and the soluble collagenase-hydrolyzed peptides (HJCp) were stored for further treatments. The pellet of collagenase digestion was considered as not-hydrolysable material. Commercial calf skin collagen (Sigma) was used as control and subjected to the same sequential hydrolysis procedure.

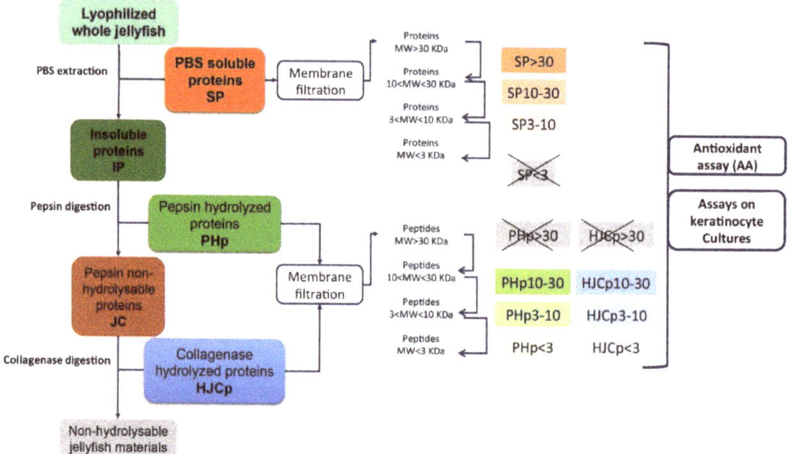

Figure 9. Flow diagram showing the various steps of extraction, hydrolysis and fractionation of the proteins from *Rhizostoma pulmo*.

Soluble proteins derived from PBS extraction (SP), pepsin hydrolyzed peptides (PHp) and hydrolyzed collagen peptides (HJCp) were subjected to fractioning by membrane filtration.

3.4. Proteins Separation by Membrane Filtration

All the obtained fractions (SP, PHp and HJCp) were separated by membrane filtration in fractions containing peptides with different molecular weight ranges. All the steps were performed at 4 °C. Each sample was filtered using Amicon® Ultra 30K device (Merck) by centrifugation at $4000\times g$ to almost total filtration, the retentate contained compounds with MW higher than 30 kDa. The filtrates (containing compounds less than 30 kDa) were further fractionated using Amicon® Ultra 10K device (Merck) by centrifugation at $4000\times g$ to obtain the 10–30 kDa fraction in the retentate. Finally, the filtrates containing compounds lower than 10 kDa were centrifuged using Amicon® Ultra 3K device (Merck) at $4000\times g$ to obtain in the retentate fractions with 10 < MW < 3 kDa and MW < 3 kDa. Each sample was analyzed for protein content, antioxidant activity and cell culture test.

3.5. Protein Content

Total protein content was estimate by Bradford assay [103]. The assay was modified and adapted to round bottom 96-well microplate for Infinite M200, quad4 monochromator™ detection system (Tecan, Männedorf, Switzerland) using bovine serum albumin (BSA) as a standard.

3.6. Antioxidant Activity

The antioxidant activity was evaluated by Trolox Equivalent Antioxidant Capacity (TEAC) method adapted for 96-well microplates and Infinite M200 (Tecan, Männedorf, Switzerland), using the radical cation ABTS•+ and Trolox (Hoffman-La Roche) as standard [104,105]. Briefly: 10 µL of each sample was added to 200 µL of ABTS•+ solution, were stirred and the absorbance at 734 nm was read at 6 min [11,106]. Trolox was used as standard and was assayed under the same conditions of the samples. Results were expressed as nmol of Trolox Equivalents per mg of contained proteins (nmol TE/mg protein).

3.7. Proteins SDS-PAGE Analysis

Total jellyfish proteins and polypeptides fractions obtained from soluble proteins extracted with PBS were analyzed by SDS-PAGE. A FastCast premixed acrylamide solution 12% was used to prepared gels and "All Blue Precision Plus Protein Standard" (Biorad) was used as molecular weight marker. In order to visualize protein bands, gels were both analyzed by stain-free system with high sensitivity imagined using ChemiDoc™ MP Imaging System (Biorad) and stained with Coomassie Brilliant Blue G-250 (Bio-Rad Protein Assay).

3.8. HEKa Cell Culture

Human epidermal keratinocytes, isolated from adult skin (HEKa) were obtained from Cascade BiologicsTM (Gibco®) and routinely grown in EpiLife® medium with 60 µM calcium (GIBCO) as described in Leone et al. [10]. Trypan blue dye exclusion and automated counting method by Countess™ was used for routinely cell viability assay and live cell counting. For all experiments, 0.15×10^6 cells/well (75000 cells/mL) were incubated in flat bottom 96-well microplates.

3.9. Cell Treatments and Oxidative Stress Induction with H_2O_2

All jellyfish protein fractions were diluted in EpiLife® culture medium to reach a final concentration on the cells ranging from 0.05 and 20 µg/mL of proteins/peptides. Soon after dilution, the jellyfish samples were added to cells grown for 24 h in 96-well microplates at 37 °C with 5% CO_2 (Thermo Forma direct heat CO_2 incubator). Controls were included in each experiment and in each microplate with medium only (without cells), cells with only medium, and cells with the vehicle

(PBS or digestion buffers), at the same final concentration as in the cells treated with the jellyfish samples. For each independent experiment, each treatment, namely each sample, each control and each concentration, was replicated in five technical replicates. Microplates were then incubated for 24 h at 37 °C with 5% CO_2 (Thermo Forma direct heat CO_2 incubator).

3.9.1. Cell Treatments with Heat-Denatured Protein

Aliquots of soluble extracted proteins (SP) and the sub-fraction SP > 30 were also heat-denatured by heating at 100 °C for 10 min (Heat Treatment 100 °C) in a water bath, cooled, diluted in EpiLife® culture medium and administrated to the cells.

3.9.2. Cell Treatments with H_2O_2

In the experiments for antioxidant activity assay, HEKa cells (0.15×10^6 cells/well) were grown for 24 h to reach 80% of confluence in flat bottom 96-well microplates, and then were treated with the collagen peptides fractions from jellyfish and from bovine collagen at the same concentrations. Two controls with medium and vehicle were also included. After 24 h, 100 µL medium contained H_2O_2, at the final concentration of 0.1 mM were supplied and cells were incubated for 1 h at 37 °C with 5% CO_2, as reported in Figure 7. Cell viability was assayed by MTS assay soon after the 1 h of treatment.

3.10. Cell Viability Assay

MTS Cell viability test was used to establish the effects of the extracted jellyfish compounds. MTS assay was performed using CellTiter 96® AQueous One Solution Reagent (Promega) according to the manufacturer's instructions. 20 µL of CellTiter 96® AQueous One Solution Reagent were added to each well, the microplates were incubated for 90 min at 37 °C with 5% CO_2 (Thermo Forma direct heat CO_2 incubator) and the absorbance was read at 490 nm with Infinite M200. Data were expressed as percentage of the respective controls.

3.11. Statistical Analysis

Statistical analyses were performed by Graphpad Prism 6.0. An unpaired Student's *t*-test was used to compare two groups; analysis of variance (ANOVA) and Dunnett's *post hoc*-test was applied to compare control with all other treatments, instead a Bonferroni post-test was applied to analyze data in oxidative stress experiments. Differences were considered statistically significant for values $p < 0.05$. All assays were replicated different time ($n = 6$) and data are represented as mean ± standard deviation (SD). The half maximal inhibitory concentration (IC_{50}) for fractions was calculated using the same program Graphpad Prism 6.0.

Supplementary Materials: The following are available online at http://www.mdpi.com/1660-3397/17/2/134/s1, Video S1: The sampling site of Rhizostoma pulmo jellyfish at Marina di Ginosa, Ionian Sea, Italy (@Antonella Leone CNR-ISPA).

Author Contributions: Conceptualization, data analysis, methodology, writing and review manuscript, funding acquisition, project administration, A.L.; performing experiments, data analysis, writing initial draft preparation, S.D.D.; performing experiments, data analysis, G.D.R.; set up first experiments, M.P.; manuscript review, S.P.

Funding: This work was supported by the project GoJelly—A gelatinous solution to plastic pollution—under grant agreement no. 774499 European Union's Horizon 2020 programme H2020-EU.3.2.5, BG-07-2017—Blue green innovation for clean coasts and seas.

Acknowledgments: The Authors thanks Leone D'Amico for his valuable technical support and Fabio Minerva, Associazione Master Wave, for the high professional support during jellyfish sampling.

Conflicts of Interest: The authors declare no conflict of interest.

References

1.   Omori, M.; Nakano, E. Jellyfish fisheries in southeast Asia. *Hydrobiologia* **2001**, *451*, 19–26. [CrossRef]

2. Purcell, J.E. Climate effects on formation of jellyfish and ctenophore blooms: A review. *J. Mar. Biol. Assoc. UK* **2005**, *85*, 461–476. [CrossRef]

3. Purcell, J.E.; Uye, S.I.; Lo, W.T. Anthropogenic causes of jellyfish blooms and their direct consequences for humans: A review. *MEPS* **2007**, *350*, 153–174. [CrossRef]

4. Dong, Z.; Liu, D.; Keesing, J.K. Jellyfish blooms in China: Dominant species, causes and consequences. *Mar. Pollut. Bull.* **2010**, *60*, 954–963. [CrossRef] [PubMed]

5. De Donno, A.; Idolo, A.; Bagordo, F.; Grassi, T.; Leomanni, A.; Serio, F.; Guido, M.; Canitano, M.; Zampardi, S.; Boero, F.; et al. Impact of stinging jellyfish proliferations along south Italian coasts: Human health hazards, treatment and social costs. *Int. J. Environ. Res. Public Health* **2014**, *11*, 2488–2503. [CrossRef] [PubMed]

6. Lucas, C.H.; Gelcich, S.; Uye, S.I. Living with jellyfish: Management and adaptation strategies. In *Jellyfish Blooms*; Pitt, K.A., Lucas, C.H., Eds.; Springer: Dordrecht, The Netherlands, 2014; Chapter 6; p. 129.

7. Doyle, T.K.; Hays, G.C.; Harrod, C.; Houghton, J.D.R. Ecological and societal benefits of jellyfish. In *Jellyfish Blooms*; Pitt, K.A., Lucas, C.H., Eds.; Springer Science + Business Media: Dordrecht, The Netherlands, 2014; Chapter 5; pp. 105–127.

8. Graham, W.M.; Gelcich, S.; Robinson, K.L.; Duarte, C.M.; Brotz, L.; Purcell, J.E.; Madin, L.P.; Mianzan, H.; Sutherland, K.R.; Uye, S.; et al. Linking human well-being and jellyfish: Ecosystem services, impacts, and societal responses. *Front Ecol. Environ.* **2014**, *12*, 515–523. [CrossRef]

9. Hsieh, Y.H.P.; Leong, F.M.; Rudloe, J. Jellyfish as food. *Hydrobiologia* **2001**, *451*, 11–17. [CrossRef]

10. Leone, A.; Lecci, R.M.; Durante, M.; Piraino, S. Extract from the zooxanthellate jellyfish *Cotylorhiza tuberculata* modulates gap junction intercellular communication in human cell cultures. *Mar. Drugs* **2013**, *11*, 1728–1762. [CrossRef] [PubMed]

11. Leone, A.; Lecci, R.M.; Durante, M.; Meli, F.; Piraino, S. The Bright Side of Gelatinous Blooms: Nutraceutical Value and Antioxidant Properties of Three Mediterranean Jellyfish (Scyphozoa) *Mar. Drugs* **2015**, *13*, 4654–4681. [CrossRef] [PubMed]

12. Hsieh, Y.H.P.; Rudloe, J. Potential of utilizing jellyfish as food in Western countries. *Trends Food Sci. Technol.* **1995**, *5*, 225–229. [CrossRef]

13. Rocha, J.; Peixe, L.; Gomes, N.C.M.; Calado, R. Cnidarians as a source of new Marine bioactive compounds—An overview of the last decade and future steps for bioprospecting *Mar. Drugs* **2011**, *9*, 1860–1886. [CrossRef] [PubMed]

14. D'Amico, P.; Leone, A.; Giusti, A.; Armani, A. Jellyfish and Humans: Not just negative interactions. In *Jellyfish: Ecology, Distribution Patterns and Human Interactions*; Mariottini, G.L., Ed.; Nova Science Publishers: Hauppauge, NY, USA, 2017; pp. 331–352.

15. Leone, A.; Lecci, R.M.; Milisenda, G.; Piraino, S. Mediterranean jellyfish as novel food: Effects of thermal processing on antioxidant, phenolic and protein contents. *Eur. Food Res. Technol.* **2019**. [CrossRef]

16. Nagai, H.; Takuwa, K.; Nakao, M.; Ito, E.; Miyake, M.; Noda, M.; Nakajima, T. Novel proteinaceous toxins from the box jellyfish (sea wasp) *Carybdea rastoni*. *Biochem. Biophys. Res. Commun.* **2000**, *275*, 582–588. [CrossRef] [PubMed]

17. Sánchez-Rodríguez, J.; Torrens, E.; Segura-Puertas, L. Partial purification and characterization of a novel neurotoxin and three cytolysins from box jellyfish (*Carybdea marsupialis*) nematocyst venom. *Arch. Toxicol.* **2006**, *80*, 163–168. [CrossRef] [PubMed]

18. Brinkman, D.L.; Burnell, J.N. Identification, cloning and sequencing of two major venom proteins from the box jellyfish, *Chironex fleckeri*. *Toxicon* **2007**, *50*, 850–860. [CrossRef] [PubMed]

19. Kang, C.; Munawir, A.; Cha, M.; Sohn, E.T.; Lee, H.; Kim, J.S.; Yoon, W.D.; Lim, D.; Kim, E. Cytotoxicity and hemolytic activity of jellyfish *Nemopilema nomurai* (Scyphozoa: Rhizostomeae) venom. *Comp. Biochem. Physiol. C Toxicol. Pharmacol.* **2009**, *150*, 85–90. [CrossRef] [PubMed]

20. Brinkman, D.L.; Konstantakopoulos, N.; McInerney, B.V.; Mulvenna, J.; Seymour, J.E.; Isbister, G.K.; Hodgson, W.C. *Chironex fleckeri* (box jellyfish) venom proteins: Expansion of a cnidarian toxin family that elicits variable cytolytic and cardiovascular effects. *J. Biol. Chem.* **2014**, *289*, 4798–4812. [CrossRef] [PubMed]

21. Ramasamy, S.; Isbister, G.K.; Seymour, J.E.; Hodgson, W.C. Pharmacologically distinct cardiovascular effects of box jellyfish (*Chironex leckeri*) venom and a tentacle-only extract in rats. *Toxicol. Lett.* **2005**, *155*, 219–226. [CrossRef] [PubMed]

22. Cuypers, E.; Yanagihara, A.; Karlsson, E.; Tytgat, J. Jellyfish and other cnidarian envenomations cause pain by affecting TRPV1 channels. *FEBS Lett.* **2006**, *580*, 5728–5732. [CrossRef] [PubMed]

23. Lazcanopérez, F.; Arellano, R.O.; Garay, E.; Arreguínespinosa, R.; Sánchezrodríguez, J. Electrophysiological activity of a neurotoxic fraction from the venom of box jellyfish *Carybdea marsupialis*. *Comp. Biochem. Physiol. Part C Toxicol. Pharmacol.* **2017**, *191*, 177–182. [CrossRef] [PubMed]

24. Ayed, Y.; Bousabbeh, M.; Mabrouk, H.B.; Morjen, M.; Marrakchi, N.; Bacha, H. Impairment of the cell-to-matrix adhesion and cytotoxicity induced by the Mediterranean jellyfish *Pelagia noctiluca* venom and its fractions in cultured glioblastoma cells. *Lipids Health Dis.* **2012**, *28*, 11–84. [CrossRef] [PubMed]

25. Ding, J.F.; Li, Y.Y.; Xu, J.J.; Su, X.R.; Gao, X.; Yue, F.P. Study on effect of jellyfish collagen hydrolysate on anti-fatigue and anti-oxidation. *Food Hydrocolloids* **2011**, *25*, 1350–1353. [CrossRef]

26. Kawabata, T.; Lindsay, D.J.; Kitamura, M.; Konishi, S.; Nishikawa, J.; Nishida, S.; Kamio, M.; Nagai, H. Evaluation of the bioactivities of water-soluble extracts from twelve deep-sea jellyfish species. *Fish Sci.* **2013**, *79*, 487–494. [CrossRef]

27. Ha, S.H.; Jin, F.; Kwak, C.H.; Abekura, F.; Park, J.Y.; Park, N.G.; Chang, Y.C.; Lee, Y.C.; Chung, T.W.; Ha, K.T.; et al. Jellyfish extract induces apoptotic cell death through the p38 pathway and cell cycle arrest in chronic myelogenous leukemia K562 cells. *Peer J.* **2017**, *5*, e2895. [CrossRef] [PubMed]

28. Yu, H.; Liu, X.; Xing, R.; Liu, S.; Li, C.; Li, P. Radical scavenging activity of protein from tentacles of jellyfish *Rhopilema esculenta*. *Bioorg. Med. Chem. Lett.* **2005**, *15*, 2659–2664. [CrossRef] [PubMed]

29. Yu, H.; Liu, X.; Xing, R.; Liu, S.; Gao, Z.; Wang, P. *In vitro* determination of antioxidant activity of proteins from jellyfish *Rhopilema esculentum*. *Food Chem.* **2006**, *95*, 123–130. [CrossRef]

30. Harada, K.; Maeda, T.; Hasegawa, Y.; Tokunaga, T.; Ogawa, S.; Fukuda, K.; Nagatsuka, N.; Nagao, K.; Ueno, S. Antioxidant activity of the giant jellyfish *Nemopilema nomurai* measured by the oxygen radical absorbance capacity and hydroxyl radical averting capacity methods. *Mol. Med. Rep.* **2011**, *4*, 919–922. [PubMed]

31. Ruan, Z.; Liu, G.; Guo, Y.; Zhou, Y.; Wang, Q.; Chang, Y.; Wang, B.; Zheng, J.; Zhang, L. First report of a thioredoxin homologue in jellyfish: Molecular cloning, expression and antioxidant activity of CcTrx1 from *Cyanea capillata*. *PLoS ONE* **2014**, *9*, e97509. [CrossRef] [PubMed]

32. Ovchinnikova, T.V.; Balandin, S.V.; Aleshina, G.M.; Tagaev, A.A.; Leonova, Y.F.; Krasnodembsky, E.D.; Menshenin, A.V.; Kokryakov, V.N. Aurelin, a novel antimicrobial peptide from jellyfish *Aurelia aurita* with structural features of defensins and channel-blocking toxins. *Biochem. Biophys. Res. Commun.* **2006**, *348*, 514–523. [CrossRef] [PubMed]

33. Lucas, C.H. Biochemical composition of the mesopelagic coronate jellyfish *Periphylla periphylla* from the Gulf of Mexico. *J. Mar. Biol. Assoc. UK* **2009**, *89*, 77–81. [CrossRef]

34. Khong, N.M.; Yusoff, F.M.; Jamilah, B.; Basri, M.; Maznah, I.; Chan, K.W.; Nishikawa, J. Nutritional composition and total collagen content of three commercially important edible jellyfish. *Food Chem.* **2016**, *196*, 953–960. [CrossRef] [PubMed]

35. Sarmadi, B.H.; Ismail, A. Antioxidative peptides from food proteins: A review. *Peptides* **2010**, *31*, 1949–1956. [CrossRef] [PubMed]

36. Rizzello, C.G.; Tagliazucchi, D.; Babini, E.; Rutella, G.S.; Saa, D.L.T.; Gianotti, A. Bioactive peptides from vegetable food matrices: Research trends and novel biotechnologies for synthesis and recovery. *J. Funct. Foods* **2016**, *27*, 549–569. [CrossRef]

37. Zhuang, Y.; Hou, H.; Zhao, X.; Zhang, Z.; Li, B. Effects of collagen and collagen hydrolysate from jellyfish (*Rhopilema esculentum*) on mice skin photoaging induced by UV irradiation. *J. Food Sci.* **2009**, *74*, 183–188. [CrossRef] [PubMed]

38. Fan, J.; Zhuang, Y.; Li, B. Effects of collagen and collagen hydrolysate from jellyfish umbrella on histological and immunity changes of mice photoaging. *Nutrients* **2013**, *5*, 223–233. [CrossRef] [PubMed]

39. Aluko, R.E. Bioactive peptides. In *Functional Foods and Nutraceuticals*; Food Science Text Series; Aluko, R.E., Ed.; Springer: New York, NY, USA, 2013; pp. 37–61.

40. Harman, D. Aging: A theory based on free-radical and radiation-chemistry. *J. Gerontol.* **1956**, *11*, 298–300. [CrossRef] [PubMed]

41. Harman, D. The biologic clock: The mitochondria? *J. Am. Geriatr. Soc.* **1972**, *20*, 145–147. [CrossRef] [PubMed]

42. Lapointe, J.; Hekimi, S. When a theory of aging ages badly. *Cell. Mol. Life Sci.* **2010**, *67*, 1–8. [CrossRef] [PubMed]

43. Gülçin, İ.; Şat, İ.G.; Beydemir, Ş.; Elmastaş, M.; Küfrevioğlu, Ö.İ. Comparison of antioxidant activity of clove (*Eugenia caryophylata Thunb*) buds and lavender (*Lavandula stoechas* L.). *Food Chem.* **2004**, *87*, 393–400. [CrossRef]

44. Adebiyi, A.P.; Adebiyi, A.O.; Yamashita, J.; Ogawa, T.; Muramoto, K. Purification and characterization of antioxidative peptides derived from rice bran protein hydrolysates. *Eur. Food Res. Technol.* **2009**, *228*, 553–563. [CrossRef]

45. De Domenico, S.; Giudetti, A. Nutraceutical intervention in ageing brain. *JGG* **2017**, *65*, 79–92.

46. Lorenzo, J.M.; Munekata, P.E.S.; Gómez, B.; Barba, F.J.; Mora, L.; Pérez-Santaescolástica, C.; Toldrá, F. Bioactive peptides as natural antioxidants in food products—A review. *Trends Food Sci Tech.* **2018**, *79*, 136–147. [CrossRef]

47. Shahidi, F.; Han, X.Q.; Synowiecki, J. Production and characteristics of protein hydrolysates from capelin (*Mallotus villosus*). *Food Chem.* **1995**, *53*, 285–293. [CrossRef]

48. Tong, L.M.; Sasaki, S.; Mc Clements, D.J.; Decker, E.A. Mechanisms of the antioxidant activity of a high molecular weight fraction of whey. *J Agric. Food Chem.* **2000**, *48*, 1473–1478. [CrossRef] [PubMed]

49. Moure, A.; Domínguez, H.; Parajó, J.C. Antioxidant properties of ultrafiltration-recovered soy protein fraction from industrial effluents and their hydrolysates. *Process Biochem.* **2006**, *41*, 447–456. [CrossRef]

50. Klompong, V.; Benjakul, S.; Kantachote, D.; Kantachote, D.; Shahidi, F. Antioxidative activity and functional properties of protein hydrolysate of yellow stripe trevally (*Selaroides Leptolepis*) as influenced by the degree of hydrolysis and enzyme type. *Food Chem.* **2007**, *102*, 1317–1327. [CrossRef]

51. Suarez-Jimenez, G.M.; Burgos-Hernandez, A.; Ezquerra-Brauer, J.M. Bioactive peptides and depsipeptides with anticancer potential: Sources from marine animals. *Mar. Drugs* **2012**, *10*, 963–986. [CrossRef] [PubMed]

52. Mendis, E.; Rajapakse, N.; Kim, S.K. Antioxidant properties of a radical-scavenging peptide purified from enzymatically prepared fish skin gelatin hydrolysate. *J. Agric. Food Chem.* **2005**, *53*, 581–587. [CrossRef] [PubMed]

53. Rajapaksea, N.; Mendisa, E.; Byunb, H.G.; Kim, S.K. Purification and *in vitro* antioxidative effects of giant squid muscle peptides on free radical-mediated oxidative systems. *J. Nutr. Biochem.* **2005**, *16*, 562–569. [CrossRef] [PubMed]

54. Wang, B.; Li, L.; Chi, C.F.; Ma, J.H.; Luo, H.Y.; Xu, J.F. Purification and characterization of a novel antioxidant peptide derived from blue mussel (*Mytilus edulis*) protein hydrolysate. *Food Chem.* **2013**, *138*, 1713–1719. [CrossRef] [PubMed]

55. Chi, C.F.; Hu, F.Y.; Wang, B.; Li, T.; Ding, G.F. Antioxidant and anticancer peptides from the protein hydrolysate of blood clam (*Tegillarca granosa*) muscle. *J. Funct. Foods* **2015**, *15*, 301–313. [CrossRef]

56. Yang, X.R.; Zhang, L.; Ding, D.G.; Chi, C.F.; Wang, B.; Huo, J.C. Preparation, identification, and activity evaluation of eight antioxidant peptides from protein hydrolysate of hairtail (*Trichiurus japonicae*) muscle. *Mar. Drugs* **2019**, *17*, 23. [CrossRef] [PubMed]

57. Thiansilakul, Y.; Benjakul, S.; Shahidi, F. Antioxidative activity of protein hydrolysate from round scad muscle using alcalase and flavourzyme. *J. Food Biochem.* **2007**, *31*, 266–287. [CrossRef]

58. Chai, T.T.; Law, Y.C.; Wong, F.C.; Kim, S.K. Enzyme-assisted discovery of antioxidant peptides from edible marine invertebrates: A Review. *Mar. Drugs* **2017**, *15*, 42. [CrossRef] [PubMed]

59. Ngo, D.H.; Kim, S.K. Marine bioactive peptides as potential antioxidants. *Curr. Protein Pept. Sci.* **2013**, *14*, 189–198. [CrossRef] [PubMed]

60. Sila, A.; Bougatef, A. Antioxidant peptides from marine by-products: Isolation, identification and application in food systems. A review. *J. Funct. Foods.* **2016**, *21*, 10–26. [CrossRef]

61. Addad, S.; Exposito, J.Y.; Faye, C.; Ricard-Blum, S.; Lethias, C. Isolation, characterization and biological evaluation of jellyfish collagen for use in biomedical applications. *Mar. Drugs* **2011**, *9*, 967–983. [CrossRef] [PubMed]

62. Ramshaw, J.A.M.; Peng, Y.Y.; Glattauer, V.; Werkmeister, J.A. Collagens as biomaterials. *J. Mater. Sci. Mater. Med.* **2009**, *20*, S3–S8. [CrossRef] [PubMed]

63. Exposito, J.Y.; Valcourt, U.; Cluzel, C.; Lethias, C. The fibrillar collagen family. *Int. J. Mol. Sci.* **2010**, *11*, 407–426. [CrossRef] [PubMed]

64. Jankangram, W.; Chooluck, S.; Pomthong, B. Comparison of the properties of collagen extracted from dried jellyfish and dried squid. *Afr. J. Biotechnol.* **2016**, *15*, 642–648.

65. Barzideh, Z.; Latiff, A.A.; Gan, C.Y.; Abedin, M.; Alias, A.K. ACE inhibitory and antioxidant activities of collagen hydrolysates from the ribbon jellyfish (*Chrysaora* sp.). *Food Technol. Biotechnol.* **2014**, *52*, 495–504. [CrossRef] [PubMed]

66. Sugahara, T.; Ueno, M.; Goto, Y.; Shiraishi, R.; Doi, M.; Akiyama, K. Immunostimulation effect of jellyfish collagen. *Biosci. Biotechnol. Biochem.* **2006**, *70*, 2131–2137. [CrossRef] [PubMed]

67. Macrì, S. *Nuove Osservazioni Intorno la Storia Naturale del Polmone Marino Degli Antichi*; Biblioteca Regia Monacensis: Munich, Germany, 1778. (In Italian)

68. Li, J.; Li, Q.; Li, J.; Zhou, B. Peptides derived from *Rhopilema esculentum* hydrolysate exhibit angiotensin converting enzyme (ACE) inhibitory and antioxidant abilities. *Molecules* **2014**, *19*, 13587–13602. [CrossRef] [PubMed]

69. Ducklow, H.W.; Mitchell, R. Composition of mucus released by coral reef coelenterates. *Limnol. Oceanogr.* **1979**, *24*, 706–714. [CrossRef]

70. Masuda, A.; Baba, T.; Dohmae, N.; Yamamura, M.; Wada, H.; Ushida, K. Mucin (qniumucin), a glycoprotein from jellyfish, and determination of its main chain structure. *J. Nat. Prod.* **2007**, *70*, 1089–1092. [CrossRef] [PubMed]

71. Liu, W.; Mo, F.; Jiang, G.; Liang, H.; Ma, C.; Li, T.; Zhang, L.; Xiong, L.; Mariottini, G.L.; Zhang, J.; et al. Stress-induced mucus secretion and its composition by a combination of proteomics and metabolomics of the jellyfish *Aurelia coerulea*. *Mar. Drugs* **2018**, *16*, 341. [CrossRef] [PubMed]

72. Ladner, C.L.; Yang, J.; Turner, R.J.; Edwards, R.A. Visible fluorescent detection of proteins in polyacrylamide gels without staining. *Anal. Biochem.* **2004**, *326*, 13–20. [CrossRef] [PubMed]

73. Posch, A.; Kohn, J.; Oh, K.; Hammond, M.; Liu, N. V3 stain-free workflow for a practical, convenient, and reliable total protein loading control in Western Blotting. *J. Vis. Exp.* **2013**, *30*, 50948. [CrossRef] [PubMed]

74. Cheng, X.; Shao, Z.; Li, C.; Yu, L.; Ali Raja, M.; Liu, C. Isolation, characterization and evaluation of collagen from jellyfish *Rhopilema esculentum* Kishinouye for use in hemostatic applications. *PLoS ONE* **2017**, *12*, e0169731. [CrossRef] [PubMed]

75. Grienke, U.; Silke, J.; Tasdemir, D. Bioactive compounds from marine mussels and their effects on human health. *Food Chem.* **2014**, *142*, 48–60. [CrossRef] [PubMed]

76. Venkatesan, J.; Anil, S.; Kim, S.-K.; Shim, M.S. Marine fish proteins and peptides for cosmeceuticals: A Review. *Mar. Drugs* **2017**, *15*, 143. [CrossRef] [PubMed]

77. Liu, R.; Xing, L.; Fu, Q.; Zhou, G.H.; Zhang, W.G. A review of antioxidant peptides derived from meat muscle and by-products. *Antioxidants* **2016**, *5*, 32. [CrossRef] [PubMed]

78. Miliauskasa, G.; Venskutonisa, P.R.; Beek, T.A. Screening of radical scavenging activity of some medicinal and aromatic plant extracts. *Food Chem.* **2004**, *85*, 231–237. [CrossRef]

79. Lee, K.J.; Oh, Y.C.; Cho, W.K.; Ma, J.Y. Antioxidant and anti-inflammatory activity determination of one hundred kinds of pure chemical compounds using offline and online screening HPLC assay. *Evid.-Based Complement. Altern. Med.* **2015**, *2015*, 165457. [CrossRef] [PubMed]

80. Li, R.; Yang, Z.S.; Sun, Y.; Li, L.; Wang, J.B.; Ding, G. Purification and antioxidant property of antioxidative oligopeptide from short-necked clam (*Ruditapes philippinarum*) hydrolysate *in vitro*. *J. Aquat. Food Prod. Technol.* **2015**, *24*, 556–565. [CrossRef]

81. Zhang, M.; Liu, W.; Li, G. Isolation and characterization of collagens from the skin of largefin longbarbel catfish (*Mystus macropterus*). *Food Chem.* **2009**, *115*, 826–831. [CrossRef]

82. Blanco, M.; Vázquez, J.A.; Pérez-Martín, R.I.; Sotelo, C.G. Hydrolysates of fish skin collagen: An pportunity for valorizing fish industry byproducts. *Mar Drugs* **2017**, *15*, 131. [CrossRef] [PubMed]

83. Nagai, T.; Yamashita, E.; Taniguchi, K.; Kanamori, N.; Suzuki, N. Isolation and characterisation of collagen from the outer skin waste material of cuttlefish (*Sepia lycidas*). *Food Chem.* **2001**, *72*, 425–429. [CrossRef]

84. Skierka, E.; Sadowska, M. The influence of different acids and pepsin on the extractability of collagen from the skin of baltic cod (*Gadus morhua*). *Food Chem.* **2007**, *105*, 1302–1306. [CrossRef]

85. Silva, T.H.; Moreira-Silva, J.; Marques, A.L.P.; Domingues, A.; Bayon, Y.; Reis, R.L. Marine origin collagens and its potential applications. *Mar. Drugs* **2014**, *12*, 5881–5901. [CrossRef] [PubMed]

86. Qian, J.; Ito, S.; Satoh, J.; Geng, H.; Tanaka, K.; Hattori, S.; Kojima, K.; Takita, T.; Yasukawa, K. The cleavage site preference of the porcine pepsin on the N-terminal α1 chain of bovine type I collagen: A focal analysis with mass spectrometry. *Biosci. Biotechnol. Biochem.* **2017**, *81*, 514–522. [CrossRef] [PubMed]

87. Jongjareonrak, A.; Benjakul, S.; Visessanguan, W.; Tanaka, M. Isolation and characterization of collagen from bigeye snapper (*Priacanthus macracanthus*) skin. *J. Sci. Food Agric.* **2005**, *85*, 1203–1210. [CrossRef]

88. Siow, H.L.; Gan, C.Y. Extraction of antioxidative and antihypertensive bioactive peptides from *Parkia speciosa* seeds. *Food Chem.* **2013**, *141*, 3435–3442. [CrossRef] [PubMed]

89. Brusca, R.C.; Brusca, G.J. *Invertebrates*, 2nd ed.; Sinauer Associates, Inc.: Sunderland, MA, USA, 2003; p. 615.

90. Cariello, L.; Romano, G.; Spagnuolo, A.; Zanetti, L. Isolation and partial characterization of rhizolysin, a high molecular weight protein with hemolytic activity, from the jellyfish *Rhizostoma pulmo*. *Toxicon* **1988**, *26*, 1057–1065. [CrossRef]

91. Rastogi, A.; Sarkar, A.; Chakrabarty, D. Partial purification and identification of a metalloproteinase with anticoagulant activity from *Rhizostoma pulmo* (Barrel Jellyfish). *Toxicon* **2017**, *132*, 29–39. [CrossRef] [PubMed]

92. Allavena, A.; Mariottini, G.L.; Carli, A.M.; Contini, S.; Martelli, A. *In vitro* evaluation of the cytotoxic, hemolytic and clastogenic activities of *Rhizostoma pulmo* toxin(s). *Toxicon* **1998**, *36*, 933–936. [CrossRef]

93. Rastogi, A.; Biswas, S.; Sarkar, A.; Chakrabarty, D. Anticoagulant activity of Moon jellyfish (*Aurelia aurita*) tentacle extract. *Toxicon* **2012**, *60*, 719–723. [CrossRef] [PubMed]

94. Chung, J.J.; Ratnapala, L.A.; Cooke, I.M.; Yanagihara, A.A. Partial purification and characterization of a hemolysin (CAH1) from Hawaiian box jellyfish (*Carybdea alata*) venom. *Toxicon* **2001**, *39*, 981–990. [CrossRef]

95. Nagai, H.; Takuwa, K.; Nakao, M.; Sakamoto, B.; Crow, G.L.; Nakajima, T. Isolation and characterization of a novel protein toxin from the Hawaiian box jellyfish (sea wasp) *Carybdea alata*. *Biochem. Biophys. Res. Commun.* **2000**, *275*, 589–594. [CrossRef] [PubMed]

96. Radwan, F.F.Y.; Gershwin, L.A.; Burnett, J.W. Toxinological studies on the nematocyst venom of *Chrysaora achlyos*. *Toxicon* **2000**, *38*, 1581–1591. [CrossRef]

97. Rottini, G.; Gusmani, L.; Parovel, E.; Avian, M.; Patriarca, P. Purification and properties of a cytolytic toxin in venom of the jellyfish *Carybdea marsupialis*. *Toxicon* **1995**, *33*, 315–326. [CrossRef]

98. Gusmani, L.; Avian, M.; Galil, B.; Patriarca, P.; Rottini, G. Biologically active polypeptides in the venom of the jellyfish *Rhopilema nomadica*. *Toxicon* **1997**, *35*, 637–648. [CrossRef]

99. Bae, S.K.; Lee, H.; Heo, Y.; Pyo, M.J.; Choudhary, I.; Han, C.H.; Yoon, W.D.; Kang, C.; Kim, E. *In vitro* characterization of jellyfish venom fibrin(ogen)olytic enzymes from *Nemopilema nomurai*. *J. Venom. Anim. Toxins Incl. Trop. Dis.* **2017**, *23*, 35. [CrossRef] [PubMed]

100. Li, G.Y.; Fukunaga, S.; Takenouchi, K.; Nakamura, F. Comparative study of the physiological properties of collagen, gelatin and collagen hydrolysate as cosmetic materials. *Int. J. Cosmet. Sci.* **2005**, *27*, 101–106. [CrossRef] [PubMed]

101. Park, S.Y.; Lim, H.K.; Lee, S.; Hwang, H.C.; Cho, S.K.; Cho, M. Pepsin-solubilised collagen (PSC) from Red Sea cucumber (*Stichopus japonicus*) regulates cell cycle and the fibronectin synthesis in HaCaT cell migration. *Food Chem.* **2012**, *132*, 487–492. [CrossRef] [PubMed]

102. Chen, R.H.; Hsu, C.N.; Chung, M.Y.; Tsai, W.L.; Liu, C.H. Effect of different concentrations of collagen, ceramides, n-acetyl glucosamine, or their mixture on enhancing the proliferation of keratinocytes, fibroblasts and the secretion of collagen and/or the expression of mRNA of type I collagen. *J. Food Drug Anal.* **2008**, *16*, 66–74.

103. Bradford, M.M. A rapid and sensitive method for the quantitation of microgram quantities of protein utilizing the principle of protein-dye binding. *Anal. Biochem.* **1976**, *72*, 248–254. [CrossRef]

104. Rice-Evans, C.A.; Miller, N.J.; Bolwell, P.G.; Bramley, P.M.; Pridham, J.B. The relative antioxidant activities of plant-derived polyphenolic flavonoids. *Free Radic. Res.* **1995**, *22*, 375–383. [CrossRef] [PubMed]

105. Re, R.; Pellegrini, N.; Proteggente, A.; Pannala, A.; Yang, M.; Rice-Evans, C. Antioxidant activity applying an improved ABTS radical cation decolorization assay. *Free Radic. Biol. Med.* **1999**, *26*, 1231–1237. [CrossRef]

106. Longo, C.; Leo, L.; Leone, A. Carotenoids, fatty acid composition and heat stability of supercritical carbon dioxide-extracted-oleoresins. *Int. J. Mol. Sci.* **2012**, *13*, 4233–4254. [CrossRef] [PubMed]

Article

The Holo-Transcriptome of the Zoantharian *Protopalythoa variabilis* (Cnidaria: Anthozoa): A Plentiful Source of Enzymes for Potential Application in Green Chemistry, Industrial and Pharmaceutical Biotechnology

Jean-Étienne R. L. Morlighem [1,2], Chen Huang [3], Qiwen Liao [3], Paula Braga Gomes [4], Carlos Daniel Pérez [5,*], Álvaro Rossan de Brandão Prieto-da-Silva [6], Simon Ming-Yuen Lee [3] and Gandhi Rádis-Baptista [1,2,*]

[1] Northeast Biotechnology Network (RENORBIO), Post-Graduation Program in Biotechnology, Federal University of Ceará, Fortaleza 60440-900, Brazil; jean.etienne.brasil@gmail.com

[2] Laboratory of Biochemistry and Biotechnology, Institute for Marine Sciences, Federal University of Ceará, Fortaleza 60165-081, Brazil

[3] State Key Laboratory of Quality Research in Chinese Medicine and Institute of Chinese Medical Sciences, University of Macau, Macau 519020, China; huangchen11@yahoo.com (C.H.); liaoqw2007@126.com (Q.L.); simonlee@umac.mo (S.M.-Y.L.)

[4] Department of Biology, Federal Rural University of Pernambuco, Recife 52171-900, Brazil; pgomes@db.ufrpe.br

[5] Academic Center in Vitória, Federal University of Pernambuco, Vitória de Santo Antão 50670-901, Pernambuco, Brazil

[6] Laboratory of Genetics, Butantan Institute, São Paulo 05503-900, Brazil; alvaro.prieto@butantan.gov.br

* Correspondence: cdperez@ufpe.br (C.D.P.); gandhi.radis@ufc.br (G.R.-B.); Tel.: +55-(81)-3114-4101 (C.D.P.); +55-(85)-3085-7007 (G.R.-B.)

Received: 1 May 2018; Accepted: 8 June 2018; Published: 13 June 2018

Abstract: Marine invertebrates, such as sponges, tunicates and cnidarians (zoantharians and scleractinian corals), form functional assemblages, known as holobionts, with numerous microbes. This type of species-specific symbiotic association can be a repository of myriad valuable low molecular weight organic compounds, bioactive peptides and enzymes. The zoantharian *Protopalythoa variabilis* (Cnidaria: Anthozoa) is one such example of a marine holobiont that inhabits the coastal reefs of the tropical Atlantic coast and is an interesting source of secondary metabolites and biologically active polypeptides. In the present study, we analyzed the entire holo-transcriptome of *P. variabilis*, looking for enzyme precursors expressed in the zoantharian-microbiota assemblage that are potentially useful as industrial biocatalysts and biopharmaceuticals. In addition to hundreds of predicted enzymes that fit into the classes of hydrolases, oxidoreductases and transferases that were found, novel enzyme precursors with multiple activities in single structures and enzymes with incomplete Enzyme Commission numbers were revealed. Our results indicated the predictive expression of thirteen multifunctional enzymes and 694 enzyme sequences with partially characterized activities, distributed in 23 sub-subclasses. These predicted enzyme structures and activities can prospectively be harnessed for applications in diverse areas of industrial and pharmaceutical biotechnology.

Keywords: Zoanthidea; holo-transcriptome; cnidarian transcriptome; marine enzyme; marine biocatalyst; marine biotechnology; pharmaceutical biotechnology

---

1. Introduction

The worldwide demand for useful enzymes is continuously increasing in various industries, including the pharmaceutical sector, green chemistry, fine chemicals and basic and applied biomedical research. The market for enzymes that are more catalytically efficient than currently used enzymes, more environmentally friendly, and have potential use as drugs in the pharmaceutical industry, is foreseen to reach over USD 17 billion, by 2024 [1]. Additionally, the global market for enzymes that have applications in molecular biology and analytical kits is estimated to reach approximately USD 12.5 billion by 2021 [2]. In green chemistry, biocatalysts are used in environmentally benign chemical synthesis, in particular, halogenation and transamination, and the future of enzyme-catalyzed reactions is foreseen to rely on the enantio (selective) synthesis and kinetic resolution of chemicals, preferentially from renewable sources [3]. The development of eco-friendly and greener biocatalyzed processes aims to achieve better environmental factor values, i.e., mass of waste per unit of product [4]. Therefore, enzymes with improved properties, including physical and chemical stability, high conversion rates, change of substrate specificity, stereoselectivity, independence of cofactors for activity, ability to produce new chemicals, possibility to assemble multi-enzyme complex, among others, are desirable. Enzymes with these attributes and properties can be prospected by screening microbiomes [5,6] and/or by in vitro mutagenesis followed by direct evolution [7,8].

With the advent of "omics" sciences, the search for novel enzymes has progressed primarily by means of high throughput metagenomics assays of uncultured environmental samples [9]. In addition, the untapped biodiversity of distinct marine environments has become a hot target source for a wide variety of bioactive compounds, particularly, enzymes, that are halotolerant, thermostable and adapted to a range of pressures and substrate selectivity [10]. Some recent examples of marine-derived enzymes include the new flavin-dependent halogenase from a marine sponge metagenome [11] and several new α-amylases isolated from a sea anemone microbial community [12]. In two recent articles, the potential of marine biomes, particularly microbiomes, was emphasized, especially with regard to marine enzymes [13,14]. The successful bioprospection of natural compounds from marine species has been reported primarily from invertebrates of the phyla Porifera and Cnidaria [15]. Most of these compounds, which were secondary metabolites that were made into chemotherapeutics or drug-leads, were originally isolated from symbiotic microorganisms, rather than their host, in the holobiont assemblage [16]. Sponges, cnidarians, tunicates and other marine invertebrates can harbor a great diversity of microbial symbionts [17–19]. Despite the microbial origin of many compounds derived from the holobiont assemblage, the own coral tissues comprise unique resources of diverse chemicals with distinct pharmacological activities, such as anti-inflammatory and anti-proliferative activities [20–22]. Therefore, prospecting biopharmaceuticals from unusual marine species, as sources of unique enzymes, focusing particularly on improved and novel biocatalysts, is also warranted.

Our local, potential of discovery, biodiverse sites, i.e., the Brazilian large marine ecosystems—namely, the Brazilian Shelves, which are rich in coral reefs and are marine biodiversity hotspots, have been largely under-studied, particularly concerning to the discovery of novel enzymes and even natural products [15,23]. Recently, we conducted whole RNA sequencing of the anthozoan *Protopalythoa variabilis* (Cnidaria: Anthozoa) and described a repertoire of bioactive peptides with cnidarian toxin features in its transcriptome [24]. *P. variabilis* is a zoantharian species that preferentially grows in the shallow, warm water of the Atlantic coast and is present in abundance along the coastal reefs of the Brazilian Northeast [25]. Colonies of *P. variabilis*, like other anthozoans, harbor symbiotic zooxanthellae and a consortium of other microorganisms [26–28]. In the present work, we investigated the expressed enzymatic content of the *P. variabilis* holo-transcriptome. This species of zoantharian belongs to the phylum Cnidaria, one of the first groups that diverged from the Bilateria and is positioned at the base of Metazoa. The holo-transcriptomic analysis of *P. variabilis* extends the prospection of marine organisms for biotechnological studies and biopharmaceutical applications, which are less numerous than those of terrestrial and microbial origin [29]. In contrast to other analytical strategies reported in most recent articles, we focused our investigation on the entire transcriptome of *P. variabilis*, from a holobiont perspective, searching for enzyme precursors expressed in the zoantharian-microbiota assemblage that resulted in the prediction of numerous enzyme sequences relevant to biotechnology and green chemistry. These putative enzymes include oxidoreductases, transferases, hydrolases, lyases, isomerases and ligases, which have potential applications in several industrial fields, such as the production of pharmaceutics and fine chemicals, bioconversion and biopolymers, and green chemistry, to mention a few. Moreover, as exemplified in the present work, a single zoantharian species can be viewed as a species-specific repository of a unique collection of marine enzymes.

2. Results and Discussion

2.1. Biodiversity in the P. variabilis Holobiont

Coral reefs are niches for different life forms, ranging from small fishes and crustaceans to associated and endosymbiotic microbial communities. Unraveling the biodiversity of a zoantharian holobiont is interesting not only from the ecological point of view, but also essential for the comprehension of the interconnected metabolic pathways, which ultimately depend on the symbiotic interactions and their enzymatic activities. To verify the overall diversity of life forms in the *P. variabilis* assemblage, we used three gene sequences commonly accepted for barcoding in species identification: the mitochondrial 16S rRNA gene for the taxonomic identification of bacteria and archaea, the mitochondrial cytochrome C oxidase subunit I (COI) gene for animals, algae, and dinoflagellates, and RuBisCo (rbcL) gene for plant and microalgae. The results of the species identification in the holobiont are shown in Figure 1A and the species are listed in Supplementary Table S1. In addition to the presence of the most common dinoflagellate algae of the genus *Symbiodinium*, involved in mutualistic symbiosis with cnidarians, the largest majority of the holobiont community identified is composed of uncultured species of cyanobateria and proteobacteria. Interestingly, four COI sequences found in the *P. variabilis* holobiont transcriptome had their best hits against terrestrial flying insects (Endopterygota), seemingly suggesting the interaction of *P. variabilis* with a group of organisms lacking known genetic information that belong to the phylum Arthropoda.

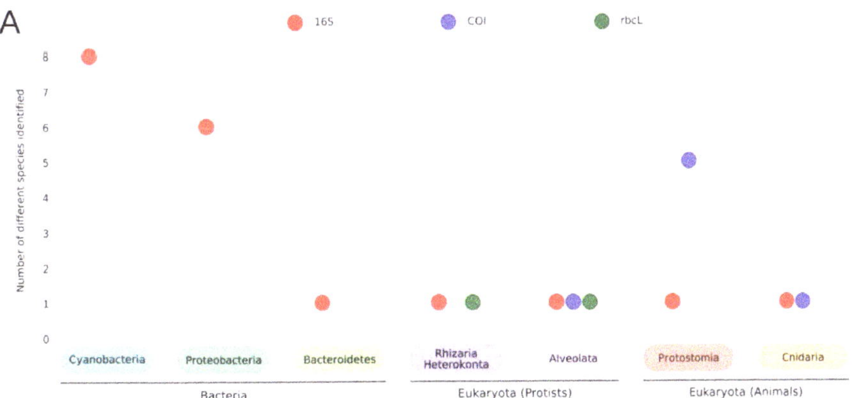

Figure 1. *Cont.*

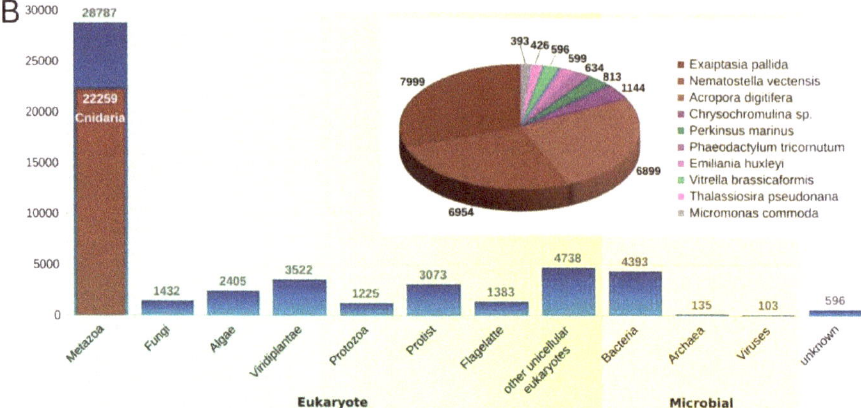

Figure 1. Biodiversity composition and taxonomic classification of unigenes from the *P. variabilis* holo-transcriptome. (**A**) Number of different species identified using the 16S rRNA, COI, and rbcL barcode sequences. (**B**) Taxonomic classification of unigenes from the *P. variabilis* holo-transcriptome after BLASTx analysis. Inserted Box: the distribution of top-hit species in the *P. variabilis* holobiont with the cnidarian species in dark red, the haptophytes and Stramenopiles in purple, the alveolates in green, and the green algae in gray.

2.2. Annotation and Functional Prediction

Approximately 40% of the unigenes (51,792 sequences) identified in this study matched to entries in at least one of the four public protein databases, corresponding to "predicted or "annotated" hits (89% of all contigs) and "hypothetical" or "putative hits" (11%).

Supplementary Figures S1 and S2 present the complete statistical analysis of the sequence annotations. Compared to other *de novo* transcriptome assembly studies of cnidarian species, the initial level of BLAST annotations observed here was in the expected range [30–33]. Based on the selected BLASTx matches for the annotations, the taxonomic distribution of species from which the predicted protein sequences originated is depicted in Figure 1B. As can be observed, the three most representative species belong to the phylum Cnidaria, class Anthozoa, subclass Hexacorallia (namely, *Exaiptasia pallida*, *Nematostella vectensis*, and *Acropora digitifera*), which together accounted for 21,852 matched contigs in the annotation process for the *P. variabilis* holobiont transcripts (~42% of all annotated transcripts).

2.2.1. Functional Classification of *P. variabilis* Predicted Enzymes

After a GO-slim reduction, a total of 29,866 unigenes (~58% of all annotated transcripts) were classified within 35 different ontological categories (Supplementary Figure S3A). The functional annotation of the *P. variabilis* holo-transcriptome, which returned GO terms for almost 30,000 unigenes, was substantially higher than that observed in the majority of previous studies but collectively consistent regarding the gene distribution within the categories of cellular or metabolic processes and functions.

2.2.2. Assignment of Metabolic Pathways of the *P. variabilis* Predicted Enzymes

The KEGG pathway mapping analysis of the annotated unigenes retrieved 135 pathway maps (Supplementary Figure S3B), which were similar to the GO annotations in the category of "Biological Process". However, the coverage differed greatly (Supplementary Table S2). These maps are available in Supplementary File S1.

From these analyses, a relatively high number of contigs were determined to encode enzymes involved in the metabolism of terpenoids and polyketides (Supplementary Table S2). This finding is interesting from the viewpoint of drug discovery, since numerous important bioactive secondary metabolites used in therapy, such as taxol, steroids and macrolide antibiotics, prostaglandins and tetracyclines, are synthesized in these biosynthetic routes. Indeed, the C-15 epimer of prostaglandin A2 (PGA2) and related structures have been identified and isolated from the Caribbean coral *Plexaura homomalla*, which produces and accumulates such compound [34].

2.3. Biotechnologically Relevant Enzymes from the P. variabilis Holo-Transcriptome

The identification of industrially relevant enzyme biocatalysts from environmental samples has been concretized by means of metagenomic analysis [9,35,36], a combination of metagenomics and metaproteomics technologies [37], and the combined application of transcriptomics and proteomics [38]. We based our search on analyzing the *P. variabilis* holo-transcriptome, providing an insight into the unexplored enzymes and related biosynthesis pathways expressed in this zoantharian-microbiota assemblage. Among all predicted proteins, 771 predicted enzymes classes with complete Enzyme Commission (EC) numbers were identified from the *P. variabilis* holo-transcriptome, representing a cumulative number of 6571 unigenes, with 606 of the predicted enzymes classes (5385 unigenes) mapped to KEGG pathways (Supplementary Table S3). Breaking down the enzymes by classes, 22% of them were oxidoreductases—EC 1 (170 enzymes for 806 unigenes), 35% were transferases—EC 2 (266 enzymes for 1208 unigenes), 21% were hydrolases—EC 3 (162 enzymes for 3792 unigenes), 10% were lyases—EC 4 (78 enzymes for 280 unigenes), 5% were isomerases—EC 5 (38 enzymes for 247 unigenes)), and 7% were ligases—EC 6 (57 enzymes for 238 unigenes).

Based on these results, we focused our study on enzymes with recognized relevance in chemical and pharmaceutical industries, as well as for green chemistry. For instance, transaminases (or aminotransferases) are stereoselective, cofactor-recycling enzymes that catalyze the transfer or exchange of an amino group from an amino-containing substrate to an acceptor molecule, resulting in the synthesis of chiral amino acids and amines. According to Supplementary Table S3, seven contigs encoding transaminases retrieved from the *P. variabilis* holo-transcriptome mapped mainly to antibiotic and amino acid biosynthesis pathways. The search for and development of optimized and high-performance transaminases that exhibit catalytic polyvalence (activity with a wide range of substrates), reactional stability and the possibility to form multi-enzyme complexes, are in demand for applications in green organic chemistry and the production of fine chemicals, food additives and pharmaceuticals [39]. In the following sections, additional selected examples of predicted enzymatic activities from the *P. variabilis* holobiont are presented, classified according to the major commercial and industrial application and discussed.

2.3.1. Relevance in the Treatment of Rare Diseases and Other Biomedical and Clinical Applications

Several classes of hydrolases, such as galactosidases and proteases, are used in therapy and the preparation of biopharmaceuticals. For instance, fibrinolytic enzymes are used clinically as thrombolytic agents to treat myocardial infarction, asparaginase and arginine deaminase are used to treat leukemia and solid tumors, and α- and β-galactosidases are used to treat lysosomal storage disease [40]. Moreover, naïve or structure-guided directly evolved glycohydrolases have been tested in trials to remove sugar residues from the surface of erythrocytes to obtain universal blood [41]. Supplementary Table S4 presents, a list of predicted *P. variabilis* enzymes for which counterparts are used in the treatment of rare metabolic diseases, cancer, and for experimental blood production and organ transplantation. Several putative enzymes in the *P. variabilis* holo-transcriptome are homologous to therapeutic enzymes currently used to treat lysosomal storage disorders (LSDs). LSDs encompass a large number of metabolic diseases, primarily characterized by a lack of hydrolases and defects in the degradation of glycoproteins, glycolipid, glycosaminoglycan and glycogen [42]. Importantly, replacement therapy with human recombinant enzymes has been proven to be effective treatment in clinical and pre-clinical settings [40,42,43].

In the fields of green chemistry, industrial and pharmaceutical biotechnology, glucosidases have been used to prepare glucopolymers of polyvinyl saccharides, such as poly(glucosyl)-acrylates, which function as drug carrier systems and non-ionic polymeric surfactants. For example, Kloosterman and colleagues (2014) [44] utilized β-glucosidase to synthesize the monomers 4-(β-glucosyloxy)-butyl acrylate, 2-(β-glucosyloxy)-ethyl acrylate and methacrylate, as a means to avoid multiple reaction steps, thereby precluding the formation of undesirable isomers. β-glucosidases were also predicted from the *P. variabilis* holo-transcriptome (Supplementary Table S4).

Two important therapeutic enzymes used in cancer therapy are L-asparaginase and arginine deaminase, for which putative homologues were also identified in this study (Supplementary Table S4). L-asparaginase catalyzes the conversion of L-asparagine into L-aspartate, preventing cancer cell survival in patients with lymphoblastic leukemia, while arginine deaminase catalytically removes, by deamination, excess adenosine molecules in the circulation of patients with cancer, thereby reducing the toxicity to the immune system caused by elevated adenosine levels [45]. In the preparation of universal O-type blood, the selective removal of α-GalNAc and α-Gal residues of the A and B oligosaccharide antigens on the surface of red blood cells can be achieved by α-galactosidases and α-N-acetylgalactosaminidases, respectively [46]. Eleven sequences from the *P. variabilis* holo-transcriptome mapped to a putative β-N-acetylhexosaminidase (EC: 3.2.1.52). One of the eleven sequences is closely related to the clade Cnidaria, whereas another belongs to the protist-algae group, and three others are in the archaea-bacteria clade. The last one, Unigene28224, fits in between the protist-algae and archaea-bacteria groups (Supplementary Figure S4A). Despite the divergence of the sequences, the proton donor glutamic acid residues in the catalytic site were found to be conserved across all *P. variabilis* sequences (Supplementary Figure S4B).

Many enzymes and therapeutic proteins used in clinical and experimental clinical trials exist in PEGylated forms, in which serum stability and half-life are usually increased, while the immunological response is decreased. The PEGylation of proteins can potentially be achieved by biocatalysis using transglutaminases, which carry out an acyl transfer reaction [47]. As noted on Supplementary Table S4, a transglutaminase (EC 2.3.2.13) was found that may be used to catalyze the covalent transfer of the PEG moiety to therapeutic enzymes and proteins. Transglutaminases are also useful in other biotech industries, such as food processing, biopolymer production and leather and wool treatment [48].

2.3.2. Relevance in Colorant, Aromas, Flavor, Fragrance, Cosmetic, and Hygienic Industries

Another group of polymer-degrading hydrolases, for which homologues are used in the fine chemicals industry, comprises glycosidases, alpha-amylase, polygalacturonase, beta-glucosidase, 1,4-alpha-glucosidase and cellulase (Supplementary Table S5). Moreover, in the industry of fine chemicals, oxidoreductases are widely used biocatalysts, and oxygenases (mono- and dioxygenases and peroxidases) are important enzymes for the modification of terpenoids. In corals, a diverse array of diterpenoids has been reported [49], with activities including anti-inflammatory, antifouling and antifeedant, anti-infective (antimicrobial, antiviral, anti-parasite), anticancer and cytotoxic effects. Thus, different species of marine cnidarian holobionts appear to be interesting sources not only for terpenoids themselves but also for enzymes involved in their biosynthesis. The holo-transcriptome of *P. variabilis* has revealed some of these predicted enzyme sequences.

2.3.3. Relevance in Agrochemical, Food and Feed Industries

In the *P. variabilis* holo-transcriptome, many predicted enzymatic activities were also identified that may be relevant to the food industry (Supplementary Table S6). Hydrolases and oxidoreductases have emerged as preferred biocatalysts on an industrial scale for the generation of chirality and enantio (selective) kinetic resolutions of chemicals, especially from renewable sources [50].

From an ecological viewpoint, the expression of chitinases and enzymes related to chitin degradation, including chitodextrinase, in the *P. variabilis* holobiont is suggestive of a complimentary mechanism of self-defense in zoantharians, equipping the organisms to cope with disease-causing agents. Chitinases have known anti-fungal [51] and algicidal properties, contributing to nutrient cycling in marine environments [49]. In pharmaceutical biotechnology, chitinases are useful for preparing chitosan composites for numerous applications, ranging from biosensors, tissue engineering and drug delivery systems to nanoarrays and lab-on-chip devices [52]. Moreover, the coupled reactions of *N*-acetylhexosaminidase (a chitinase) and deacetylases can produce, a valuable nutraceutical supplement, D-Glucosamine, with high yields from polymeric chitin in a proof-of-concept environmentally friendly dual-catalysis process [53]. Interestingly, a predicted deacetylase (*N*-acetyl-D-glucosamine-6-phosphate deacetylase, EC 3.5.1.25) was found in the holo-transcriptome of *P. variabilis* (Supplementary Table S3).

Again, from an ecological viewpoint, the presence of transcriptional precursors encoding enzymes for the biosynthesis of herbicide intermediates, in the *P. variabilis* holo-transcriptome, can be seen as a possible way for this anthozoan species to restrain macroalgae overgrowth, as seen in corals [54]. Hence, considering the marine environment in which these enzymes presumably work, the possibility to obtain naturally evolved salt-tolerant biocatalysts for industrial bioprocesses is high.

2.3.4. Relevance in Bioconversion and Biopolymer Synthesis

In the *P. variabilis* holo-transcriptome, some, but not all, industrially useful putative glycosidases that could potentially be applied to the processing of lignocellulose were found (Supplementary Table S7). Moreover, enzymatic activities involved in the production of precursors of biopolymers, such as recyclable bioplastic, were also observed. Examples of such enzymes include, 3-oxoacyl-ACP reductase (E.C. 1.1.1.100), acetyl-CoA C-acyltransferase (2.3.1.16) and enoyl-CoA hydratase (4.2.1.17), which catalyzes the biosynthesis of polyhydroxyalkanate (PHA) precursors. The predicted enzymes epoxide hydrolase (3.3.2.9), nitrile hydratase (4.2.1.84) and γ-glutamyltransferase (2.3.2.13) are important for bioconversion in the renewable energy industry and for the synthesis of fine chemicals, in green chemistry and bioremediation, as will be discussed later. Finally, the peroxidases identified in this zoantharian holo-transcriptome could be further evaluated for applications in the conversion and biosynthesis of phenol- and vinyl-containing polymers, accordingly to the current use of enzymes of these catalytic classes.

2.3.5. Relevance in the Cleaning and Recovery Industries

The identification of enzymatic activities in a given microbiome is useful for estimating the capacity of the microbes to recover an environmental site. The disclosed enzymatic activities, serve as indicators of the bioremediation potential for a given set of contaminants, and they also indicate potential enzymatic catalysts that may be isolated for downstream processing. In this regard, among the enzymes listed in Supplementary Table S8, a very interesting enzyme is glutathione γ-glutamylcysteinyltransferase (E.C. 2.3.15.2), also known as phytochelatin synthase, which catalyzes the synthesis of phytochelatins. Phytochelatins are cysteine-rich peptides responsible for the chelation and sequestration of essential microelements (e.g., copper and zinc) and toxic heavy metals (e.g., cadmium, lead and mercury). Members of the phytochelatin family are biosynthesized from glutathione and are structurally characterized by *n* repetitions of the γ-GluCys dipeptide followed by a terminal Gly, i.e., (γ-GluCys)*n*-Gly [55]. The genes encoding enzymes for phytochelatin synthesis have a recognized wide phylogenetic distribution, indicating the importance of coping with metal and metalloid (arsenic, selenium and silver) detoxification across species [55,56]. Thus, the expression of phytochelatin synthase in the *P. variabilis* holo-transcriptome could be related to detoxification of heavy metals and metal homeostasis in zoantharians. From the biotechnological point of view, immobilized phytochelatin synthase has been utilized to prepare highly stable cadmium-sulfur (CdS) nanocrystals of tunable sizes with optical and electrical properties [57]. Prepared in this way, nanocrystals are useful in the fabrication of semiconductor quantum dots (QDs) and for application as fluorescent tags in biological systems for molecular imaging. Another potential technological use for phytochelatin synthases is related to bioremediation; engineered bacteria have been designed that overexpress a mutated enzyme in combination with a cadmium protein transporter, resulting in Cd accumulation [58].

Last but not least, in this group of enzymes relevant to the recovery and processing industry, epoxide hydrolase and nitrile hydratase homologous were found. Epoxide hydrolases (EC 3.3.2.3) are cofactor-independent biocatalysts that convert epoxides to the corresponding diols, and epoxide hydrolases of microbial origin are useful for producing enantioselective intermediates with vicinal diols in the synthesis of fine chemicals and pharmaceuticals, such as (*S*)-Ibuprofen, (*R*)-Mevanolactone and (*R*)-Eliprodil [59,60]. Nitrile hydratases are Fe- or Co-type metalloenzymes that convert nitriles (organo-cyanides) into industrially valuable amides, such as acrylamides, from acrylonitriles. Tons of acrylamide are used as coagulators, soil conditioners and additives in the paper industry, as well as adhesives, paint components and agents for petroleum recovery [61,62]. Additionally, wild-type and engineered nitrile hydratases that act on myriad cyanide-containing substrates can be applied in bioremediation, wastewater treatment and even for the development of cyanide biosensors [63]. These examples provide a glimpse of the biotechnologically significant enzymes found in the holo-transcriptome of *P. variabilis* that may be used as biocatalysts with improved activities and selectivity over those currently in use. Moreover, one can speculate on the eco-physiological role of these enzymes in the cnidarian-holobiont assemblage.

2.3.6. Relevance in Molecular Biology and Analytical Applications

In Supplementary Table S9, homologs of enzymes useful for manipulating nucleic acids found in the *P. variabilis* holo-transcriptome are listed. Among these, two enzymes—deoxyribonuclease I and *N*-acetylglucosaminidase—have also been used to treat cystic fibrosis and lysosomal storage disease, respectively.

In recombinant DNA technology, deoxyribonuclease (DNase) I catalyzes the hydrolysis of phosphodiester bonds and cleaves double-stranded (ds) and single-stranded (ss) DNA in a sequence-nonspecific manner. DNase I is used, for instance, to make "nick translations" and in DNase I footprinting—a technique used to study the interaction of ligands (proteins and drugs) with DNA [64]. Other predicted endonucleases found in the *P. variabilis* holo-transcriptome include DNase II and IV; DNase II is a lysosomal "acidic" DNase that preferentially cleaves ssDNA, whereas DNase IV (flap endonuclease-1) is a multifunctional enzyme that cleaves 5′-ssDNA flaps of DNA or RNA. DNase IV has been investigated as a marker of disease risk, since it is involved in DNA metabolism, genomic stability and apoptosis [65].

Several types of predicted ribonucleases were also found, such as ribonucleases (RNases) H and T (Supplementary Table S9). RNase H is a monomeric enzyme that degrades RNA in RNA:DNA heteroduplexes and is useful for the synthesis of complementary DNA (cDNA); RNase III works in multicomponent assemblies to bind and cleave dsRNAs, especially in the processing of dsRNA maturation and the decay of coding and noncoding RNAs, such as miRNAs and siRNAs [66].

DNA-dependent DNA polymerases catalyze the biosynthesis of polydeoxyribonucleotides; diverse applications for DNA polymerases exist, including nucleotide sequencing, in vitro synthesis of the second cDNA strand, DNA amplification and the preparation of DNA hybridization probes [67]. RNA polymerases (RNA-dependent DNA polymerases) are enzymes that transcribe gene sequences into the corresponding RNAs. RNA polymerases are used to prepare hybridization probes and in vitro-transcribed RNA to direct a high-level of expression of cloned genes, as well as to produce capture reagents for RNA-binding proteins and as antisense probes [68]. In the *P. variabilis* zoantharian holo-transcriptome, predicted and structurally conserved RNA-dependent RNA polymerase (reverse transcriptase) sequences were retrieved. Because reverse transcriptase catalyzes the synthesis of DNA from an RNA template, the uses for this type of enzyme include the preparation of cDNA and the molecular design of inhibitors of retrovirus replication [69]. DNA and RNA ligases catalyze the formation of $3' \to 5'$ phosphodiester bonds in nucleic acid (DNA and RNA) molecules. These enzymes have a range of uses in recombinant DNA technology, from elongation or circularization of dsDNA, in the case of DNA ligases, to $3'$-labeling of RNA, DNA $5'$-tailing of DNA and production of elongated molecules, e.g., in cDNA cloning, in the case of RNA ligases [70]. Transcripts encoding both types of ligases were found to be expressed in the zoantharian holo-transcriptome in this study.

2.4. Prediction of Enzymes with Two or More Activities

In searching for different types of enhanced and unique marine biocatalysts, we wondered whether the *P. variabilis* holobiont transcriptome contains enzymes with multiple activities. A widely known example of a dual catalytic enzyme is RuBisCO (ribulose-1,5-bisphosphate carboxylase/oxygenase, EC 4.1.1.39), which catalyzes both CO_2 fixation in the dark phase of photosynthesis and carbon oxidation in the process of photorespiration [71]. Another interestingly example of a dual catalytic enzyme from plants is the enzyme hydroxycinnamoyl-Coenzyme A:quinate hydroxycinnamoyl transferase (HQT, EC 2.3.1.99), which catalyzes the formation of the strong phenolic antioxidants, chlorogenic and dicaffeoylquinic acids, which are useful as phytonutrients in foods and as pharmaceuticals. It has been demonstrated that in addition to the transesterification of caffeoyl-CoA with quinic acid to produce chlorogenic acid, HQT can form dicaffeoylquinic acid via its chlorogenate:chlorogenate transferase activity [72]. In biotechnology, multi-enzyme systems have been designed with different assembly strategies to mimic natural enzyme complexes and pathways, with the aim of improving catalytic efficiency [73]. We devised an annotation iterative process that resulted in the prediction of 13 putative enzymes with two to three enzymatic activities (Table 1). Six of these enzymes have activities that are partially identified and are related to more than one metabolic pathway (CL12403.contig2, CL2444.contig1, unigene12818, unigene14615, unigene32504, as well as unigene33780), and another four enzymes with dual activities were positioned in a single metabolic pathway. Enzymes that were predicted to possess dual catalytic activity are also shown in Supplementary Figure S5; these were found as a result of our stringent analysis. Some putative dual-activity enzymes might have been missed; however, this strategy proved to be effective, as it included an initial convenient and rapid data mining and screening approach of bi-functional biocatalysts from this zoantharian holo-transcriptome.

Table 1. List of predicted multi-functional enzymes from the *P. variabilis* holo-transcriptome with dual catalytic activities.

Unigene	ECs	Activities	Substrate	Product
CL12403.Contig1	1.5.1.20	methylenetetrahydrofolate reductase (NAD(P)H)	5-methyltetrahydrofolate	5,10-methylenetetrahydrofolate
	2.1.2.1	glycine hydroxymethyltransferase	5,10-methylenetetrahydrofolate	tetrahydrofolate
CL12403.Contig2	1.5.1.20	methylenetetrahydrofolate reductase (NAD(P)H)	5-methyltetrahydrofolate	5,10-methylenetetrahydrofolate
	2	transferase	?	?
CL2444.Contig1	4.2.1.11	phosphopyruvate hydratase	2-phospho-D-glycerate	phosphoenolpyruvate
	5.3.1.1	triose-phosphate isomerase	D-glyceraldehyde 3-phosphate	glycerone phosphate
Unigene12818	1.3.99.1	succinate dehydrogenase	succinate	fumarate
	1.8.4	oxidoreductase, acting on a sulfur group of donors	?	?
	5.3.4.1	protein disulfide isomerase	-S-S- bonds rearrangement	none
Unigene14615	2.4.2.30	NAD ADP-ribosyltransferase	NAD+	nicotinamide
	2.4.2.31	NAD(P)+protein-arginine ADP-ribosyltransferase	NAD+ & protein-L-arginine	nicotinamide & protein-omega-N-(ADP-D-ribosyl)-L-arginine
Unigene28009	2.5.1.21	farnesyl-diphosphate farnesyltransferase	(2E,6E)-farnesyl diphosphate	squalene
	1.14	oxidoreductase	?	?
Unigene32504	2.1.1	O-methyltransferase	?	?
	2.6.1.42	branched-chain-amino-acid transaminase	L-leucine	4-methyl-2-oxopentanoate
Unigene33780	2.7.7.9	UTP:glucose-1-phosphate uridylyltransferase	alpha-D-glucose 1-phosphate	UDP-glucose
	5.3.1.1	triose-phosphate isomerase	D-glyceraldehyde 3-phosphate	glycerone phosphate
Unigene34807	3.4.21	serine-type endopeptidase	?	?
	3.4.24	metalloendopeptidase	?	?
Unigene38918	2.7.2.3	phosphoglycerate kinase	3-phospho-D-glycerate	3-phospho-D-glyceroyl phosphate
	4.2.1.11	phosphopyruvate hydratase	2-phospho-D-glycerate	phosphoenolpyruvate
Unigene52468	2.1.1	methyltransferase	?	?
	2.5.1	transferase, transferring alkyl or aryl groups	?	?
Unigene9562	2.1.2.11	3-methyl-2-oxobutanoate hydroxymethyltransferase	5,10-methylenetetrahydrofolate & 3-methyl-2-oxobutanoate	tetrahydrofolate & 2-dehydropantoate
	6.3.2.1	pantoate-beta-alanine ligase	(R)-pantoate	(R)-pantothenate
Unigene9804	2.5.1.21	farnesyl-diphosphate farnesyltransferase	(2E,6E)-farnesyl diphosphate	squalene
	5.3.3.2	isopentenyl-diphosphate delta-isomerase	isopentenyl diphosphate	dimethylallyl diphosphate

2.5. Annotation of Novel Predicted Enzyme Sequences with Partial EC Number

In addition to the prediction of enzymes with activities that are fully characterized, groups of enzymes that are of general interest comprise expressed sequences that are completely new. These, as analyzed herein, could not be mapped to a specific and detailed catalytic reaction, i.e., they comprise enzymatic precursors with an EC number lacking the fourth categorization numbers. Indeed, several predicted enzymes were found with a sequence similarity that were reasonably close to be classified within a known sub-subclass (EC with at least three numbers), but, still, distinct enough to be completely mapped in given reactional group of characterized enzymes. Predicted precursors that fit in this category could hypothetically point to isozymes with already described reactions but working on different substrates, with distinct kinetic parameters, in distinct catalytic conditions or, even, comprising a totally novel catalytic reaction. Based on these facts, from the *P. variabilis* holo-transcriptome, additional 694 predicted enzyme sequences were found with incomplete EC numbers (unknown fourth serial digit), distributed into 23 sub-subclasses (Table 2), with a large majority related to the class of hydrolases. In this study, based on these preceding findings, we focused our further analysis on two sub-subclasses that *P. variabilis* sequences are grouped in distinct new clades, representing new structures, namely, cysteine dioxygenases (EC:1.13.11.20) and carboxypeptidases A, B, A2, and U (ECs:3.4.17.1, 3.4.17.2, 3.4.17.15, 3.4.17.20). I Initially, all predicted *P. variabilis* sequences mapped to a sub-subclass were evaluated by phylogenetic inference with corresponding counterparts representing each species for each enzyme in this sub-subclass. It was found that eleven *P. variabilis* sequences were related to the cysteine dioxygenases while forming an out-clade, and five sequences with the carboxypeptidases A, B, A2, and U forming a distinct clade all together. Afterwards, a second tree was inferred to confirm that, even if they are related, the *P. variabilis* sequences form a distinct clade, showing their uniqueness (Figures 2 and 3). Cysteine dioxygenase is a key enzyme in the synthesis of taurine, an important compound, product of cysteine metabolism, that is used in functional foods, as well as in pharmaceutical and cosmetic industries. Several patents are granted to the production of taurine by fermentation methods with transgenic microorganisms. Carboxypeptidases are generally applied either in research or in the pharmaceutical industries, but also found some application in food industries, as exemplified by the use of the carboxylase A in baking industry [74].

Table 2. List of predicted enzymes with incomplete Enzyme Commission numbers.

Class	Sub-Subclass EC Number	Sub-Subclass Principal Enzyme Type(s)	Sub-Subclass Known Enzyme Entries	*P. variabilis* Number Sequences
Oxidoreductases	1.13.11	dioxygenase	80	26
	1.14.11	dioxygenase, hydroxylase, demethylase	56	9
	1.14.12	dioxygenase	23	1
	1.14.13	monooxygenase, hydroxylase	235	1
	1.14.16	monooxygenase	7	3
	1.14.17	monooxygenase	3	13
	1.14.19	desaturase	51	5
Transferases	2.7.10	protein-tyrosine kinase	2	72
	2.7.11	protein-serine/threonine kinase	33	100
Hydrolases	3.1.13	exoribonuclease	5	3
	3.1.21	endodeoxyribonuclease	9	1
	3.4.11	aminopeptidase	23	25
	3.4.13	dipeptidase	12	7
	3.4.15	peptidyl-dipeptidase	4	1
	3.4.16	serine-type carboxypeptidase	4	10
	3.4.17	metallocarboxypeptidase	20	25
	3.4.19	omega peptidase	12	9
	3.4.21	serine endopeptidase	100	136
	3.4.22	cysteine endopeptidase	58	20
	3.4.23	aspartic endopeptidase	41	70
	3.4.24	metalloendopeptidase	83	120
	3.4.25	threonine endopeptidase	2	34
Lyases	4.1.99	carbon-carbon lyases	16	3

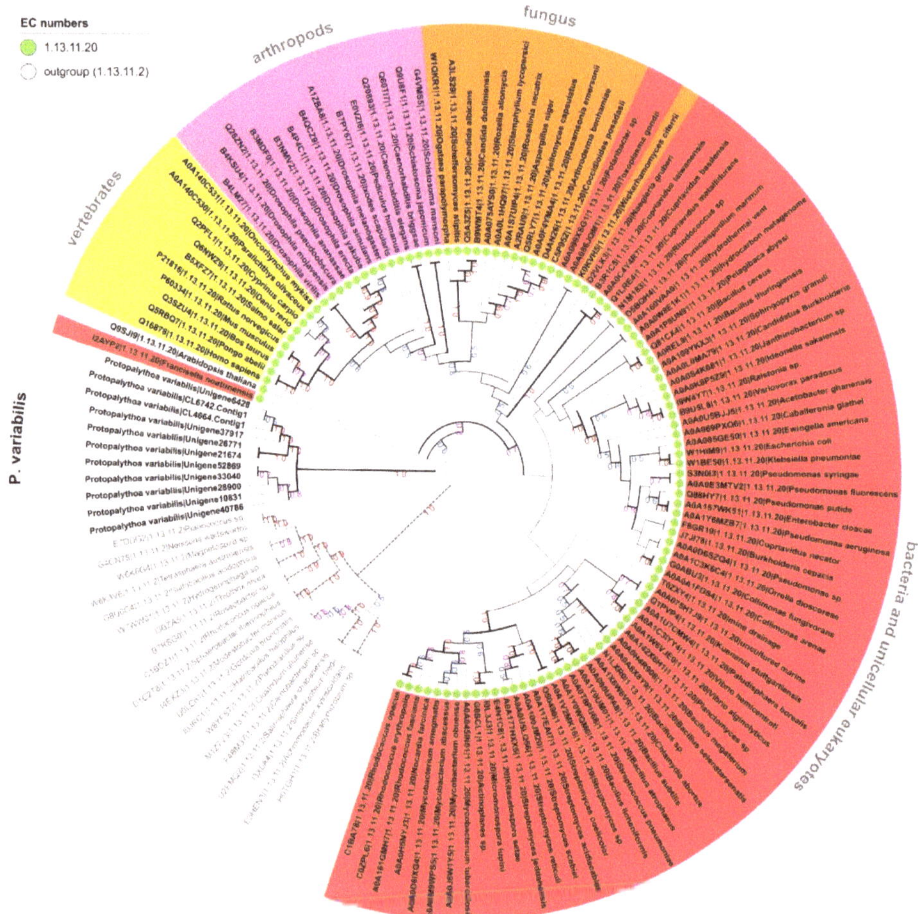

Figure 2. Cladogram depiction of the *P. variabilis* predicted sequences related to cysteine dioxygenase (EC:1.13.11.20) members. Tree based on the distance between the protein sequences of cysteine dioxygenase (EC:1.13.11.20) from 106 species and eleven *P. variabilis* predicted enzymes with an incompletely annotated EC:1.13.11 activity. Twenty catechol 2,3-dioxygenase (EC:1.13.11.2) sequences were used as outgroup. Only bootstrap values greater than 50% are shown at the branch points, in blue, purple, or red color for values comprised between 50–69%, 70–89%, and 90–100% respectively. Enzyme activities are indicated at the name base by circles colored as indicated in the legend.table.

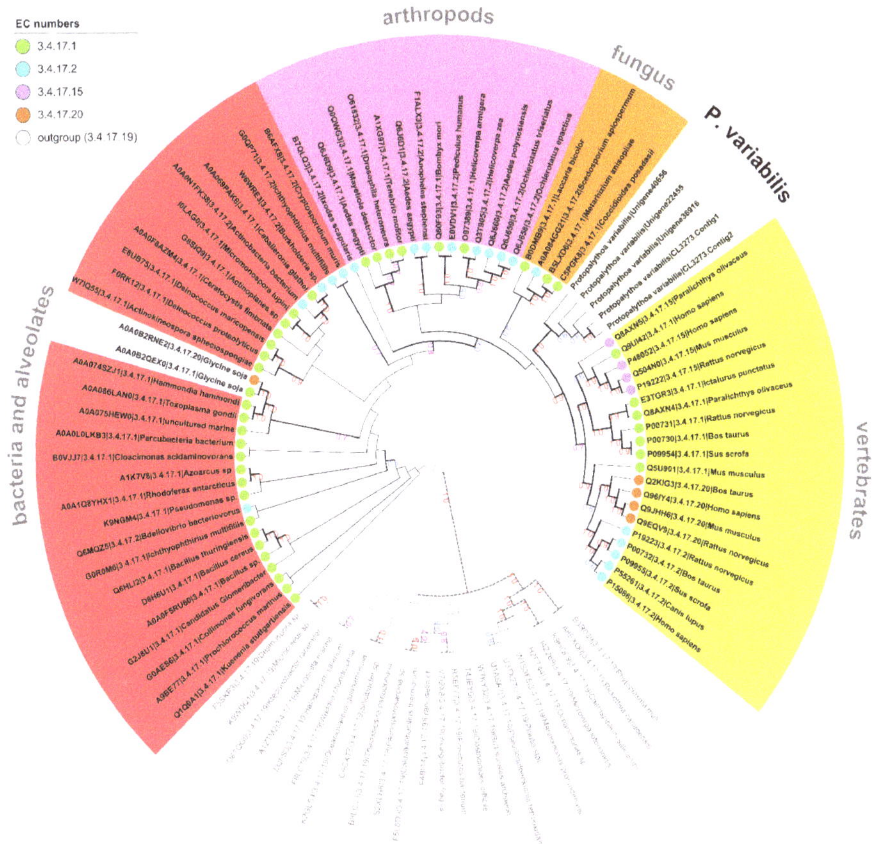

Figure 3. Cladogram depiction of the *P. variabilis* predicted sequences related to carboxypeptidase A, B, A2, and U (EC:3.4.17.1, 2, 15, and 20) members. Tree based on the distance between the protein sequences of carboxypeptidase A, B, A2, and U (EC:3.4.17.1, 2, 15, and 20) from 40, 19, 4, and 5 species respectively and five *P. variabilis* predicted enzymes with an incompletely annotated EC:3.4.17 activity. Sequences from 24 carboxypeptidase Taq (EC:3.4.17.19) sequences were used as outgroup. Tree legend same as in Figure 2.

3. Materials and Methods

3.1. Biological Sample

Specimens of *Protopalythoa variabilis* (Duerden, 1898) were collected in coastal reefs of Porto de Galinhas, Pernambuco, Brazil (8°30′20″ S, 35°00′34″ W) during low tide. A voucher specimen (GPA 181) was identified by us and was kept at the cnidarian collection of the Anthozoan Research Group (GPA) at the Academic Center of Vitória, Federal University of Pernambuco (Brazil). This species has been mentioned by some authors as *Palythoa variabilis*, in reason of a proposed synonymy for the genera *Palythoa* and *Protopalythoa* [75]. However, the issue of distinctive genera is not completely solved yet (see, for example, [76]), despite new molecular phylogenetic approach, based on the universal target-enrichment baits, has been recently developed to help resolve long-standing controversial relationships in the class Anthozoa [77]. Thus, until an extensive revision of the group is not definitively resolved with morphological and molecular data precisely combined, with inclusion of species of both genera, the binomial nomenclature *Protopalythoa variabilis* is used herein, as for decades.

3.2. RNA Library Construction and Origin of Zoantharian RNA Sequences

The RNA isolation, library preparation and the transcriptome assembly of the *P. variabilis* holobiont were performed as described in one of our previous articles [24]. RNA sequencing was performed using the Illumina HiSeq 2500 platform. Reads were cleaned up before the de novo transcriptome assembly using the Trinity method for transcriptome reconstruction [78]. This Transcriptome Shotgun Assembly (TSA) project was deposited in DDBJ/EMBL/GenBank under the accession GCVI00000000, associated with the BioProject PRJNA279783 and biosample SAMN03450566. The statistics of the RNA sequencing and contig assembling are summarized in the Supplementary Table S1.

3.3. Assessment of the Biodiversity Composition of the P. variabilis Holobiont

P. variabilis transcripts with a high similarity to 16S rRNA, COI, and rbcL sequences were identified using BLASTn with an E-value limit of 1E-40. The closest related species were characterized by homology search of these transcripts against the NCBI nr database. Only the first hits in concordance with the selected gene markers aforementioned were retained for species composition identification in the holobiont assemblage.

3.4. Sequence Annotations for Enzyme Precursors in the P. variabilis Holobiont

The unigenes from the *P. variabilis* transcriptome were investigated for structural enzyme homology using BLASTx (BLAST+ suite, version 2.5.0) [79] with a fixed E-value of 1E-5 against four public protein databases: the NCBI non-redundant (nr) protein database (accessed from October to November 2016), the Clusters of Orthologous Groups (COG) database, version 2003–2014, the UniProtKB/Swiss-Prot database, downloaded in October 2016), and the EuKaryotic Orthologous Groups (KOG) database, version 2003. Basic statistics of the sequence annotations are described in Supplementary Figures S1 and S2. Species information of the selected annotations was extracted from the BLAST output files to discern the taxonomic distribution.

3.5. Gene Ontology, Enzyme Codes and KEGG Pathway Assignments

The Blast2GO software, version 4.0.2 [80], was used for the subsequent steps under default parameters to carry out the InterProScan protein domain analysis, followed by the Gene Ontology (GO) annotation. The annotations were then subjected to a generic GO-slim reduction, prior to the mapping of the Enzyme Commission codes (EC) and KEGG pathways. The GO annotations chart was plotted using WEGO [81], while further KEGG pathways information was retrieved through the KEGG BRITE hierarchies site [82].

3.6. Sequence Alignment and Phylogenetic Inference

Multiple sequence alignments of predicted enzyme sequences were performed using MUSCLE v3.8.31 [83] or Kalign v2.9.0b2 [84], depending on whether the enzyme domain/motif was a single structural unit or multiple-domain repeats, respectively. The phylogenetic tree was inferred using MEGA 7.0.26 [85] based on the LG+G+I model, with a bootstrap test of 500 replicates and edited using FigTree v 1.4.3. The amino acid sequence identities and similarities were determined using Jalview 2.9.0b2 [86].

3.7. Prediction of Enzymes with Two or More Activities

An alternative annotation strategy was used to identify potential enzymes with two or more enzymatic activities. To favor the multiple domains (hits) discovery, we first performed a BLASTx search against the UniProtKB/Swiss-Prot database, keeping an E-value of 1E-5 with the additional "culling_limit" option set to "1". The selected set was reduced to only sequences with at least two non-overlapping hits on the same frame. During a second round of selection, only sequences with predicted protein products containing regions found by BLAST were kept. Finally, an InterProScan protein domain analysis [87] was performed to map each sequence to its GO annotation and its associated EC codes and KEGG pathways using the external EC2GO [88] and KEGG2GO mapping databases [89,90]. Only sequences with two or more ECs were kept, and data on the enzyme's substrate(s) and product(s) were retrieved from KEGG Enzyme [82].

3.8. Analysis of Predicted Enzyme with Partial EC Number

Results of the annotation generated during the Blast2GO analysis were used to extract a list of the transcripts with a partial EC number. For each of these sequences, the corresponding protein sequence is predicted and subsequently submitted to InterProScan 5 via RESTful service 80 to validate the presence of a catalytic domain corresponding to the partial EC. These steps were automatized using a python script, depending on the external EC2GO mapping database [88]. To perform the observed distance analysis between the predicted *P. variabilis* and the known enzyme sequences, the available collections of sequences corresponding to each EC of interest were retrieved from the BRENDA website (www.brenda-enzymes.org, release 2018.1) [91] and sequences with a length shorter or higher to two population standard deviation were removed. The multiple sequence alignment was done using Kalign v2.04 77, and the analysis was computed with MEGA 7.0.26 78 using the Neighbor-Joining method and the p-distance method, removing missing ambiguous positions for each sequence pair, and with a bootstrap test of 500 replicates. The cladograms were edited in the Interactive Tree Of Life website [92].

4. Conclusions

Marine invertebrates with associated microbiota form complex holobiont assemblages, which are attractive sources of biologically active organic compounds and (poly-)peptides, including enzymes. Marine enzymes have a high potential to be applied in green organic synthesis and in pharmaceutical and industrial biotechnology. The search for improved biocatalysts can be carried out using different strategies, such as screening a huge number of environmental samples, pursuing enzyme engineering, mining genomic and proteomic data, or a combination of more than one approach. The process of data mining transcriptomes has some advantages over genomic analysis; the most obvious advantage is that only enzymes that are expressed in a given environmental context are retrieved, including enzymes not completely characterized and with unknow enzyme-substrate specificity. This is particularly advantageous in the case of the marine assemblages of microbionts that form species-specific holobionts, from which the purification of enzymes with a high yield may be a concern. Thus, once identified, the cloning and the recombinant production of desirable marine biocatalysts can be structure-guided and based on the nature of expressed transcripts.

According to data reported in this work, the zoantharian *P. variabilis* expresses a variety of putative enzymes that could potentially be converted into biotechnologically useful biocatalysts and biopharmaceuticals. This holo-transcriptomic data demonstrates that a single holobiont assemblage comprises a unique repository of relevant biotechnological enzymes. Finally, the integrative analyses of this holo-transcriptome point to a valuable marine resource for the discovery of improved enzymes with applications in green chemistry, industrial and pharmaceutical biotechnology.

Supplementary Materials: The following are available online at http://www.mdpi.com/1660-3397/16/6/207/s1. Figure S1: Summary of the unigenes mapped to the public databases. Figure S2: Characteristics of the sequence homology search results. Figure S3: GO and KEGG pathways assignments. Figure S4: Maximum Likelihood (ML) phylogenetic tree of the predicted *P. variabilis* beta-*N*-acetylhexosaminidase (3.2.1.52) and their closest homologous sequences. Figure S5: Prediction of enzymes with two activities closely positioned in given metabolic pathway. Table S1: Species from the BLAST search of the barcode sequences from the *P. variabilis* holobiont transcriptome against the NCBI nr database. Table S2: KEGG pathways mapping summary. Table S3: List of predicted enzymes in *P. variabilis*. Table S4: List of enzymatic activities with relevance in treatment of rare diseases and other pharmaceutical fine chemicals predicted in *Protopalythoa variabilis* holo-transcriptome. Table S5: List of enzymatic activities with relevance in colorant, aromas, flavor, fragrance, cosmetic and hygienic industries predicted in *Protopalythoa variabilis* holo-transcriptome. Table S6: List of enzymatic activities with relevance in agrochemical, and food and feed industries predicted in *Protopalythoa variabilis* holo-transcriptome. Table S7: List of enzymatic activities with relevance in bioconversion and biopolymer synthesis predicted in *Protopalythoa variabilis* holo-transcriptome. Table S8: List of enzymatic activities with relevance in other industries predicted in *Protopalythoa variabilis* holo-transcriptome. Table S9: List of enzymatic activities with relevance in molecular biology and analytical applications predicted in *Protopalythoa variabilis* holo-transcriptome. Table S10: RNA-sequencing and assembling statistics. File S1: KEGG pathways.

Author Contributions: Conceived and designed the experiments: J.-É.R.L.M. and G.R.-B. Performed the experiments: J.-É.R.L.M., C.H. and Q.L. Analyzed the data: J.-É.R.L.M., A.R.B.P.-S. and G.R.-B. Contributed reagents, materials and/or analysis tools: S.M.-Y.L. and G.R.-B. Field trip missions, species identification and deposit collection: P.B.G. and C.D.P. wrote the manuscript with the contribution from all other authors: J.-É.R.L.M. and G.R.-B. All authors have read and approved the final article.

Funding: Financial support for G.R.-B. and research at the Institute for Marine Sciences, Federal University of Ceará, was provided by the Brazilian National Council for Scientific and Technological Development, CNPq (Marine Biotechnology Network—Proc. 408835/2013-3), the Ministry of Science, Technology and Innovation (MCTI), Brasília, DF, Brazil. At the University Federal of Pernambuco, financial support was CNPq (Proc. 408934/2013-1) granted to C.D.P. Research at the University of Macau was supported by grants from the Science and Technology Development Fund (FDCT) of Macao SAR and Research Committee, University of Macau, Macau, China.

Acknowledgments: Our gratitude to the Program on Toxinology (Issue 2010), the Coordination for the Improvement of Higher Education Personnel (CAPES), the Ministry of Education of the Federal Government of Brazil, Brasília, DF, Brazil. J-ÉRM was a doctoral fellowship recipient from CAPES.

Conflicts of Interest: The authors declare no conflict of interest. The funders had no role in the design of the study; in the collection, analyses, or interpretation of data; in the writing of the manuscript, and in the decision to publish the results.

References

1. Enzymes Market by Type (Industrial, Specialty), by Product (Carbohydrases, Proteases, Lipases, Polymerases & Nucleases), by Application (Food & Beverages, Detergents, Animal Feed, Textile, Paper & Pulp, Nutraceutical, Personal Care & Cosmetics, Wastewater, Research & Biotechnology, Diagnostics, Biocatalyst) and Segment Forecasts to 2024. Available online: https://www.grandviewresearch.com/industry-analysis/enzymes-industry (accessed on 13 June 2018).

2. Molecular Biology Enzymes and Kits & Reagents Market Analysis by Product (Kits and Reagents, Enzymes), by Application (PCR, Sequencing, Cloning, Epigenetics), by End User, and Segment Forecasts, 2018–2025. Available online: https://www.grandviewresearch.com/industry-analysis/molecular-biology-enzymes-kits-reagents-market (accessed on 13 June 2018).

3. Itoh, T.; Hanefeld, U. Enzyme catalysis in organic synthesis. *Green Chem.* **2017**, *19*, 331–332. [CrossRef]

4. Wenda, S.; Illner, S.; Mell, A.; Kragl, U. Industrial biotechnology-the future of green chemistry? *Green Chem.* **2011**, *13*, 3007–3047. [CrossRef]

5. Yamaguchi, S. The quest for industrial enzymes from microorganisms. *Biosci. Biotechnol. Biochem.* **2017**, *81*, 54–58. [CrossRef] [PubMed]

6. Shimizu, S.; Ogawa, J.; Kataoka, M.; Kobayashi, M. Screening of novel microbial enzymes for the production of biologically and chemically useful compounds. *Adv. Biochem. Eng. Biotechnol.* **1997**, *58*, 45–87. [PubMed]

7. Wang, J.B.; Li, G.; Reetz, M.T. Enzymatic site-selectivity enabled by structure-guided directed evolution. *Chem. Commun.* **2017**, *53*, 3916–3928. [CrossRef] [PubMed]

8. Hibbert, E.G.; Dalby, P.A. Directed evolution strategies for improved enzymatic performance. *Microb. Cell Fact.* **2005**, *4*, 29. [CrossRef] [PubMed]

9. Madhavan, A.; Sindhu, R.; Parameswaran, B.; Sukumaran, R.K.; Pandey, A. Metagenome Analysis: A Powerful Tool for Enzyme Bioprospecting. *Appl. Biochem. Biotechnol.* **2017**, *183*, 636–651. [CrossRef] [PubMed]

10. Trincone, A. Marine biocatalysts: Enzymatic features and applications. *Mar. Drugs* **2011**, *9*, 478–499. [CrossRef] [PubMed]

11. Smith, D.R.M.; Uria, A.R.; Helfrich, E.J.N.; Milbredt, D.; van Pee, K.H.; Piel, J.; Goss, R.J.M. An Unusual Flavin-Dependent Halogenase from the Metagenome of the Marine Sponge *Theonella swinhoei* WA. *ACS Chem. Biol.* **2017**, *12*, 1281–1287. [CrossRef] [PubMed]

12. Sarian, F.D.; Janecek, S.; Pijning, T.; Ihsanawati; Nurachman, Z.; Radjasa, O.K.; Dijkhuizen, L.; Natalia, D.; van der Maarel, M.J. A new group of glycoside hydrolase family 13 alpha-amylases with an aberrant catalytic triad. *Sci. Rep.* **2017**, *7*, 44230. [CrossRef] [PubMed]

13. Lima, R.N.; Porto, A.L. Recent Advances in Marine Enzymes for Biotechnological Processes. *Adv. Food Nutr. Res.* **2016**, *78*, 153–192. [PubMed]

14. Trincone, A. Enzymatic Processes in Marine Biotechnology. *Mar. Drugs* **2017**, *15*, 93. [CrossRef] [PubMed]

15. Leal, M.C.; Puga, J.; Serodio, J.; Gomes, N.C.; Calado, R. Trends in the discovery of new marine natural products from invertebrates over the last two decades—Where and what are we bioprospecting? *PLoS ONE* **2012**, *7*, e30580. [CrossRef] [PubMed]

16. Lindequist, U. Marine-Derived Pharmaceuticals—Challenges and Opportunities. *Biomol. Ther.* **2016**, *24*, 561–571. [CrossRef] [PubMed]

17. Ryu, T.; Seridi, L.; Moitinho-Silva, L.; Oates, M.; Liew, Y.J.; Mavromatis, C.; Wang, X.; Haywood, A.; Lafi, F.F.; Kupresanin, M.; et al. Hologenome analysis of two marine sponges with different microbiomes. *BMC Genomics* **2016**, *17*, 158. [CrossRef] [PubMed]

18. Macintyre, L.; Zhang, T.; Viegelmann, C.; Martinez, I.J.; Cheng, C.; Dowdells, C.; Abdelmohsen, U.R.; Gernert, C.; Hentschel, U.; Edrada-Ebel, R. Metabolomic tools for secondary metabolite discovery from marine microbial symbionts. *Mar. Drugs* **2014**, *12*, 3416–3448. [CrossRef] [PubMed]

19. Dunlap, W.C.; Battershill, C.N.; Liptrot, C.H.; Cobb, R.E.; Bourne, D.G.; Jaspars, M.; Long, P.F.; Newman, D.J. Biomedicinals from the phytosymbionts of marine invertebrates: A molecular approach. *Methods* **2007**, *42*, 358–376. [CrossRef] [PubMed]

20. Huang, C.Y.; Sung, P.J.; Uvarani, C.; Su, J.H.; Lu, M.C.; Hwang, T.L.; Dai, C.F.; Wu, S.L.; Sheu, J.H. Glaucumolides A and B, Biscembranoids with New Structural Type from a Cultured Soft Coral *Sarcophyton glaucum*. *Sci. Rep.* **2015**, *5*, 15624. [CrossRef] [PubMed]

21. Huang, C.Y.; Su, J.H.; Liaw, C.C.; Sung, P.J.; Chiang, P.L.; Hwang, T.L.; Dai, C.F.; Sheu, J.H. Bioactive Steroids with Methyl Ester Group in the Side Chain from a Reef Soft Coral *Sinularia brassica* Cultured in a Tank. *Mar. Drugs* **2017**, *15*, 280. [CrossRef] [PubMed]

22. Hsiao, T.H.; Sung, C.S.; Lan, Y.H.; Wang, Y.C.; Lu, M.C.; Wen, Z.H.; Wu, Y.C.; Sung, P.J. New Anti-Inflammatory Cembranes from the Cultured Soft Coral *Nephthea columnaris*. *Mar. Drugs* **2015**, *13*, 3443–3453. [CrossRef] [PubMed]

23. Ferrer, M.; Martinez-Martinez, M.; Bargiela, R.; Streit, W.R.; Golyshina, O.V.; Golyshin, P.N. Estimating the success of enzyme bioprospecting through metagenomics: Current status and future trends. *Microb. Biotechnol.* **2016**, *9*, 22–34. [CrossRef] [PubMed]

24. Huang, C.; Morlighem, J.R.; Zhou, H.; Lima, E.P.; Gomes, P.B.; Cai, J.; Lou, I.; Perez, C.D.; Lee, S.M.; Radis-Baptista, G. The Transcriptome of the Zoanthid *Protopalythoa variabilis* (Cnidaria, Anthozoa) Predicts a Basal Repertoire of Toxin-like and Venom-Auxiliary Polypeptides. *Genome Biol. Evol.* **2016**, *8*, 3045–3064. [CrossRef] [PubMed]

25. Rabelo, E.F.; Soares, M.D.O.; Bezerra, L.E.A.; Matthews-Cascon, H. Distribution pattern of zoanthids (Cnidaria: Zoantharia) on a tropical reef. *Mar. Biol. Res.* **2015**, *11*, 584–592. [CrossRef]

26. Rabelo, E.; Rocha, L.; Colares, G.; Araújo Bomfim, T.; Lúcia Rodrigues Nogueira, V.; Katzenberger, M.; Matthews-Cascon, H.; Melo, V. *Symbiodinium Diversity Associated with Zoanthids (Cnidaria: Hexacorallia) in Northeastern Brazil*; Springer: Dordrecht, The Netherlands, 2015.

27. Sun, W.; Zhang, F.; He, L.; Li, Z. Pyrosequencing reveals diverse microbial community associated with the zoanthid *Palythoa australiae* from the South China Sea. *Microb. Ecol.* **2014**, *67*, 942–950. [CrossRef] [PubMed]

28. Chu, N.D.; Vollmer, S.V. Caribbean corals house shared and host-specific microbial symbionts over time and space. *Environ. Microbiol. Rep.* **2016**, *8*, 493–500. [CrossRef] [PubMed]

29. Newman, D.J.; Cragg, G.M.; Snader, K.M. The influence of natural products upon drug discovery. *Nat. Prod. Rep.* **2000**, *17*, 215–234. [CrossRef] [PubMed]

30. Urbarova, I.; Karlsen, B.O.; Okkenhaug, S.; Seternes, O.M.; Johansen, S.D.; Emblem, A. Digital marine bioprospecting: Mining new neurotoxin drug candidates from the transcriptomes of cold-water sea anemones. *Mar. Drugs* **2012**, *10*, 2265–2279. [CrossRef] [PubMed]

31. Anderson, D.A.; Walz, M.E.; Weil, E.; Tonellato, P.; Smith, M.C. RNA-Seq of the Caribbean reef-building coral *Orbicella faveolata* (*Scleractinia-Merulinidae*) under bleaching and disease stress expands models of coral innate immunity. *PeerJ* **2016**, *4*, e1616. [CrossRef] [PubMed]

32. Ponce, D.; Brinkman, D.L.; Potriquet, J.; Mulvenna, J. Tentacle Transcriptome and Venom Proteome of the Pacific Sea Nettle, *Chrysaora fuscescens* (Cnidaria: Scyphozoa). *Toxins* **2016**, *8*, 102. [CrossRef] [PubMed]

33. Hasegawa, Y.; Watanabe, T.; Takazawa, M.; Ohara, O.; Kubota, S. De Novo Assembly of the Transcriptome of *Turritopsis*, a Jellyfish that Repeatedly Rejuvenates. *Zool. Sci.* **2016**, *33*, 366–371. [CrossRef] [PubMed]

34. Valmsen, K.; Jarving, I.; Boeglin, W.E.; Varvas, K.; Koljak, R.; Pehk, T.; Brash, A.R.; Samel, N. The origin of 15R-prostaglandins in the Caribbean coral *Plexaura homomalla*: Molecular cloning and expression of a novel cyclooxygenase. *Proc. Natl. Acad. Sci. USA* **2001**, *98*, 7700–7705. [CrossRef] [PubMed]

35. Uria, A.R.; Zilda, D.S. Metagenomics-Guided Mining of Commercially Useful Biocatalysts from Marine Microorganisms. *Adv. Food Nutr. Res.* **2016**, *78*, 1–26. [PubMed]

36. Popovic, A.; Tchigvintsev, A.; Tran, H.; Chernikova, T.N.; Golyshina, O.V.; Yakimov, M.M.; Golyshin, P.N.; Yakunin, A.F. Metagenomics as a Tool for Enzyme Discovery: Hydrolytic Enzymes from Marine-Related Metagenomes. *Adv. Exp. Med. Biol.* **2015**, *883*, 1–20. [PubMed]

37. Sukul, P.; Schakermann, S.; Bandow, J.E.; Kusnezowa, A.; Nowrousian, M.; Leichert, L.I. Simple discovery of bacterial biocatalysts from environmental samples through functional metaproteomics. *Microbiome* **2017**, *5*, 28. [CrossRef] [PubMed]

38. Sturmberger, L.; Wallace, P.W.; Glieder, A.; Birner-Gruenberger, R. Synergism of proteomics and mRNA sequencing for enzyme discovery. *J. Biotechnol.* **2016**, *235*, 132–138. [CrossRef] [PubMed]

39. Guo, F.; Berglund, P. Transaminase biocatalysis: Optimization and application. *Green Chem.* **2017**, *19*, 333–360. [CrossRef]

40. McGrath, B.M.; Walsh, G. *Directory of Therapeutic Enzymes*; CRC Press, Taylor & Francis Group: Boca Raton, FL, USA, 2005.

41. Sun, Z.; Ilie, A.; Reetz, M.T. Towards the Production of Universal Blood by Structure-Guided Directed Evolution of Glycoside Hydrolases. *Angew. Chem.* **2015**, *54*, 9158–9160. [CrossRef] [PubMed]

42. Greiner-Tollersrud, O.K.; Berg, T. Lysosomal Storage Disorders. In *Madame Curie Bioscience Database [Internet]*; Landes Bioscience: Austin, TX, USA, 2013.

43. Desnick, R.J.; Schuchman, E.H. Enzyme replacement therapy for lysosomal diseases: Lessons from 20 years of experience and remaining challenges. *Annu. Rev. Genomics Hum. Genet.* **2012**, *13*, 307–335. [CrossRef] [PubMed]

44. Kloosterman, W.M.J.; Roest, S.; Priatna, S.R.; Stavila, E.; Loos, K. Chemo-enzymatic synthesis route to poly(glucosyl-acrylates) using glucosidase from almonds. *Green Chem.* **2014**, *16*, 1837–1846. [CrossRef]

45. Vellard, M. The enzyme as drug: Application of enzymes as pharmaceuticals. *Curr. Opin. Biotechnol.* **2003**, *14*, 444–450. [CrossRef]

46. Olsson, M.L.; Clausen, H. Modifying the red cell surface: Towards an ABO-universal blood supply. *Br. J. Haematol.* **2008**, *140*, 3–12. [CrossRef] [PubMed]

47. Dozier, J.K.; Distefano, M.D. Site-Specific PEGylation of Therapeutic Proteins. *Int. J. Mol. Sci.* **2015**, *16*, 25831–25864. [CrossRef] [PubMed]

48. Zhang, D.; Zhu, Y.; Chen, J. Microbial transglutaminase production: Understanding the mechanism. *Biotechnol. Genet. Eng. Rev.* **2010**, *26*, 205–222. [CrossRef] [PubMed]

49. Li, Y.; Lei, X.; Zhu, H.; Zhang, H.; Guan, C.; Chen, Z.; Zheng, W.; Fu, L.; Zheng, T. Chitinase producing bacteria with direct algicidal activity on marine diatoms. *Sci. Rep.* **2016**, *6*, 21984. [CrossRef] [PubMed]

50. Hollmann, F.; Arends, I.W.C.E.; Holtmann, D. Enzymatic reductions for the chemist. *Green Chem.* **2011**, *13*, 2285–2314. [CrossRef]

51. Han, Y.; Yang, B.; Zhang, F.; Miao, X.; Li, Z. Characterization of antifungal chitinase from marine *Streptomyces* sp. DA11 associated with South China Sea sponge *Craniella australiensis*. *Mar. Biotechnol.* **2009**, *11*, 132–140. [CrossRef] [PubMed]

52. Hein, S.; Wang, K.; Stevens, W.F.; Kjems, J. Chitosan composites for biomedical applications: Status, challenges and perspectives. *Mater. Sci. Technol.* **2008**, *24*, 1053–1061. [CrossRef]

53. Lyu, Y.; Laborda, P.; Huang, K.; Cai, Z.P.; Wang, M.; Aimin, L.; Doherty, C.; Liu, L.; Flitsch, S.; Voglmeir, J. Highly Efficient and Selective Biocatalytic Production of Glucosamine from Chitin. *Green Chem.* **2016**, *19*, 527–535.

54. Shearer, T.L.; Snell, T.W.; Hay, M.E. Gene expression of corals in response to macroalgal competitors. *PLoS ONE* **2014**, *9*, e114525. [CrossRef] [PubMed]

55. Cobbett, C.; Goldsbrough, P. Phytochelatins and metallothioneins: Roles in heavy metal detoxification and homeostasis. *Annu. Rev. Plant Biol.* **2002**, *53*, 159–182. [CrossRef] [PubMed]

56. Bundy, J.G.; Kille, P. Metabolites and metals in Metazoa—What role do phytochelatins play in animals? *Metallomics* **2014**, *6*, 1576–1582. [CrossRef] [PubMed]

57. Liu, F.; Kang, S.H.; Lee, Y.-I.; Choa, Y.-H.; Mulchandani, A.; Myung, N.V.; Chen, W. Enzyme mediated synthesis of phytochelatin-capped CdS nanocrystals. *Appl. Phys. Lett.* **2010**, *97*, 123703. [CrossRef]

58. Kang, S.H.; Singh, S.; Kim, J.Y.; Lee, W.; Mulchandani, A.; Chen, W. Bacteria metabolically engineered for enhanced phytochelatin production and cadmium accumulation. *Appl. Environ. Microbiol.* **2007**, *73*, 6317–6320. [CrossRef] [PubMed]

59. Archelas, A.; Furstoss, R. Synthetic applications of epoxide hydrolases. *Curr. Opin. Chem. Biol.* **2001**, *5*, 112–119. [CrossRef]

60. Van Loo, B.; Kingma, J.; Arand, M.; Wubbolts, M.G.; Janssen, D.B. Diversity and biocatalytic potential of epoxide hydrolases identified by genome analysis. *Appl. Environ. Microbiol.* **2006**, *72*, 2905–2917. [CrossRef] [PubMed]

61. Yamada, H.; Shimizu, S.; Kobayashi, M. Hydratases involved in nitrile conversion: Screening, characterization and application. *Chem. Rec.* **2001**, *1*, 152–161. [CrossRef] [PubMed]

62. Yamada, H.; Kobayashi, M. Nitrile hydratase and its application to industrial production of acrylamide. *Biosci. Biotechnol. Biochem.* **1996**, *60*, 1391–1400. [CrossRef] [PubMed]

63. Park, J.M.; Trevor Sewell, B.; Benedik, M.J. Cyanide bioremediation: The potential of engineered nitrilases. *Appl. Microbiol. Biotechnol.* **2017**, *101*, 3029–3042. [CrossRef] [PubMed]

64. Eun, H.-M. 3-Nucleases. In *Enzymology Primer for Recombinant DNA Technology*; Academic Press: San Diego, CA, USA, 1996; pp. 145–232.

65. Kim, I.; Hur, N.W.; Shin, H.D.; Park, B.L.; Cheong, H.S.; Bae, S.C. Associations of DNase IV polymorphisms with autoantibodies in patients with systemic lupus erythematosus. *Rheumatology* **2008**, *47*, 996–999. [CrossRef] [PubMed]

66. Nicholson, A.W. Ribonuclease III mechanisms of double-stranded RNA cleavage. *Wiley Interdiscip. Rev. RNA* **2014**, *5*, 31–48. [CrossRef] [PubMed]

67. Eun, H.-M. 6-DNA Polymerases. In *Enzymology Primer for Recombinant DNA Technology*; Academic Press: San Diego, CA, USA, 1996; pp. 345–489.

68. Eun, H.-M. 7-RNA Polymerases. In *Enzymology Primer for Recombinant DNA Technology*; Academic Press: San Diego, CA, USA, 1996; pp. 491–565.

69. Bachman, J. Reverse-transcription PCR (RT-PCR). *Methods Enzymol.* **2013**, *530*, 67–74. [PubMed]

70. Eun, H.-M. 2-Ligases. In *Enzymology Primer for Recombinant DNA Technology*; Academic Press: San Diego, CA, USA, 1996; pp. 109–144.

71. Spreitzer, R.J.; Salvucci, M.E. Rubisco: Structure, regulatory interactions, and possibilities for a better enzyme. *Annu. Rev. Plant Biol.* **2002**, *53*, 449–475. [CrossRef] [PubMed]

72. Moglia, A.; Lanteri, S.; Comino, C.; Hill, L.; Knevitt, D.; Cagliero, C.; Rubiolo, P.; Bornemann, S.; Martin, C. Dual catalytic activity of hydroxycinnamoyl-coenzyme A quinate transferase from tomato allows it to moonlight in the synthesis of both mono- and dicaffeoylquinic acids. *Plant Physiol.* **2014**, *166*, 1777–1787. [CrossRef] [PubMed]

73. Wang, S.Z.; Zhang, Y.H.; Ren, H.; Wang, Y.L.; Jiang, W.; Fang, B.S. Strategies and perspectives of assembling multi-enzyme systems. *Crit. Rev. Biotechnol.* **2017**, *37*, 1024–1037. [CrossRef] [PubMed]

74. Kupski, L.; Queiroz, M.I.; Badiale-Furlong, E. Application of carboxypeptidase A to a baking process to mitigate contamination of wheat flour by ochratoxin A. *Process Biochem.* **2018**, *64*, 248–254. [CrossRef]

75. Reimer, J.D.; Ono, S.; Takishita, K.; Tsukahara, J.; Maruyama, T. Molecular evidence suggesting species in the zoanthid genera *Palythoa* and *Protopalythoa* (Anthozoa: Hexacorallia) are congeneric. *Zool. Sci.* **2006**, *23*, 87–94. [CrossRef] [PubMed]

76. Swain, T.D. Revisiting the phylogeny of Zoanthidea (Cnidaria: Anthozoa): Staggered alignment of hypervariable sequences improves species tree inference. *Mol. Phyl. Evol.* **2018**, *118*, 1–12. [CrossRef] [PubMed]

77. Quattrini, A.M.; Faircloth, B.C.; Duenas, L.F.; Bridge, T.C.L.; Brugler, M.R.; Calixto-Botia, I.F.; DeLeo, D.M.; Foret, S.; Herrera, S.; Lee, S.M.Y.; et al. Universal target-enrichment baits for anthozoan (Cnidaria) phylogenomics: New approaches to long-standing problems. *Mol. Ecol. Resour.* **2018**, *18*, 281–295. [CrossRef] [PubMed]

78. Haas, B.J.; Papanicolaou, A.; Yassour, M.; Grabherr, M.; Blood, P.D.; Bowden, J.; Couger, M.B.; Eccles, D.; Li, B.; Lieber, M.; et al. De novo transcript sequence reconstruction from RNA-seq using the Trinity platform for reference generation and analysis. *Nat. Protocols* **2013**, *8*, 1494–1512. [CrossRef] [PubMed]

79. Camacho, C.; Coulouris, G.; Avagyan, V.; Ma, N.; Papadopoulos, J.; Bealer, K.; Madden, T.L. BLAST+: Architecture and applications. *BMC Bioinform.* **2009**, *10*, 421. [CrossRef] [PubMed]

80. Conesa, A.; Gotz, S.; Garcia-Gomez, J.M.; Terol, J.; Talon, M.; Robles, M. Blast2GO: A universal tool for annotation, visualization and analysis in functional genomics research. *Bioinformatics* **2005**, *21*, 3674–3676. [CrossRef] [PubMed]

81. Ye, J.; Fang, L.; Zheng, H.; Zhang, Y.; Chen, J.; Zhang, Z.; Wang, J.; Li, S.; Li, R.; Bolund, L.; et al. WEGO: A web tool for plotting GO annotations. *Nucleic Acids Res.* **2006**, *34*, W293–W297. [CrossRef] [PubMed]

82. Kanehisa, M.; Sato, Y.; Kawashima, M.; Furumichi, M.; Tanabe, M. KEGG as a reference resource for gene and protein annotation. *Nucleic Acids Res.* **2016**, *44*, D457–D462. [CrossRef] [PubMed]

83. Edgar, R.C. MUSCLE: A multiple sequence alignment method with reduced time and space complexity. *BMC Bioinform.* **2004**, *5*, 113. [CrossRef] [PubMed]

84. Lassmann, T.; Frings, O.; Sonnhammer, E.L. Kalign2: High-performance multiple alignment of protein and nucleotide sequences allowing external features. *Nucleic Acids Res.* **2009**, *37*, 858–865. [CrossRef] [PubMed]

85. Kumar, S.; Stecher, G.; Tamura, K. MEGA7: Molecular Evolutionary Genetics Analysis Version 7.0 for Bigger Datasets. *Mol. Biol. Evol.* **2016**, *33*, 1870–1874. [CrossRef] [PubMed]

86. Waterhouse, A.M.; Procter, J.B.; Martin, D.M.; Clamp, M.; Barton, G.J. Jalview Version 2—A multiple sequence alignment editor and analysis workbench. *Bioinformatics* **2009**, *25*, 1189–1191. [CrossRef] [PubMed]

87. Jones, P.; Binns, D.; Chang, H.Y.; Fraser, M.; Li, W.; McAnulla, C.; McWilliam, H.; Maslen, J.; Mitchell, A.; Nuka, G.; et al. InterProScan 5: Genome-scale protein function classification. *Bioinformatics* **2014**, *30*, 1236–1240. [CrossRef] [PubMed]

88. Hill, D.P.; Davis, A.P.; Richardson, J.E.; Corradi, J.P.; Ringwald, M.; Eppig, J.T.; Blake, J.A. PROGRAM DESCRIPTION: Strategies for Biological Annotation of Mammalian Systems: Implementing Gene Ontologies in Mouse Genome Informatics. *Genomics* **2001**, *74*, 121–128. [CrossRef] [PubMed]

89. Ashburner, M.; Ball, C.A.; Blake, J.A.; Botstein, D.; Butler, H.; Cherry, J.M.; Davis, A.P.; Dolinski, K.; Dwight, S.S.; Eppig, J.T.; et al. Gene Ontology: Tool for the unification of biology. *Nat. Genet.* **2000**, *25*, 25. [CrossRef] [PubMed]

90. The Gene Ontology Consortium. Expansion of the Gene Ontology knowledgebase and resources. *Nucleic Acids Res.* **2017**, *45*, D331–D338.

91. Placzek, S.; Schomburg, I.; Chang, A.; Jeske, L.; Ulbrich, M.; Tillack, J.; Schomburg, D. BRENDA in 2017: New perspectives and new tools in BRENDA. *Nucleic Acids Res.* **2017**, *45*, D380–D388. [CrossRef] [PubMed]

92. Letunic, I.; Bork, P. Interactive tree of life (iTOL) v3: An online tool for the display and annotation of phylogenetic and other trees. *Nucleic Acids Res.* **2016**, *44*, W242–W245. [CrossRef] [PubMed]

Article

Stress-Induced Mucus Secretion and Its Composition by a Combination of Proteomics and Metabolomics of the Jellyfish *Aurelia coerulea*

Wenwen Liu [1,2,†], Fengfeng Mo [3,†], Guixian Jiang [4,†], Hongyu Liang [1,2], Chaoqun Ma [2], Tong Li [5], Lulu Zhang [2], Liyan Xiong [6], Gian Luigi Mariottini [7], Jing Zhang [1,*] and Liang Xiao [2,*]

[1] College of Traditional Chinese Medicine, Jilin Agricultural University, Changchun 130118, China; lww0325@hotmail.com (W.L.); lianghongyu0819@hotmail.com (H.L.)

[2] Department of Marine Biotechnology, Faculty of Naval Medicine, Second Military Medical University, Shanghai 200433, China; maria940501@outlook.com (C.M.); zhanglulu2822@outlook.com (L.Z.)

[3] Department of Ship Hygiene, Faculty of Navy Medicine, Second Military Medical University, Shanghai 200433, China; fengfengmo@hotmail.com

[4] Clinical Medicine, Grade 2015, Second Military Medical University, Shanghai 200433, China; jiangdaxian@hotmail.com

[5] School of Marine Sciences, Ningbo University, Ningbo 315211, China; 18317128586@163.com

[6] Department of Traditional Chinese Medicine Identification, School of Pharmacy, Second Military Medical University, Shanghai 200433, China; xlyiverson@126.com

[7] Department of Earth, Environment and Life Sciences (DISTAV), University of Genova, Viale Benedetto XV 5, I-16132 Genova, Italy; Gian.Luigi.Mariottini@unige.it

* Correspondence: zhjing0701@163.com (J.Z.); hormat830713@hotmail.com (L.X.); Tel.: +86-13353144693 (J.Z.); +86-15921431590 (L.X.)

† These authors contributed equally to this work.

Received: 24 August 2018; Accepted: 9 September 2018; Published: 18 September 2018

Abstract: Background: Jellyfish respond quickly to external stress that stimulates mucus secretion as a defense. Neither the composition of secreted mucus nor the process of secretion are well understood. Methods: *Aurelia coerulea* jellyfish were stimulated by removing them from environmental seawater. Secreted mucus and tissue samples were then collected within 60 min, and analyzed by a combination of proteomics and metabolomics using liquid chromatography coupled with tandem mass spectrometry (LC-MS/MS) and ultra-performance liquid chromatography/quadrupole time-of-flight mass spectrometry (UPLC-QTOF-MS/MS), respectively. Results: Two phases of sample collection displayed a quick decrease in volume, followed by a gradual increase. A total of 2421 and 1208 proteins were identified in tissue homogenate and secreted mucus, respectively. Gene Ontology (GO) analysis showed that the mucus-enriched proteins are mainly located in extracellular or membrane-associated regions, while the tissue-enriched proteins are distributed throughout intracellular compartments. Tryptamine, among 16 different metabolites, increased with the largest-fold change value of 7.8 in mucus, which is consistent with its involvement in the Kyoto Encyclopedia of Genes and Genomes (KEGG) pathway 'tryptophan metabolism'. We identified 11 metalloproteinases, four serpins, three superoxide dismutases and three complements, and their presence was speculated to be related to self-protective defense. Conclusions: Our results provide a composition profile of proteins and metabolites in stress-induced mucus and tissue homogenate of *A. coerulea*. This provides insight for the ongoing endeavors to discover novel bioactive compounds. The large increase of tryptamine in mucus may indicate a strong stress response when jellyfish were taken out of seawater and the active self-protective components such as enzymes, serpins and complements potentially play a key role in innate immunity of jellyfish.

Keywords: jellyfish; *Aurelia coerulea*; mucus; proteomics; metabolomics

---

1. Introduction

In recent decades, jellyfish blooms have become an important issue in coastal areas worldwide. These blooms are likely related to issues such as overfishing, global warming and eutrophication [1]. Blooming jellyfish consume fish eggs, crush captured fish, clog or destroy fish nets and block power-plant intakes, leading to disruption of marine ecosystems and thereby causing significant economic losses. Moreover, with massive increases of jellyfish blooms in recent years in coastal areas, the number of victims stung by jellyfish, including swimmers, fishermen and divers, has consequently increased [2,3]. Contact with jellyfish tentacles can trigger millions of nematocysts to pierce the skin and inject venom, causing responses ranging from no effect or local pain, to a series of severe systemic manifestations such as cardiovascular collapse, liver dysfunction, renal failure, and even death [4–6]. It is widely reported that the comprehensive toxicities of jellyfish venoms are attributed to numerous active components exerting hemolytic, cardiovascular, muscular, neural, antioxidant and cytotoxic effects [7,8].

An often reported scenario is that children touch or pick up a dying or dead jellyfish while playing on the beach and are stung. The usual speculation is that the poisoning comes from the nematocyst venom released when the tentacles are touched. Interestingly, however, almost all fishermen interviewed described to us that even the 'residual seawater' on their fishing nets after trawling through a jellyfish bloom when in contact with their bodies would cause serious cutaneous pain and local swelling. Residual seawater from a jellyfish bloom is reported to be very sticky, most likely derived from a mixture of the secreted mucus [9–12] and the nematocyst venom of the jellyfish. In the laboratory we have also observed that jellyfish secreted mucus as a defense resulting from external stimulation of being gently shaken or stirred. Moreover, because the gelatinous body of jellyfish is very fragile and easy to autolyze, it is reasonable that a beached jellyfish would also have started to autolyze when touched or picked up by children. Therefore, the envenomation of children by beached jellyfish would likely be due to a mixture of nematocyst venom, secreted mucus and autolyzed tissue fluid.

Like most other aquatic organisms, the surfaces of jellyfish are covered by a thin layer of mucus that originates from the epidermal cells. It is greatly affected when jellyfish are processed or when interfered with [13]. Some species of jellyfish can produce nets of mucus or release blobs to trap food particles [14]. Meanwhile, skin mucus from other aquatic organisms, like fish, contain a variety of immunity-related factors including lectins, lysozymes, calmodulin, immunoglobulins, complement, C-reactive proteins, proteolytic enzymes and anti-microbial peptides [15–17]. Although there are few papers reported, it is reasonable to speculate that jellyfish mucus is a rich library of active components for predation by adhesion and digestion, as well as modulators of innate immunity against triggers such as physical damage, microbial invasion and pollutants [14,18–21].

Aurelia coerulea (*Aurelia* sp.1) is a species of moon jellyfish found in the coastal waters of Chinese seas [22,23]. Compared with other marine jellyfish, *A. coerulea* is of low toxicity and can be maturely reproduced in an artificial environment, thus facilitating mucus collection [13]. Our interest in *A. coerulea* is focused on exploring its stress-induced mucus secretion and its composition by a combination of proteomics and metabolomics. Consequently, we aimed to provide insight into the protein and metabolite composition of stress-induced mucus and tissue homogenate to facilitate a better understanding of the process of stress-induced mucus secretion, as well as its involvement in innate immunity, along with the discovery of novel bioactive compounds.

2. Results

2.1. Stress-Induced Mucus Secretion and Autolysis of A. coerulea

Jellyfish are able to respond quickly to external environmental stimuli, although they have limited movement ability. We have previously noted the active secretion of jellyfish mucus induced by external stimulation, in that the surrounding seawater turns sticky when disturbed. In this study, we first checked the quantity-time relationship of stress-induced mucus secretion as well as the autolysis that rapidly occurs in dying jellyfish.

External stress was performed by removing *A. coerulea* from the environmental seawater and, as expected, the sticky liquid samples were largely secreted [13] and collected every 10 min for a total of 1 h (Figure 1). Two obvious phases in volume collection were displayed, whereby the volume decreases to a minimum at 30 min, followed by a gradual increase within 60 min (Figure 1A). However, protein concentration of each sample is positively associated with the time (Figure 1B), which is further confirmed by sodium dodecyl sulfate–polyacrylamide gel electrophoresis (SDS-PAGE) (Figure 1C). Proteins in mucus are mainly distributed in three concentrated molecular weight ranges—100–250 kDa, 50–100 kDa and 37–50 kDa—while proteins in tissue are more dispersed. A gentle trough of the curve for protein quantity (mg/kg) of each sample is shown at 30 min (Figure 1D). Meanwhile, obvious crevices in the umbrella part indicate that jellyfish autolysis starts at or even earlier than 30 min. Therefore, the decreases of mucus volume (Figure 1A) and protein amount (Figure 1D) in the first 30 min imply an adaption to the stress while the increase of mucus volume (Figure 1A) and protein amount (Figure 1D) in the latter 30 min is probably due to jellyfish autolysis. Interestingly, straight line correlations ($R^2 > 0.99$) for both mucus volume (mL/kg, Figure 1E) and protein quantity (mg/kg, Figure 1F) with time indicate a continuous release of proteins through two different mechanisms—i.e., stress-induced mucus secretion followed by jellyfish autolysis—without clear boundaries. The 20 min sample is less influenced by both residual seawater and jellyfish autolysis, and is therefore the sample selected for proteomics and metabolomics.

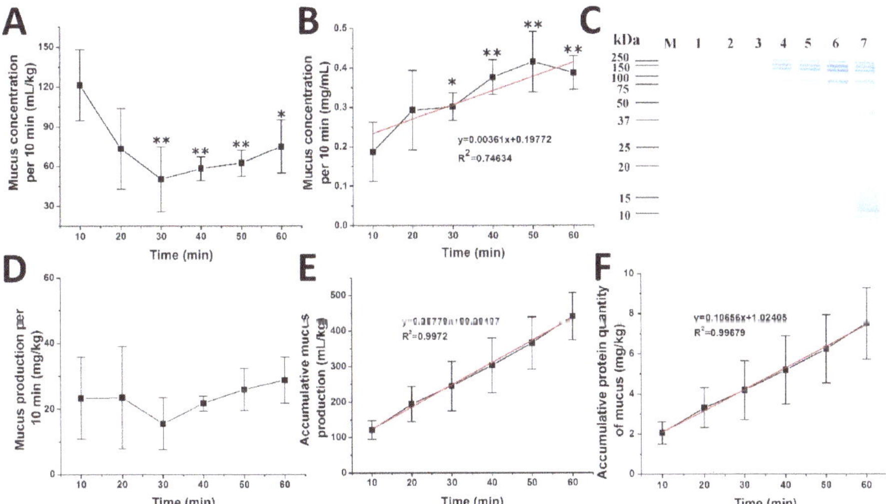

Figure 1. Stress-induced mucus secretion and autolysis of *A. coerulea*. (**A**) Mucus volume of each sample/10 min; (**B**) Protein concentration of each sample/10 min was determined by the Bradford method; (**C**) SDS-PAGE of collected mucus and tissue homogenate of *A. coerulea*. M: protein molecular size marker; Lanes 1–6: mucus samples of *A. coerulea* collected in 60 min; Lane 7: jellyfish tissue homogenate; (**D**) Protein quantity (mg/kg)/10 min; (**E**) Accumulative volume of mucus (mL/kg)/60 min; (**F**) Accumulative protein quantity of mucus (mg/kg)/60 min. Mean ± SD (*n* = 4) is shown. * $p < 0.05$ and ** $p < 0.01$ indicate a significance difference as compared to the control.

2.2. Proteomic Comparison of Secreted Mucus and Tissue Homogenate

All proteomics raw MS data were aligned to obtain peptide sequence information and matched to proteins from our previously constructed transcriptomic database for *A. coerulea*. A total of 2729 proteins from 10,560 peptides were identified by LC-MS/MS, where 2421 proteins from 8866 peptides were matched in tissue and 1208 proteins from 4148 peptides were matched in mucus. A Venn diagram shows that 1438 and 225 proteins are separately located in tissue and mucus, respectively. Proteins numbering 183, 523, and 267 are elevated, lowered or unchanged, respectively, among the 983 overlapped proteins in mucus when compared to those in tissue (Figure 2A). This profile is further supported by the quantitative ratio histogram of the overlapped proteins between the two groups where the $\log_2$(FC) (fold change) value from mucus (numerator) vs. tissue (denominator) values were distributed from −6 to +6, with a peak located at around −2, rather than 0 (Figure 2B). Although protein number of secreted mucus is far less than that of tissue homogenate, two proteomic indexes including amino acid (AA) (Figure 2E) and molecular weight (MW) distributions (Figure 2F) are the same for the two groups. The distribution curves of four other indexes, including peptide count (Figure 2C), protein sequence coverage (Figure 2D), electric point (Figure 2G) and exponentially modified protein abundance index (emPAI) (Figure 2H), are slightly shifted to the right in mucus when compared to those in tissue. These proteomic analysis results indicate that proteins identified in secreted mucus were successfully separated to a high level of purity and that they were independent of those identified in tissue homogenate.

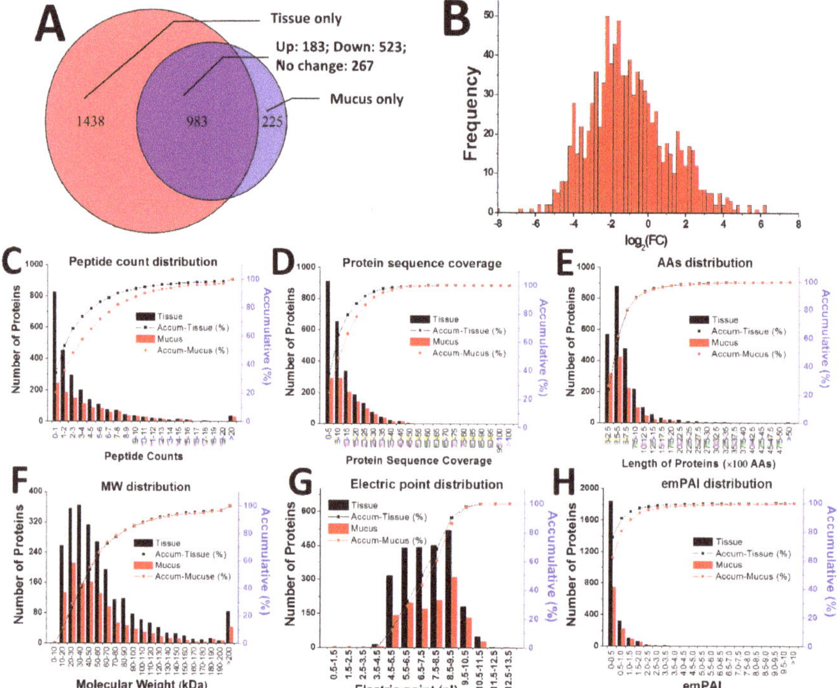

Figure 2. Proteomic comparison of secreted mucus and tissue homogenate. (**A**) Venn diagram of protein composition in mucus and tissue. There were 2421 proteins identified in tissue and 1208 identified in mucus. 1438 and 225 proteins were only found in tissue and mucus, respectively. Of these 983 overlapping proteins in both groups, 183 were found at elevated levels in mucus, while 523 were at lower levels and 267 were at consistent levels when compared to those in tissue. (**B**) Histogram of quantitative ratio of the overlapped proteins between the two groups. The $\log_2$(FC) (fold change) value from mucus (numerator) vs. tissue (denominator) is distributed mainly between −6 to +6, with a peak located at around −2 instead of 0. Six indexes, including peptide count distribution (**C**), protein sequence coverage (**D**), AA distribution (**E**), MW distribution (**F**), electric point distribution (**G**), and emPAI distribution (**H**), are compared for tissue and mucus.

2.3. Gene Ontology Analysis

The Gene Ontology (GO) project provides an ontology of defined terms describing the characteristics of genes and their products in any organism [24]. It covers three domains: biological process (BP), cellular component (CC), and molecular function (MF). We categorized all identified proteins according to the levels of protein expression in secreted mucus and tissue homogenate. The distributing tendencies of tissue-enriched proteins (FC < 0.5), mucus-enriched proteins (FC > 2) and proteins with no change (0.5 < FC < 2) are similar, although obvious differences are seen in specific terms. Since the quantity of proteins in mucus is much less than that in tissue, we used percentage of the total identified proteins in each group as the horizontal axis, whereas the exact amount of proteins is labeled on the right side of the transverse column (Figure 3A–C).

Among the top 10 terms of BP, the ratios of proteins fall from near 50% in 'cellular process' to less than 5% in 'biological adhesion' in tissue-enriched proteins. The top three terms, 'cellular process', 'metabolic process' and 'biological regulation' in mucus-enriched proteins have much lower ratios than those in tissue-enriched proteins (Figure 3A). An intermediate distribution is seen in the proteins with no change. The largest difference between the two groups comes from the CC subcategories. Tissue-enriched proteins are mainly distributed in three intracellular locations: 'cell', 'cell part' and 'organelle', then followed by the membrane-related terms 'membrane', 'membrane part' and 'macromolecular complex'. Comparatively, the most abundant locations in mucus-enriched proteins are 'membrane' and 'membrane part'. The ratios of protein levels in the intracellular locations 'cell', 'cell part' and 'organelle' are much lower than those in tissue-enriched proteins, while the extracellular terms 'extracellular region' and 'extracellular part' show elevated percentages of proteins and larger ratios in mucus-enriched proteins when compared to those in tissue-enriched proteins (Figure 3B). The distribution profiles of MF subcategories are similar across all three groups. Interestingly, the ratios of the top two terms, 'binding' and 'catalytic activity', are close to 50%, which is significantly higher than that of other terms, with ratios of less than 10%. Although the number of 'molecular function regulator' proteins is similar between 'mucus-enriched' and 'tissue-enriched' samples, the ratio of 'molecular function regulator' proteins in mucus-enriched proteins is much higher than those in the other two groups, which may potentially be used as the molecular indicators of jellyfish stress (Figure 3C).

We turned our attention to the mucus-enriched proteins, of which the GO enrichment diagram (Figure 3D) is built with the parameters: protein number > 15, rich factor 0.2–0.8 and $-\log_{10}(p$ value) > 7. The most significant feature is that the 'extracellular region' shows the highest $-\log_{10}(p$ value) (bright red), although its 'rich factor' and 'protein quality' are not the largest. Moreover, a Venn diagram was constructed to further divide the extracellular proteins in mucus-enriched proteins into three subclasses—'extracellular region', 'extracellular matrix' and 'extracellular space'—with 23, 32 and 28 proteins, respectively, in each subclass. The 'extracellular region' and 'extracellular matrix', 'extracellular region and extracellular space', 'extracellular space and extracellular matrix' share five, eight and two proteins, respectively. Only two proteins overlap across all three subgroups (Figure 3E).

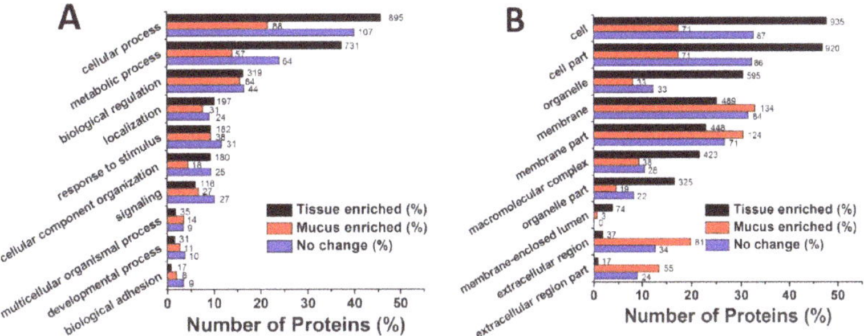

Figure 3. *Cont.*

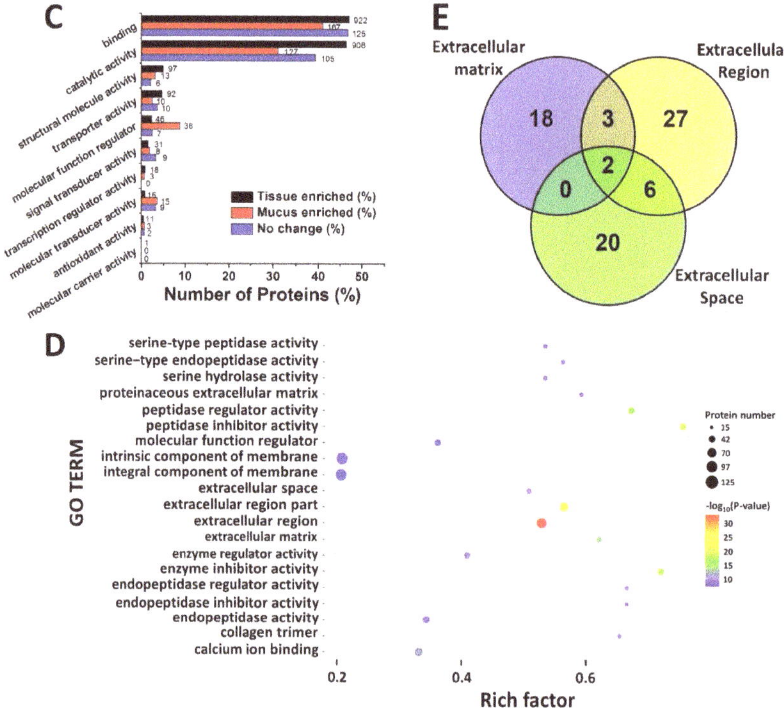

Figure 3. Comparative Gene Ontology (GO) analysis of identified proteins in tissue and mucus of *A. coerulea*. Three groups, namely 'tissue-enriched proteins', 'mucus-enriched proteins', and 'proteins with no change' are displayed. The tissue-enriched proteins or mucus-enriched proteins represent proteins exclusively and highly expressed in tissue homogenate (FC < 0.5) or secreted mucus (FC > 2). The group 'proteins with no change' implies that the proteins expressed in both mucus and tissue show no obvious difference (0.5 < FC < 2). (**A**) Biological process (BP). The horizontal axis is the ratio of proteins in the total identified proteins, whereas the vertical axis provides description of the matched GO terms. The protein numbers are labeled on the right side of each transverse column. (**B**) Cellular component (CC). (**C**) Molecular function (MF). (**D**) Diagram of GO enrichment in mucus-enriched proteins. The horizontal axis indicates the rich factor, i.e., the proportion of the number of differentially expressed proteins vs. the total number of proteins in the same GO term. The vertical ordinates represent the matched GO terms. The bubble shows the number of proteins matched in each GO term. The color represents $-\log_{10}(p$ value): Logarithmic conversion of Fisher exact test p value. (**E**) Venn diagram of the extracellular proteins in mucus-enriched proteins. Three subclasses 'extracellular matrix', 'extracellular region' and 'extracellular space' are colored by blue, yellow and green, respectively.

2.4. KEGG Pathway Analysis

KEGG (Kyoto Encyclopedia of Genes and Genomes) is a common bioinformatics tool and was utilized in this study to provide pathway mapping of identified proteins. The top 20 matched pathways among the 226 successfully mapped pathways are mainly associated with the intracellular synthesis of metabolites, as well as intracellular functions (Figure 4A) in tissue-enriched proteins. The number of proteins gradually falls from 74 in the 'Ribosome' and 72 in 'Carbon metabolism' to only one protein in each pathway. By comparison, only 81 pathways were matched in mucus-enriched proteins. There are 25 proteins matched to the 'ECM (extracellular matrix)—receptor interaction' pathway, significantly more than in other pathways (Figure 4B). The largest spot (shown in red) has a rich factor of 0.52 and represents an 'ECM-receptor interaction'. This is shown in the KEGG enrichment diagram for protein-enriched mucus (Figure 4C). Three subclasses, namely collagen, laminin and thrombospondin (THBS) are further divided from the 25 matched proteins identified in the 'ECM-receptor interaction' that are known to function in the matrix-receptor interactions (Figure 4D).

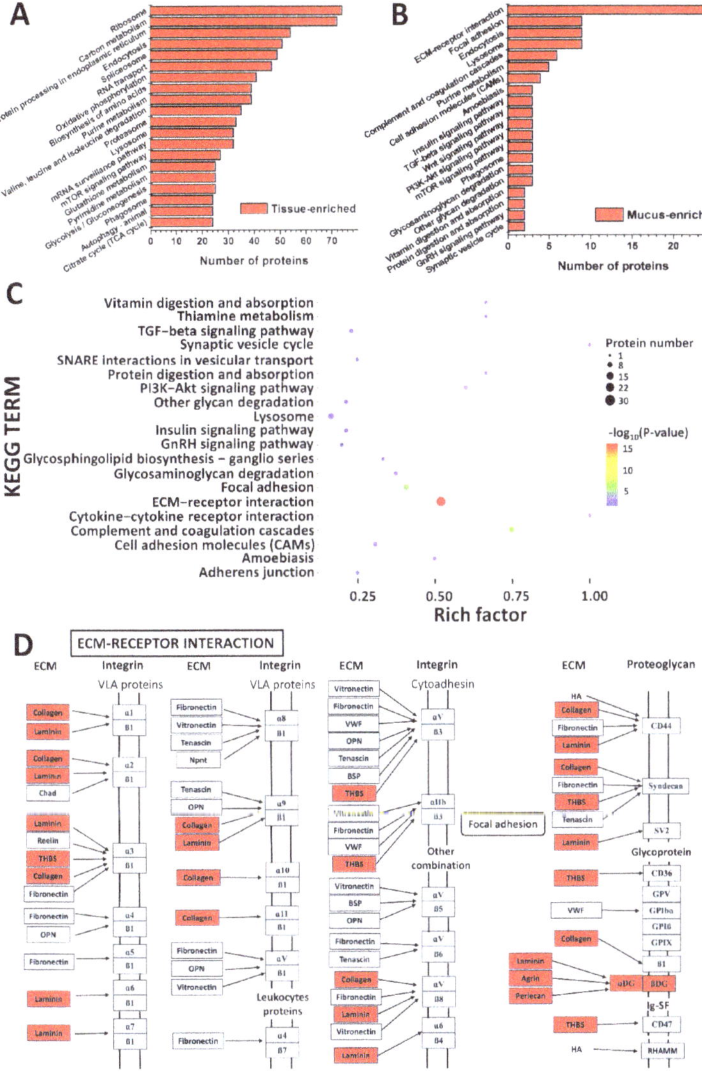

Figure 4. Kyoto Encyclopedia of Genes and Genomes (KEGG) pathway annotation of identified proteins in tissue and mucus of *A. coerulea*. (**A**) KEGG pathway annotation of tissue-enriched proteins. The horizontal axis is the number of proteins, whereas the vertical ordinates are the terms of the KEGG pathways. (**B**) KEGG pathway annotation of mucus-enriched proteins. (**C**) KEGG pathway enrichment of mucus-enriched proteins. The horizontal axis indicates rich factor and vertical ordinates are the terms of the KEGG pathways. Rich factor is the proportion of the number of differentially expressed proteins vs. the total number of proteins in the same KEGG pathway. The bubble shows the number of proteins matched in the KEGG pathway. The color represents $-\log_{10}(p$ value): Logarithmic conversion of Fisher exact test p value. (**D**) The 'ECM-receptor interaction' pathway matched from the KEGG PATHWAY database where the elevated proteins from mucus-enriched proteins are colored by red.

2.5. Metabolomics

Besides the protein or peptide components, metabolites in mucus were also determined by UPLC-QTOF MS/MS and identified from common metabolite databases by comparing the molecular weights. Principal component analysis (PCA) provides a summary of all the observations, revealing significant differences between mucus and tissue in both positive-ion (Figure 5A) and negative-ion (Figure 5C) detection modes. Similar results were obtained by orthogonal partial least-squares-discriminant analysis (OPLS-DA) and was subsequently used to determine the most significant metabolites in mucus and tissue in both positive-ion (Figure 5B) and negative-ion (Figure 5D) MS detection modes using the variable important plot (VIP) value (>1). The R2X, R2Y, and Q2Y values in the positive ion mode were 0.884, 0.999 and 0.997, respectively, whereas their values in negative ion mode were 0.712, 0.998 and 0.993, respectively, indicating a good predictive power and goodness-of-fit of OPLS-DA plots.

A total of 16 discriminating metabolites with three subgroups were obtained (Table 1) according to FC values of mucus vs. tissue. The FCs of nine metabolites including L-Glutamate, Succinylacetone, Linoleyl linolenate, Uridine, L-Proline, Inosine, Hypoxanthine, L-Valine and Guanosine are smaller than 1, representing a lower concentration in mucus than that in tissue. Five metabolites including 4-Hydroxy-L-proline, Citrulline, L-Leucine, 3-(Phosphoacylase mido)-L-alanine and L-Threonine have FC values ≈ 1, indicating similar concentrations between the two groups. The metabolite Tryptamine, a derivative of Tryptophan, is a potential neurotransmitter or neurotransmodulator, and displayed the largest FC value of 7.8. This value was significantly higher than all other values and indicates elevated enrichment in mucus over that of all other metabolites. This feature is of particular interest to us. A further KEGG scanning of Tryptamine shows that there are 16 proteins identified from proteomics and 26 mRNAs from transcriptomics (Figure 5E) in tissue homogenate. Only 1 'mucus-enriched' protein identified was matched to the downstream region of the 'Tryptophan metabolism' pathway. This indicates a strong Tryptophan/Tryptamine metabolic system in the cytosol, but not in mucus, allowing for Tryptamine synthesis or accumulation.

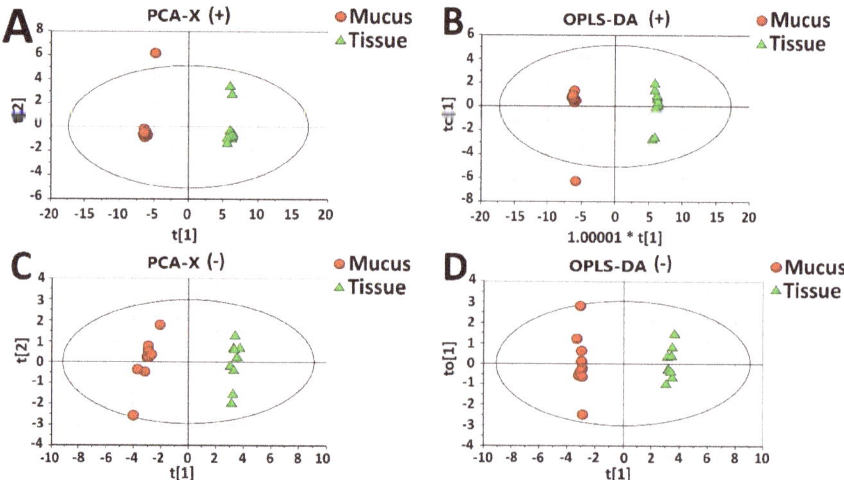

Figure 5. *Cont.*

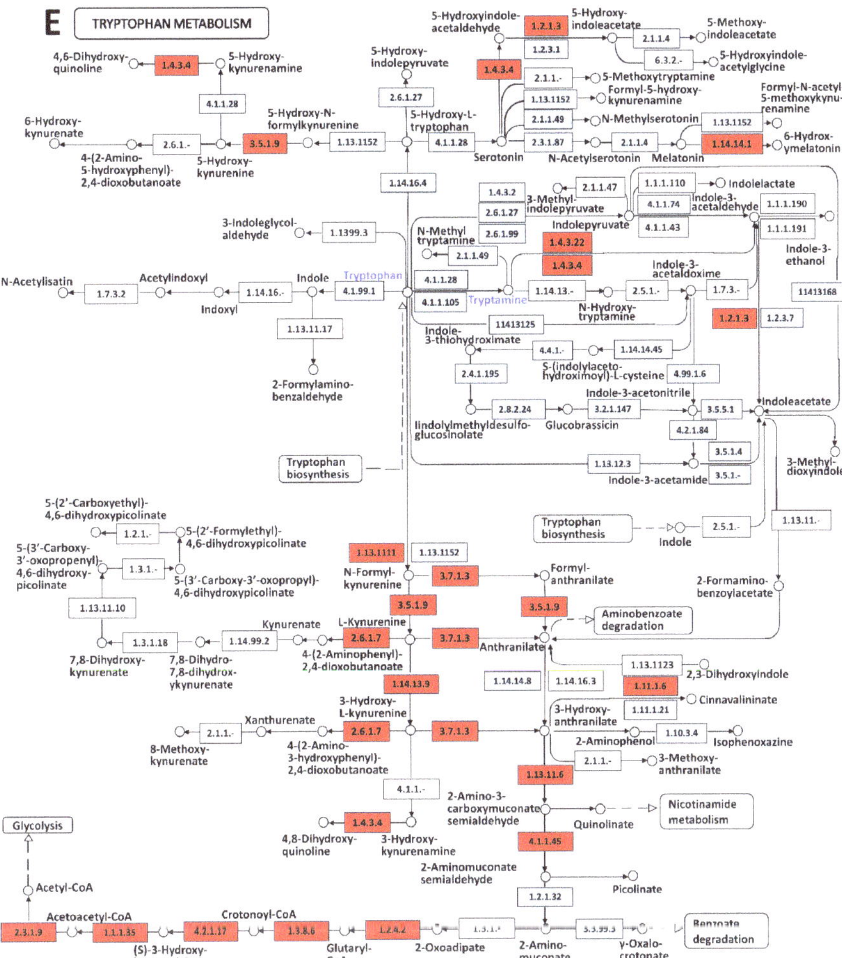

Figure 5. Plots of multivariate statistical analysis of all experimental groups in electrospray ionization (ESI) positive and negative-ion MS detection modes. The difference in substances between the two groups was screened by a variable importance plot (VIP). (**A**) Principal component analysis-X variogram (PCA-X) (+) and (**B**) Orthogonal partial least-squares-discriminant analysis (OPLS-DA) (+) scores plot of the mucus and tissue groups in the ESI positive-ion mode. Asterisk (*) indicates multiplication sign. (**C**) PCA-X (−) and (**D**) OPLS-DA (−) scores plot of the mucus and tissue groups in the ESI negative ion mode. The red plots represent the data from the secreted mucus, whereas the green plots represent data from the tissue homogenate. (**E**) The 'Tryptophan metabolism' pathway from KEGG PATHWAY database where the mRNAs from the transcriptomics of jellyfish tissue are colored by red, while the light blue and white boxes represent the background genes annotated or not in the KEGG PATHWAY database. Meanwhile, the metabolites 'Tryptophan' and 'Trytamine' are highlighted by blue.

Table 1. Metabolite difference between the secreted mucus and tissue homogenate.

Metabolite	FC	*p* Value	Related Pathway	Metabolite	FC	*p* Value	Related Pathway
Tryptamine	7.80	<0.0001	Tryptophan metabolism	Linoleyl linolenate	0.28	<0.0001	
4-Hydroxy-L-proline	1.26	<0.0001	Arginine and proline metabolism	Uridine	0.19	<0.0001	Pyrimidine metabolism
Citrulline	1.17	<0.0005	Arginine biosynthesis	L-Proline	0.13	<0.0001	Arginine and proline metabolism
L-Leucine	1.14	<0.0001	Valine, leucine and isoleucine metabolism	Inosine	0.12	<0.0001	Purine metabolism
3-(Phosphoacetylamido)-L-alanine	1.09	<0.0001	L-asparagine biosynthesis	Hypoxanthine	0.12	<0.0001	Purine metabolism
L-Threonine	0.88	0.0007	Valine, leucine and isoleucine biosynthesis	L-Valine	0.07	<0.0001	Valine, leucine and isoleucine metabolism
L-Glutamate	0.54	<0.0001	Arginine and proline metabolism	Guanosine	0.05	<0.0001	Purine metabolism
Succinylacetone	0.44	<0.0001	Tyrosine metabolism				

Note: FC values were obtained by comparing the mean concentration of each metabolite in mucus with that in tissue; FC value > 1 indicates a higher concentration in mucus, while the value < 1 indicates a lower concentration in mucus (n = 10). Metabolites analyzed based on MS/MS chromatograms.

2.6. Self-Protective Proteins

When danger or other stresses arise, a large secretion of mucus occurs as an emergent self-protective reaction. According to our GO and KEGG data, some of the mucus-enriched proteins identified relate to immunization or prevention mechanisms against harmful or invading organisms. Therefore, we have identified these as self-protective proteins of jellyfish mucus. On the whole, 11 metalloproteinases, 4 serine proteinase inhibitors, 3 superoxide dismutases (SODs) and 3 complements were successfully scanned (Table 2, Supplementary Table S1).

Table 2. Summary of self-protective proteins enriched in mucus from the jellyfish *A. coerulea*.

Accession	Swissprot Annotation	Description	Matched Species	Identify (%)
Metalloproteases				
TRINITY_DN45838_c0_g1 | m.27904	sp | Q20191	Zinc metalloproteinase nas-13	*Caenorhabditis elegans*	42.6
TRINITY_DN44124_c13_g11 | m.23323	sp | Q20191	Zinc metalloproteinase nas-13	*Caenorhabditis elegans*	34.9
TRINITY_DN35212_c0_g1 | m.10040	sp | P55115	Zinc metalloproteinase nas-15	*Caenorhabditis elegans*	30.8
TRINITY_DN45621_c1_g1 | m.27292	sp | Q8N119	Matrix metalloproteinase-21	*Mus musculus*	34.9
TRINITY_DN42325_c7_g8 | m.19392	sp | P51511	Matrix metalloproteinase-15	*Homo sapiens*	33.6
TRINITY_DN47471_c0_g1 | m.30857	sp | Q9UKF2	ADAM30	*Homo sapiens*	45.1
TRINITY_DN45145_c0_g4 | m.26001	sp | Q05910	ADAM8	*Mus musculus*	34.8
TRINITY_DN55675_c0_g2 | m.32080	sp | O75077	ADAM23	*Homo sapiens*	31.1
TRINITY_DN46375_c2_g1 | m.29588	sp | Q9UKP4	ADMTMS7	*Homo sapiens*	41.5

Table 2. *Cont.*

TRINITY_DN45686_c0_g1 l m.27481	sp l Q9P2N4	ADMTMS9	*Homo sapiens*	33.2
TRINITY_DN44955_c0_g2 l m.25457	sp l Q69Z28	ADMTMS16	*Mus musculus*	31.4
TRINITY_DN44955_c0_g1 l m.25456	sp l Q9UKP5	ADMTMS6	*Homo sapiens*	30.6
Serine protease inhibitors				
TRINITY_DN43329_c0_g6 l m.21485	sp l Q6J201	serine protease inhibitor	*Cyanea capillata*	56.4
TRINITY_DN45322_c11_g3 l m.26476	sp l Q60854	Serpin B6	*Mus musculus*	37.5
TRINITY_DN9397_c0_g3 l m.37796	sp l Q4R3G2	Serpin B6	*Macaca fascicularis*	38.8
TRINITY_DN29892_c0_g1 l m.6537	sp l Q4R3G2	Serpin B6	*Macaca fascicularis*	36.6
Superoxide dismutase				
TRINITY_DN37669_c0_g1 l m.12406	sp l Q8HXP2	Superoxide dismutase [Mn], mitochondrial	*Macaca mulatta*	75.3
TRINITY_DN36380_c0_g1 l m.11103	sp l P11428	Superoxide dismutase [Cu-Zn] 2	*Zea mays*	65.8
TRINITY_DN37490_c0_g1 l m.12223	sp l P24706	Superoxide dismutase [Cu-Zn]	*Onchocerca volvulus*	51.8
Complements				
TRINITY_DN45426_c0_g1 l m.26742	sp l Q00685	Complement C3	*Lethenteron camtschaticum*	30.7
TRINITY_DN45758_c0_g1 l m.27712	sp l Q00685	Complement C3	*Lethenteron camtschaticum*	25.9
TRINITY_DN43257_c1_g1 l m.21313	sp l P81187	Complement factor B	*Bos taurus*	23.6

2.6.1. Metalloproteinases

Metalloproteinases are enzymes characterized by a catalytic zinc ion in its active site [25]. They have been described as the toxic components responsible for the induction of tissue damage, necrosis and hemorrhage [26,27] in various venoms [28–30]. Eleven metalloproteases with four subgroups—zinc metalloproteinase, matrix metalloproteinase (MMP) [31], A disintegrin and metalloproteinase domain-containing protein (ADAM) [32,33], and A disintegrin and metalloproteinase with thrombospondin motifs (ADMTMS) [34,35]—are predicted with their similarity 30.6–42.6% in mucus-enriched proteins of *A. coerulea* (Table 2). All three predicted zinc metalloproteinases display a relatively conserved astacin-like subfamily segment with a HEXXH or even more strict HEXXHXXGXXH zinc-binding site/active site [36,37]. A model metalloproteinase (pdb ID: 3LQB) is utilized to do the sequence alignment (Figure 6A) and 3D modeling (Figure 6B). With the exception of the highly conserved HEXXHXXGXXH motif, four parallel β-sheets and three α-helixes are also structurally similar. The differences include an initial coil and turn in all three sequences, an extra α-helix and a turn in [28]DILKEVGS[35] of TRINITY_DN45838_c0_g1 l m.27904, an α-helix in [90]GPRCY[95], a changeable area with an α-helix in [139]RKYSHGQ[145] of TRINITY_DN44124_c13_g11 l m.23323, a turn in [143]PGGASTLGA[152] of TRINITY_DN35212_c0_g1 l m.10040 and an α-helix and turn in TRINITY_DN45838_c0_g1 l m.27904, an extra coil between 169 and 173 of all three predicted sequences, and a turn in [169]PEIT[183] of TRINITY_DN45838_c0_g1 l m.27904. Interestingly, two sequences—TRINITY_DN45838_c0_g1 l m.27904 and TRINITY_DN44124_c13_g11 l m.23323—aligned to the same metalloproteinase, nas-13, display a large molecular evolutional distance, and are close to XP 022799178.1 from the coral *Stylophora pistillata* and AAX09930.1 from the same jellyfish *A. coerulea*, respectively. Meanwhile, the remaining sequence, TRINITY_DN35212_c0_g1 l m.10040, is evolutionally close to a sequence JAC85096 from *Clytia hemisphaerica* (Figure 6C).

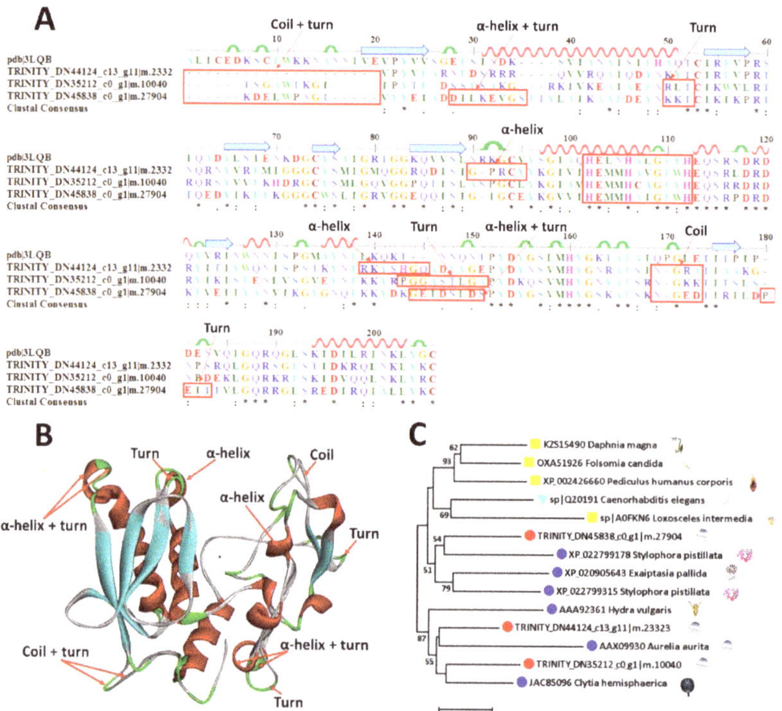

Figure 6. Sequence alignment, 3D modeling and phylogenetic analysis of putative zinc metalloproteinases from *A. coerulea*. (**A**) Three putative sequences TRINITY_DN45838_c0_g1 | m.27904, TRINITY_DN44124_c13_g11 | m.23323, and TRINITY_DN35212_c0_g1 | m.10040 in mucus-enriched proteins are aligned with a model metalloproteinase (pdb ID: 3LQB). At the bottom of columns, asterisks (*) show conserved positions, colons (:) show conserved substitutions and points (.) show non-conserved substitutions. Grey line, green bend, blue banded arrowhead and red solenoid represent coil, turn, sheet and helix, respectively. Different fragments are framed by red lines. (**B**) 3D modeling was simulated using the template metalloproteinase (pdb ID: 3LQB) by SWISS-MODEL and viewed by Discovery Studio 4.5. The colors grey, green, blue and red represent coils, turns, sheets and helices, respectively. Different fragments are indicated by red arrows. (**C**) Phylogenetic tree constructed using three putative zinc metalloproteinases and 11 other sequences from different species using MEGA 7 with the Neighbor-Joining method.

2.6.2. Serine Protease Inhibitors

Serine protease inhibitors or serpins [38] are the largest and most diverse superfamily of protease inhibitors [39], with inhibitory or non-inhibitory functions in blood coagulation, fibrinolysis, host defense, and impairment of motility of human glioblastoma cells [29,40,41]. They have widely been found in the venoms of many poisonous animals such as sea anemones, snakes, scorpions, spiders, anurans and hymenopterans [42]. Four sequences have been successfully scanned with their similarities identified at 36.6–56.4% to mucus-enriched proteins of *A. coerulea* (Table 2). Similar to zinc metalloproteinases, serpins are more similar in spatial structure than amino acid sequences simulated with a model serpin (pdb ID: 5CDZ) (Figure 7A,B). Two clusters of β-sheets are distributed in the central and tail areas, whereas 11 α-helixes are scattered in the periphery of proteins. The main structural variations contain an extended loop in [86]HPNDPLEP[93] of TRINITY_DN29892_c0_g1 | m.6537, [82]KHTAD[86] of TRINITY_DN45322_c11_g3 | m.26476 and [63]DLKEDIRGSSAFGFVGSEAALE[84] of TRINITY_DN9397_c0_g3 | m.37796, an extended coil and turn in [367]FPSLRFDE[374] of TRINITY_DN29892_c0_g1 | m.6537, and an extended coil in [368]RCAIPIPLV[376] of TRINITY_DN9397_c0_g3 | m.37796. Phylogenetic analysis shows that TRINITY_DN45322_c11_g3 | m.26476 and TRINITY_DN29892_c0_g1 | m.6537 are close to AHC98669 in *Rhipicephalus microplus* while TRINITY_DN9397_c0_g3 | m.37796 is near XP_015915455 in *Parasteatoda tepidariorum* (Figure 7C).

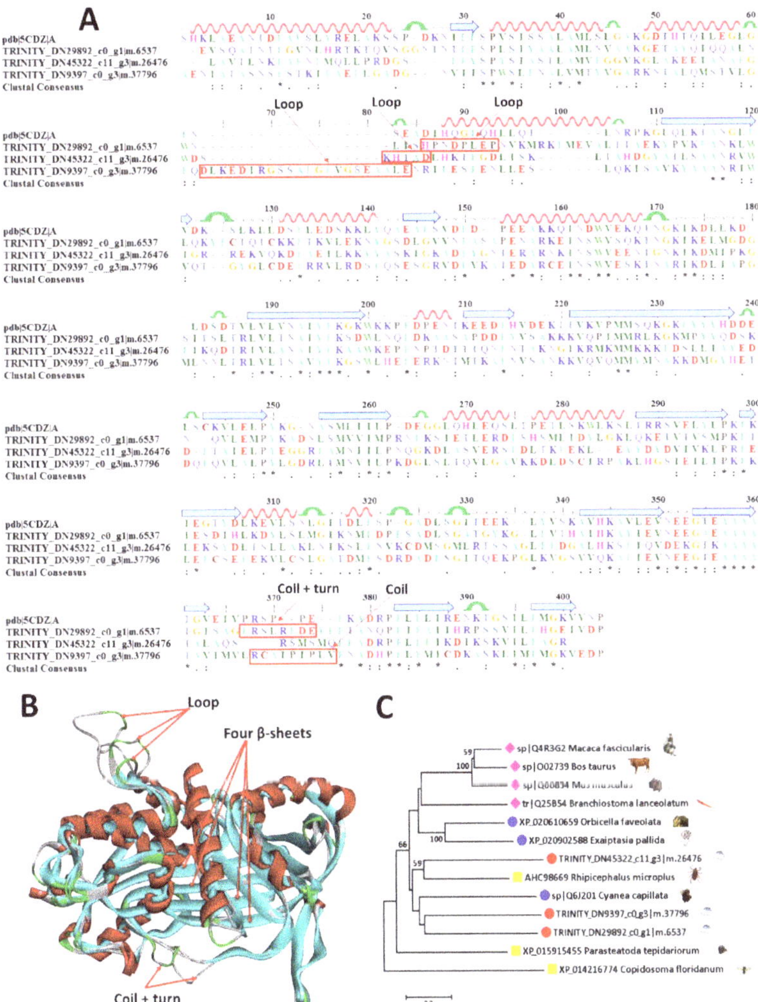

Figure 7. Sequence alignment, 3D modeling and phylogenetic analysis of the putative serpins from *A. coerulea*. (**A**) Three putative sequences TRINITY_DN45322_c11_g3 | m.26476, TRINITY_DN9397_c0_g3 | m.37796, and TRINITY_DN29892_c0_g1 | m.6537 in mucus-enriched proteins are aligned with a model serpin (pdb ID: 5CDZ). At the bottom of columns, asterisks (*) show conserved positions, colons (:) show conserved substitutions and points (.) show non-conserved substitutions. Grey line, green bend, blue banded arrowhead and red solenoid represent coil, turn, sheet and helix, respectively. Different fragments are framed by red lines. (**B**) 3D modeling was simulated using the template serpin (pdb ID: 5CDZ) by SWISS-MODEL and viewed by Discovery Studio 4.5. The colors grey, green, blue and red represent coils, turns, sheets and helices, respectively. Different fragments are indicated by red arrows. (**C**) Phylogenetic tree constructed by three putative serpins and 10 other sequences from different species using MEGA 7 with the Neighbor-Joining method.

2.6.3. Superoxide Dismutase

The serum superoxide dismutases (SODs) could effectively eliminate reactive oxygen species (ROS) and maintain the redox balance [43], thereby playing a key role in protection against oxidative tissue injury in both prokaryotes and eukaryotes [44–48]. Three sequences were successfully scanned with their similarities identified at 51.8–75.3% in mucus-enriched proteins of *A. coerulea* (Table 2). A SOD (pdb ID: 1Q0E) was selected as the model sequence (Figure 8A) and structural template (Figure 8B) of the two putative Cu/Zn-SODs. The core is constituted of two groups of parallel β-sheets that are surrounded by scattered helixes, turns and coils. The discrepancies include an extra helix-turn-helix in ^{52}DIYTRGC^{58} and an extended 130GTGAKRAGSQKTG^{142} of TRINITY_DN37490_c0_g1 | m.12223. Phylogenetic analysis shows that TRINITY_DN36380_c0_g1 | m.11103 is close to ALQ81852 in *Cyanea capillata* while TRINITY_DN37490_c0_g1 | m.12223 is near XP_022803558 in *Stylophora pistillata* (Figure 8C).

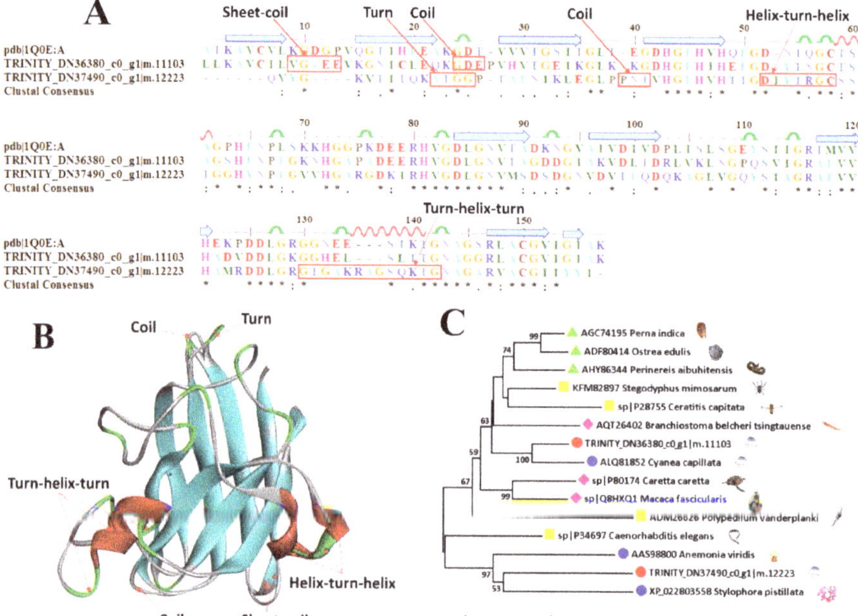

Figure 8. Sequence alignment, 3D modeling and phylogenetic analysis of the putative Cu/Zn superoxide dismutases (SODs) from *A. coerulea*. (**A**) Two putative sequences, TRINITY_DN36380_c0_g1 | m.11103 and TRINITY_DN37490_c0_g1 | m.12223, in mucus-enriched proteins are aligned with a model SOD (pdb ID: 1Q0E). At the bottom of columns, asterisks (*) show conserved positions, colons (:) show conserved substitutions and points (.) show non-conserved substitutions. Grey lines, green bends, blue-banded arrowheads and red solenoids represent coils, turns, sheets and helices, respectively. Different sections are framed by red lines. (**B**) 3D modeling was simulated using the template SOD (pdb ID: 1Q0E) by SWISS-MODEL and viewed by Discovery Studio 4.5. The colors grey, green, blue and red represent coils, turns, sheets and helices, respectively. Different sections are indicated by red arrows. (**C**) Phylogenetic tree constructed by two putative SODs in mucus-enriched proteins and 13 representative sequences from different species by MEGA 7 with the Neighbor-Joining method.

3. Discussion

3.1. Jellyfish Mucus Is a Rich Library for the Discovery of Novel Bioactive Compounds

Mucus provides a unique and multi-functional hydrogel interface between the epithelial cells and their external environment [49]. In most phyla of aquatic and terrestrial metazoans, it has exceptional properties, including elasticity, changeable rheology and an ability to self-repair, and therefore is an ideal medium for trapping and immobilizing pathogens [50]. Moreover, mucus is a rich library of bioactive components functioning against invasion or microorganisms in innate immunity. Sea star integument produces mucus with antioxidant proteins such as peroxiredoxin, catalase and SODs, and anti-microbial proteins such as lysozyme, melanotransferrin and ribosomal proteins [18]. Fish skin mucus serves as the first line of defense against pathogens and external stressors with antioxidant proteins, lectin, calmodulin, histone proteins, Cystatin-B, Apolipoprotein A1, and heat shock proteins [15]. In the human body, the protective mucus widely covers the epithelial cells on the surface of various tissues and organs such as respiratory organs, stomach, intestine, genitourinary organs, etc. [51–53]. Nasal mucus plays crucial roles in preventing microbial infections and protecting the lower airways from the unhealthy conditions of ambient air through fibrinogen, plasminogen, and complement factor C3 [54], whereas self-protective proteins from cervical mucus include serpins, phosphorylated proteins and heat-shock proteins [55].

In recent years, the large-scale outbreak of jellyfish blooms and dramatic increase of patients by jellyfish envenomation has drawn great attention in both fishery and medicine [56]. Immediate pain, redness and swelling occurs on local skin after being injured by contacting the tentacles of jellyfish, while serious systemic symptoms include multiple organ failures or even death [57]. The venom that is released from nematocyst of jellyfish tentacle possesses a large amount of toxins and other active proteins leading to these various manifestations, and is therefore naturally considered to be a rich library of bioactive compounds. Using the integrated methods of traditional liquid chromatography, amino acid sequencing and cDNA library alignment, the hemolytic proteins were firstly purified and identified from two box jellyfish, *Carybdea rastoni* [58] and *Carybdea alata* [59], in 2000; subsequently, more than ten hemolytic proteins have successfully been identified as a novel cytotoxic protein family [60–62]. By the combination of transcriptomics and proteomics, more than 170 potential toxin proteins, including metalloproteinases, an alpha-macroglobulin domain containing protein, two CRISP proteins, a turripeptide-like protease inhibitor, and particularly, nine novel members of a taxonomically restricted family of cnidarian pore forming toxins, were successfully identified from the jellyfish *Chironex fleckeri* on the basis of homology to known toxins in public sequence databases [63]. Similarly, 174 potential toxic proteins with 27 homologs to the toxins from venomous animals, including phospholipase A2, zinc metalloproteinase-disintegrin agkistin, serine protease inhibitor, plancitoxin-1, alpha-latrocrustotoxin-Lt1a, etc. were scanned in the jellyfish *Cyanea nozakii* [64].

When an envenomation happens, the skin firstly contacts the jellyfish mucus; this is also an important poisoning source described by fishers indicating that mucus also contains a large number of active components potentially deriving from nematocyst venoms. Because mucus covers the surface of tentacles, including nematocysts, any release of nematocyst venom has a certain portion of debris, thereby displaying a conventional material communication with mucus. Meanwhile, considering the long evolutionary history of jellyfish and the key role of mucus in innate immunity of other aquatic organisms, there should also be a large number of bioactive compounds to function as adhesion and in defense. In this study, we explored the stress-induced mucus secretion and its composition by a combination of proteomics and metabolites using a low-toxicity jellyfish, *A. coerulea*. We successfully scanned 1208 and 2421 proteins in mucus and tissue, respectively. Among them, 225 proteins are exclusively identified in mucus while 183 proteins are up-regulated when compared to that of the tissue. As expected, the mucus-enriched proteins possess dozens of functions as indicated by GO and KEGG analyses. Moreover, 21 self-protective proteins, such as metalloproteinases, serine proteinase inhibitors, SODs and complements were successfully scanned against potential external invasion.

3.2. Tryptamine Release Indicates an Elevated Stress of Jellyfish When Stimulated

Metabolomics is a technique for studying metabolic networks in biological systems by examining the metabolite profiles and their dynamic changes before and after stimulation or disturbance [65,66]. The number of metabolites is much less than those of genes and proteins, and small changes in gene and protein expression can be amplified at the metabolite levels [67]. In this study, we have found three groups of metabolites with lower, equal and higher expressions in mucus when compared to those in tissue by metabolomics. Because metabolites in mucus have the tendency to disperse into the surrounding seawater [68,69] and all are feasibly synthesized and secreted from jellyfish tissue, it is reasonable that metabolites with lower concentrations (L-Glutamate, Succinylacetone, Linoleyl linolenate, Uridine, L-Proline, Inosine, Hypoxanthine, L-Valine and Guanosine) in mucus are in normal conditions while those with equal concentrations (4-Hydroxy-L-proline, Citrulline, L-Leucine, 3-(Phosphoacylase mido)-L-alanine and L-Threonine) should already have been well enriched with the balance between release and diffusion.

Of particular interest, the metabolite Tryptamine displays the maximal FC value of 7.8, which is far higher and indicates much better enrichment than that of other metabolites. Tryptamine is a group of monoamine alkaloids including serotonin (5-hydroxytryptamine, 5-HT) and melatonin, as well as other compounds known for their neurotransmitter properties [70]. It derives from the amino acid-trytophan by tryptophan decarboxylase (EC 4.1.1.105 and EC 4.1.1.28), which is also named aromatic L-amino acid decarboxylase (AADC or AAAD), DOPA decarboxylase (DDC) and 5-hydroxytryptophan decarboxylase, and which plays an important role in the dopaminergic system participating in the uptake and decarboxylation of amine precursors in the peripheral tissues [71,72]. The typical tryptamine, serotonin, is one of the most important and widely studied hormones in humans and other vertebrates involved in regulation and modulation of multiple processes within the central nervous system and behavior [73]. It also plays an important role in gastrointestinal motility, vascular tone and platelet function, and has been related to various pathophysiological processes [74]. However, less research on tryptamine is reported in marine invertebrates. It is reported that cells with serotonin are concentrated at the anterior pole of hydrozoan planulae [75]. Brain serotonin levels in crayfish are reported to greatly increase when exposed to pressure [76]. Because jellyfish have a well-developed peripheral nervous system, and it is reported that tryptamine can be accumulated and released in large quantities under stress in marine invertebrates [76,77], we hypothesize that the increases of tryptamine release and mucus secretion indicate an elevated stressful response when stimulated by taking the jellyfish out of seawater. This will provide a new mechanism for jellyfish to secrete mucus although further validation is needed.

3.3. Self-Protective Proteins Play a Key Role in Innate Immunity of Jellyfish

In mucus-enriched proteins, we have successfully scanned dozens of self-protective proteins including 11 metalloproteinases, four serine protease inhibitors, three SODs, and three complements. According to the activity of metalloproteinase that degrades extracellular matrix proteins and bioactive molecules on the surface of the jellyfish *A. coerulea*, three functions are proposed as an important self-protective factor. The first function is to process the extracellular matrix such as collagen, laminin and THBS to form the main component of the mucus [78]. The second self-protective effect is to directly digest or degrade the invasive toxic component or microorganism, which can further induce the third function, transferring the external stimuli to the body by activating or deactivating the signaling through cutting or digesting the extracellular part of membrane proteins such as Her2 receptors. On the contrary, serine protease inhibitors form stable complexes with their target enzymes to control the activity of serine proteases [79], thereby playing an important role in innate immunity and environmental stability in the body [80]. On the surface of jellyfish, serine protease inhibitors are able to prevent the over-digestion of functional proteins through inhibiting the serine proteases from both jellyfish themselves and the exogenous invading pathogens. As the natural antioxidant function, SODs in jellyfish mucus should catalyze the production of O_2 and H_2O_2 from superoxide

(O_2^-), which results in less harmful reactants and protects cellular components from being oxidized by ROSs [81,82] that are feasibly from jellyfish tissue, invading microorganisms or environmental conditions such as water pollution, eutrophication, anoxia and radiation. In addition to the above, we also listed three complements. The complement system is a highly complicated defense system in the innate immunity of invertebrates, which directly participates in the lysis of pathogen cell [83]. Most studies on cnidarian complement have focused on C3 [83,84], of which the activation plays a central role and is required for all three pathways of complement activation [85]. By comparison, the complement factor B is a component of the alternative pathway of complement activation to form the pore complexes and lyse the invading micrograms.

4. Materials and Methods

4.1. A. coerulea Samples

A. coerulea were collected alive from an artificial aquafarm in Shanghai, China in May 2017. Jellyfish were transported in a 3-L plastic bag fully filled with seawater, to prevent damage from sloshing. In the laboratory all individuals were maintained in buckets of seawater at 18–22 °C.

4.2. Transcriptome Sequencing and Its Annotation

Total RNA of jellyfish *A. coerulea* was extracted with TRIzol (Life Technologies) following the manufacturer's procedure. RNA purity, concentration and integrity were evaluated using NanoDrop 2000 UV–Vis spectrophotometer (Thermo Fisher Scientific, Waltham, MA, USA), Agilent 2100 Bioanalyzer (Agilent Technologies, Santa Clara, CA, USA) and RNA 6000 Nano LabChip®Kit (Agilent Technologies, Santa Clara, CA, USA). The mRNA was enriched by polyT oligo-conjugated magnetic beads, then fragmentation buf Santa Clara, America fer was added to break the mRNA into short fragments. A single strand of cDNA was synthesized by using random hexamers as template. The double-stranded cDNA was synthesized by adding buffer, dNTPs, DNA polymerase I and RNase H. The double-stranded cDNA was purified by AMPure XP beads (Beckman, Atlanta, GA, USA). The purified double-stranded cDNA was repaired at the end, A-tail was added and sequencing junction was connected, and AMPure XP beads were used to select the fragment size. Finally, PCR was amplified and purified by AMPure XP beads, and the final cDNA library was obtained. Illumina HiSeq sequencing was carried out after qualification.

The raw data of sequencing were evaluated by FastQC (version: 0.11.2) and filtered by Trimmomatic (version: 0.36) to remove the joints as well as the sequences with low quality. Then the clean data were de novo assembled into transcripts by Trinity, which were further annotated by blast using the databases NR (National Center for Biotechnology Information non-redundant protein sequences) and Swissprot. According to the annotation results of transcripts NR and Swissprot, GO and KEGG analyses were finally performed according to the annotation of transcripts by NR and Swissprot.

4.3. Mucus Collection and Tissue Homogenate Preparation

The jellyfish *A. coerulea* were fasted for 48 h and gently washed thoroughly with sterile filtered artificial seawater before experiments. After measurement of body weight, jellyfish were gently put into a funnel with one layer of medical gauze, which was then sealed with a plastic wrap to avoid liquid evaporation. Jellyfish mucus was collected every 10 min with a 15 mL centrifuge tubes, for a total of 1 h. The debris was removed by centrifugation at 4000 rpm for 10 min and the supernatant was collected and stored at −80 °C for further experiments. Jellyfish tissue homogenate was obtained using the method of ultrasonic extraction. In every working cycle, the working time and resting time were 20 s and 1 min with a power of 200 W. The total working time was 2 min. The supernatant of jellyfish tissue homogenate was collected after centrifugation at 4000 rpm for 10 min at 4 °C, which was then stored at −80 °C for further analysis.

Four replicates of all experiments were performed, using independent batches of *A. coerulea* samples. Protein concentrations of both mucus and tissue homogenate were determined by Bradford's assay. Bovine Serum Albumin (BSA) was used to construct the standard curve.

4.4. SDS-PAGE

Jellyfish mucus and tissue homogenate were assessed by sodium dodecyl sulfate polyacrylamide gel electrophoresis (SDS-PAGE) using 5% stacking gel and 10% resolving gel. Samples were mixed with 5× Loading Buffer (v/v = 4:1) and then heated at 100 °C for 5 min prior to loading. Eight samples including six mucus (10–60 min), one tissue homogenate and the protein marker with the molecular weight scale 2–250 kDa were electrophoresed simultaneously. The selected voltages were 120 V and 180 V for stacking and resolving gel (PowerPac™ Basic, Bio-Rad, Hercules, CA, USA), respectively. To reduce the staining background as well as to improve the dye sensitivity, the PAGE gel was boiled in deionized water for 3 min and then shaken at 60–70 rev/min for 5 min in a rotary shaker. Subsequently, the PAGE gel was stained with Coomassie Brilliant Blue R-250 staining (Beyotime, Shanghai, China) at 60–70 r/min for 30 min, decolored with the deionized water overnight, and finally scanned for images (Perfection V700 Photo, Epson, Suwa, Japan).

4.5. Proteomics

4.5.1. Protein Sample Preparation

The proteins of jellyfish mucus and tissue homogenate were precipitated with four times the volume of trichloroacetic acid (TCA) at 4 °C overnight. The sediment was collected after centrifugation at 15,000× *g* for 15 min at 4 °C, further treated with a pre-cooled acetone solution for 30 min, and re-centrifuged at 15,000× *g* for 15 min at 4 °C. After freeze-drying, the precipitates were dissolved in the lysis buffer (8.0 M Urea, 1× Protease inhibitor, 100 mM Tris-HCl, pH 8.0) overnight. Subsequently, the undissolved debris was removed by centrifugation at 15,000× *g* for 15 min at 4 °C, whereas the supernatant was stored at −80 °C for further use.

After mass quantification of the pre-treated samples, 60 µg mucus or tissue homogenate were mixed with 5 µL 1 M dithiothreitol (DTT) at 37 °C for 1 h, and reacted with 20 µL 1 M indoleacetic acid (IAA) in the dark at room temperature for 1 h. After that, samples were collected by centrifugation in ultrafiltration tubes and sequentially rinsed three times with 100 µL UA solution (8 M urea, 100 mM Tris-HCl, pH 8.0) and 100 µL 50 mM NH_4HCO_3, respectively. Finally, the collected samples were digested with Trypsin (protein:trypsin = 50:1) at 37 °C for 12–16 h, and the digested proteins were lyophilized and stored at −80 °C for further use.

4.5.2. LC-MS/MS

Samples were analyzed by liquid chromatography-mass spectrometry/mass spectrometry (LC-MS/MS) using an Orbitrap Fusion Lumos mass spectrometer coupled to an Easy nLC/Ultimate 3000 nano-HPLC chromatography system (Thermo Fisher Scientific, Waltham, MA, USA). In the process of separation, the column was equilibrated with 95% buffer A (0.1% formic acid). The digested proteins were desalted with a C18 pre-column (3 mm, 100 µm × 20 mm, Thermo Scientific, USA). After loading and washing the digested proteins, the separation was performed with an analytical C18 column (1.9 mm, 150 µm × 120 mm, Thermo Scientific, USA) at a flow rate of 600 nL/min for 75 min. The mobile phase was the mixture of the buffer A (0.1% FA, H_2O) and the buffer B (0.1% formic acid (FA), 100% acetonitrile (ACN)). Elution gradient parameters were setup as follow: 0–14 min, 7–13% buffer B; 14–51 min, 13–23% buffer B; 51–68 min, 23–36% buffer B; 68–69 min, 36–100% buffer B; and 69–75 min, 100% buffer B. Mass spectrometric analyses were carried out by an automated data-dependent Tandem Mass Spectrometry (MS/MS) analysis with full scans (300–1400 *m*/*z*) that was acquired from proteins in the Orbitrap at a mass resolution of 120,000. The positive ion mode and the negative ion mode were separately employed with the spray voltage of the mass spectrometer at 2000 V and 600 V, and the spray temperature of 320 °C for peptides. Normalized collision energy was set to 35% and the stepped collision energy was 5%. Automatic gain control settings for Fourier Transform Mass Spectrometry (FTMS) survey scans were 500,000 and for FT MS/MS scans 5000. Maximum injection time was 50 ms for survey scans and 35 ms for MS/MS scans.

4.5.3. Database Search and Bioinformatics Analysis

All raw data were aligned from the *A. coerulea* transcriptomic database, which was previously built by us. Raw MS files were processed for the peptide analysis using the software Proteome Discoverer 1.4 (ver. 1.4.0.2888; Thermo Fisher Scientific, Waltham, MA, USA). The parameters used for data analysis were: enzyme = trypsin, max missed cleavages = 2, fix modifications = cysteine carbamido methylation, variable modifications = methionine oxidation, *N*-terminal acetylation, peptide mass tolerance = ±15 ppm, and fragment mass tolerance = 20 mmu. The false discovery rate (FDR) < 0.01 was selected for peptide and protein identification. Differential protein screening was performed at a threshold of 2 fold change (FC). FC ≥ 2, FC ≤ 0.5 and 0.5 < FC < 2 representing, up, down and no significant change in the data of protein expression, respectively.

The bioinformatics of Gene Ontology (http://www.geneontology.org, GO) was analyzed on the differentially expressed proteins with a 2 FC according to biological processes, cellular components and molecular functions. GO enrichment was analyzed on the differentially expressed proteins with $p \leq 0.05$ that was calculated based on a hypergeometric distribution with the default database as the background. The signal pathways were analyzed by the primary public database Kyoto Encyclopedia of Genes and Genomes (KEGG) (http://www.kegg.jp/kegg/pathway.html), and enriched by the tool Pathway Maps. *p*-values were calculated based on a hypergeometric distribution with the default KEGG database as the background. Multiple sequence alignment analysis was performed with BioEdit software (http://www.mbio.ncsu.edu/bioedit/page2.html) under default parameters. RNA sequences of *A. coerulea* were translated into protein through ORFfinder online service (https://www.ncbi.nlm.nih.gov/orffinder). Sequences used for multiple sequence alignment were collected from SwissProt (http://www.uniprot.org/uniprot) or NCBI (https://blast.ncbi.nlm.nih.gov/Blast.cgi) databases, and performed with CLUSTALW program using Bioedit software (version: 7.0.5.3). Finally, phylogenetic trees were constructed using MEGA7 using the Neighbor-Joining method, whereas the 3D modeling was carried out by the combination of the online service SWISS-MODEL (https://swissmodel.expasy.org/) and software Discovery studio 4.5.

4.6. Metabolomics

4.6.1. Sample Preparation

Jellyfish mucus and tissue homogenate were precipitated with three times the volume of methanol. The precipitates were removed by 4000 rpm centrifugation for 10 min at 4 °C after shaking of the mixture for 1 min. The clear supernatant (100 μL) was transferred to a sampling vial for UPLC-QTOF-MS/MS analysis. Meanwhile, the 100 μL aliquot of each sample was also transferred to a sampling vial and mixed as a quality control (QC) to check the stability of the system and method.

4.6.2. Metabolites Acquisition

An Agilent 6520 UPLC-QTOF MS/MS (Agilent Technologies, Santa Clara, CA, USA) was used in the study. Chromatographic separations were performed on an XSELECTTM HSS T3 column (2.1 mm × 100 mm, 2.5 μm, Waters, Milford, CT, USA) at a column oven temperature of 40 °C; 0.1% formic acid (A) and ACN (B) were used as the mobile phase. The gradient conditions were as follows: 0–2 min, 5% buffer B; 2–17 min, 5–95% buffer B; 17–20 min, 95% buffer B; 20–21 min, 95% buffer B. The post time was 6 min for column equilibration. The flow rate was maintained at 0.4 mL/min, the injection volume was 4 μL and the auto-sampler temperature was set at 4 °C. Electrospray ionization source (ESI) was set in both positive and negative-ion mode, and MS parameters were performed as follows:

the scanning range was set 50~1100 m/z, electrospray capillary voltage with 4 kV in positive mode and 3.5 kV in negative mode were used, the nebulizer pressure was set at 45 psi, nitrogen was used as drying gas with a flow rate of 11 L/min and temperature was set as 350 °C, fragment voltage was maintained at 120 V, skimmer voltage was set at 60 V, Octopole RF Peake was set at 750 V, 121.0509 Da and 922.0098 Da were used at reference masses (m/z).

4.6.3. Data Reduction and Pattern Recognitio

All data were acquired using Agilent MassHunter workstation software version B.01.04 (Agilent, Santa Clara, CA, USA). Firstly, the UPLC-QTOF MS/MS raw data were converted to mzdata files. The isotope interferences were excluded and the threshold of the absolute peak height was set at 500. The R package "xcms" was employed to generate a data matrix through peak extraction, alignment and integration, and the forma visual table including sample name, and peak indexes (m/z-Rt pairs and peak area). XCMS parameters were default settings except for the following: fwhm = 8, bw = 10 and snthersh = 5. All the ions were filtered based on the 80% rule before all of the detected ions in each sample were normalized to the sum of the peak area to obtain the relative intensity of metabolites based on MATLAB7.1 (MathWorks, Natick, MA, USA). After being normalized, ion intensities were converted to CSV (Comma-Separated Values) data and imported into the SIMCA-P program (version 12.0, Umetrics, Umea, Sweden) for principal component analysis (PCA) and orthogonal partial least-squares-discriminant analysis (OPLS-DA) after mean-centering and Pareto scaling. The parameters (R2X, R2Y, and Q2Y) showing the goodness of fit and prediction were assessed by SIMCA-P for internal validation.

In order to verify the different metabolites, first ions based on the extracted ion chromatogram (EIC) and then the extracted molecular weight with the common metabolite databases, such as the Human Metabolome Database, (http://metlin.scripps.edu) were confirmed. KEGG website was also adopted to enrich the relative pathways for these metabolites.

4.7. Statistics

All data were expressed as the mean ± standard deviation (SD). Statistically significant differences between groups were determined by one-way ANOVA and the Tukey test for multiple comparisons. All results were considered to be statistically significant at $p < 0.05$.

5. Conclusions

In this study, we have explored the stress-induced mucus secretion and its constituent composition in the jellyfish *A. coerulea* by a combination of proteomics and metabolomics. Our first conclusion is that two different but successive phases have been drawn from the initial stress to the final autolysis and death, with an obvious inflection point occurring at 30 min after removing the jellyfish from seawater. The results of proteomics using GO and KEGG analyses drew our second conclusion that the proteins in stress-induced mucus are independent (i.e., different) to those in tissue homogenate. We also identified that the mucus-enriched proteins are mainly located in the extracellular or membrane-associated region, while the tissue-enriched proteins are located in the intracellular compartment. The results of metabolomics are of particular interest; the potential neurotransmitter or neuromodulator, tryptamine, displays the maximal FC value of 7.8, a significantly elevated value among 16 other different metabolites in stress-induced mucus compared to those of tissue homogenate. This supports the hypothesis that a drastic nerve stress response, as well as a tempestuous release of neurotransmitters, occurs upon stress initiation. Finally, 11 metalloproteinases, four serine proteinase inhibitors, three SODs, three complements, and four toxin-related proteins were successfully assigned to function as self-protective components. In summary, our results provide a constituent profile of proteins and metabolites in stress-induced mucus and tissue homogenate of *A. coerulea*. This profile is, we believe, particularly important in equipping us with a better understanding of the process of stress-induced mucus secretion, as well as signaling the important role these self-protective components play in the innate immunity of jellyfish and in the ongoing discovery of novel bioactive compounds.

Supplementary Materials: The following are available online at http://www.mdpi.com/1660-3397/16/9/341/s1, Table S1: Sequences of self-protective proteins.

Author Contributions: Investigation, W.L., F.M. and G.J.; Methodology, L.X.; Project administration, L.Z.; Software, H.L., C.M. and T.L.; Writing—original draft, W.L. and L.X.; Writing—review & editing, G.L.M., J.Z. and L.X.

Funding: This work was supported by the Pujiang talent plan project (16PJ1411000) from Shanghai science and Technology Commission, Excellent youth talent program from Shanghai Municipal Commission of Health and Family Planning (2017YQ007), and the general program from the National Natural Science Foundation of China (81470518, 81770329).

Acknowledgments: The authors acknowledge the assistance of MeiGui Zhu, Na Yi and Ning Chen for determination of proteomics from the company Bangfei Bioscience in China, and the help of language refining by I. Darren Grice from Institute for Glycomics of Griffith University, Australia. We also thank Xinshu Li who generously provided us with the jellyfish *A. coerulea*.

Conflicts of Interest: The authors declare no conflict of interest.

References

1. Zoccarato, L.; Celussi, M.; Pallavicini, A.; Fonda Umani, S. *Aurelia aurita* Ephyrae Reshape a Coastal Microbial Community. *Front. Microbiol.* **2016**, *7*, 749. [CrossRef] [PubMed]
2. Li, R.; Yu, H.; Xue, W.; Yue, Y.; Liu, S.; Xing, R.; Li, P. Jellyfish venomics and venom gland transcriptomics analysis of *Stomolophus meleagris* to reveal the toxins associated with sting. *J. Proteom.* **2014**, *106*, 17–29. [CrossRef] [PubMed]
3. Wilcox, C.L.; Headlam, J.L.; Doyle, T.K.; Yanagihara, A.A. Assessing the Efficacy of First-Aid Measures in *Physalia sp.* Envenomation, Using Solution- and Blood Agarose-Based Models. *Toxins* **2017**, *9*, 149. [CrossRef] [PubMed]
4. Zhang, L.; He, Q.; Wang, Q.; Zhang, B.; Wang, B.; Xu, F.; Wang, T.; Xiao, L.; Zhang, L. Intracellular Ca^{2+} overload induced by extracellular Ca^{2+} entry plays an important role in acute heart dysfunction by tentacle extract from the jellyfish *Cyanea capillata*. *Cardiovasc. Toxicol.* **2014**, *14*, 260–274. [CrossRef] [PubMed]
5. Lazcanopérez, F.; Arellano, R.O.; Garay, E.; Arreguínespinosa, R.; Sánchezrodríguez, J. Electrophysiological activity of a neurotoxic fraction from the venom of box jellyfish *Carybdea marsupialis*. *Comp. Biochem. Physiol. Part C Toxicol. Pharmacol.* **2017**, *191*, 177–182. [CrossRef] [PubMed]
6. Sánchez-Rodríguez, J.; Torrens, E.; Segura-Puertas, L. Partial purification and characterization of a novel neurotoxin and three cytolysins from box jellyfish (*Carybdea marsupialis*) nematocyst venom. *Arch. Toxicol.* **2006**, *80*, 163–168. [CrossRef] [PubMed]
7. Ruan, Z.; Liu, G.; Guo, Y.; Zhou, Y.; Wang, Q.; Chang, Y.; Wang, B.; Zheng, J.; Zhang, L. First report of a thioredoxin homologue in jellyfish: Molecular cloning, expression and antioxidant activity of CcTrx1 from *Cyanea capillata*. *PLoS ONE* **2014**, *9*, e97509. [CrossRef] [PubMed]
8. Harada, K.; Maeda, T.; Hasegawa, Y.; Tokunaga, T.; Ogawa, S.; Fukuda, K.; Nagatsuka, N.; Nagao, K.; Ueno, S. Antioxidant activity of the giant jellyfish *Nemopilema nomurai* measured by the oxygen radical absorbance capacity and hydroxyl radical averting capacity methods. *Mol. Med. Rep.* **2011**, *4*, 919–922. [PubMed]
9. Cassoli, J.S.; Verano-Braga, T.; Oliveira, J.S.; Montandon, G.G.; Cologna, C.T.; Peigneur, S.; Pimenta, A.M.; Kjeldsen, F.; Roepstorff, P.; Tytgat, J.; et al. The proteomic profile of Stichodactyla duerdeni secretion reveals the presence of a novel O-linked glycopeptide. *J. Proteom.* **2013**, *87*, 89–102. [CrossRef] [PubMed]
10. Stabili, L.; Schirosi, R.; Parisi, M.G.; Piraino, S.; Cammarata, M. The Mucus of *Actinia equina* (Anthozoa, Cnidaria): An Unexplored Resource for Potential Applicative Purposes. *Mar. Drugs* **2015**, *13*, 5276–5296. [CrossRef] [PubMed]
11. Provan, F.; Nilsen, M.M.; Larssen, E.; Uleberg, K.E.; Sydnes, M.O.; Lyng, E.; Oysaed, K.B.; Baussant, T. An evaluation of coral lophelia pertusa mucus as an analytical matrix for environmental monitoring: A preliminary proteomic study. *J. Toxicol. Environ. Health Part A* **2016**, *79*, 647–657. [CrossRef] [PubMed]
12. Salinas, E.M.; Cebada, J.; Valdes, A.; Garateix, A.; Aneiros, A.; Alvarez, J.L. Effects of a toxin from the mucus of the Caribbean sea anemone (*Bunodosoma granulifera*) on the ionic currents of single ventricular mammalian cardiomyocytes. *Toxicon* **1997**, *35*, 1699–1709. [CrossRef]

13. Patwa, A.; Thiéry, A.; Lombard, F.; Lilley, M.K.; Boisset, C.; Bramard, J.F.; Bottero, J.Y.; Barthélémy, P. Accumulation of nanoparticles in "jellyfish" mucus: A bio-inspired route to decontamination of nano-waste. *Sci. Rep.* **2015**, *5*, 11387. [CrossRef] [PubMed]

14. Hanaoka, K.I.; Ohno, H.; Wada, N.; Ueno, S.; Goessler, W.; Kuehnelt, D.; Schlagenhaufen, C.; Kaise, T.; Irgolic, K.J. Occurrence of organo-arsenicals in jellyfishes and their mucus. *Chemosphere* **2001**, *44*, 743–749. [CrossRef]

15. Patel, D.M.; Brinchmann, M.F. Skin mucus proteins of lumpsucker (*Cyclopterus lumpus*). *Biochem. Biophys. Rep.* **2017**, *9*, 217–225. [CrossRef] [PubMed]

16. Li, S.; Jia, Z.; Li, X.; Geng, X.; Sun, J. Calmodulin is a stress and immune response gene in Chinese mitten crab *Eriocheir sinensis. Fish. Shellfish Immunol.* **2014**, *40*, 120–128. [CrossRef] [PubMed]

17. Rajan, B.; Fernandes, J.M.; Caipang, C.M.; Kiron, V.; Rombout, J.H.; Brinchmann, M.F. Proteome reference map of the skin mucus of Atlantic cod (*Gadus morhua*) revealing immune competent molecules. *Fish. Shellfish Immunol.* **2011**, *31*, 224–231. [CrossRef] [PubMed]

18. Elise, H.; Baptiste, L.; Ruddy, W.; Peter, L. An integrated transcriptomic and proteomic analysis of sea star epidermal secretions identifies proteins involved in defense and adhesion. *J. Proteom.* **2015**, *128*, 83–91.

19. Jatkar, A.A.; Brown, B.E.; Bythell, J.C.; Guppy, R.; Morris, N.J.; Pearson, J.P. Coral mucus: The properties of its constituent mucins. *Biomacromolecules* **2010**, *11*, 883. [CrossRef] [PubMed]

20. Shanks, A.L.; Graham, W.M. Chemical defense in a scyphomedusa. *Mar. Ecol. Prog. Ser.* **1988**, *45*, 81–86. [CrossRef]

21. Heeger, T.; Möller, H. Ultrastructural observations on prey capture and digestion in the scyphomedusa *Aurelia aurita. Mar. Biol.* **1987**, *96*, 391–400. [CrossRef]

22. Bonnet, D.; Molinero, J.C.; Schohn; Yahia, M.D. Seasonal changes in the population dynamics of *Aurelia aurita* in Thau lagoon. *Cah. De Biol. Mar.* **2012**, *53*, 343–347.

23. Scorrano, S.; Aglieri, G.; Boero, F.; Dawson, M.N.; Piraino, S. Unmasking Aurelia species in the Mediterranean Sea: An integrative morphometric and molecular approach. *Zool. J. Linn. Soc.* **2016**, *180*, 243–267.

24. Wang, Y.; Jiang, H.; Meng, H.; Li, J.; Yang, X.J.; Zhao, B.; Sun, Y.; Bao, T. Antidepressant Mechanism Research of Acupuncture—Insights from an Genome-wide Transcriptome Analysis of Frontal Cortex in Rats with Chronic Restraint Stress. *Evid. Based Complement. Altern. Med.* **2017**, *2017*, 1676808. [CrossRef] [PubMed]

25. Takeda, S.; Takeya, H.; Iwanaga, S. Snake venom metalloproteinases: Structure, function and relevance to the mammalian ADAM/ADAMTS family proteins. *Biochim. Biophys. Acta* **2012**, *1824*, 164–176. [CrossRef] [PubMed]

26. Cristina, S.M.; Tamires, L.S.; Vieira, S.M.; Martins, M.C.; Maria, S.F.; Cortes, F.K.; Fábio, O.; Patriarca, M.T.W.; Roberto, M.I. Interaction between TNF and BmooMP-Alpha-I, a Zinc Metalloprotease Derived from Bothrops moojeni Snake Venom, Promotes Direct Proteolysis of This Cytokine: Molecular Modeling and Docking at a Glance. *Toxins* **2016**, *8*, 223. [CrossRef]

27. Jacob-Ferreira, A.L.; Menaldo, D.L.; Sartim, M.A.; Riul, T.B.; Dias-Baruffi, M.; Sampaio, S.V. Antithrombotic activity of Batroxase, a metalloprotease from Bothrops atrox venom, in a model of venous thrombosis. *Int. J. Biol. Macromol.* **2016**, *95*, 263. [CrossRef] [PubMed]

28. Kang, C.; Han, D.Y.; Park, K.I.; Pyo, M.J.; Heo, Y.; Lee, H.; Kim, G.S.; Kim, E. Characterization and neutralization of Nemopilema nomurai (Scyphozoa: Rhizostomeae) jellyfish venom using polyclonal antibody. *Toxicon* **2014**, *86*, 116–125. [CrossRef] [PubMed]

29. Liu, G.; Zhou, Y.; Liu, D.; Wang, Q.; Ruan, Z.; He, Q.; Zhang, L. Global Transcriptome Analysis of the Tentacle of the Jellyfish Cyanea capillata Using Deep Sequencing and Expressed Sequence Tags: Insight into the Toxin- and Degenerative Disease-Related Transcripts. *PLoS ONE* **2015**, *10*, e0142680. [CrossRef] [PubMed]

30. Li, N.; Chiang, D.Y.; Wang, S.; Wang, Q.; Sun, L.; Voigt, N.; Respress, J.L.; Ather, S.; Skapura, D.G.; Jordan, V.K.; et al. Ryanodine receptor-mediated calcium leak drives progressive development of an atrial fibrillation substrate in a transgenic mouse model. *Circulation* **2014**, *129*, 1276–1285. [CrossRef] [PubMed]

31. Van Lint, P.; Libert, C. Chemokine and cytokine processing by matrix metalloproteinases and its effect on leukocyte migration and inflammation. *J. Leukocyte Biol.* **2007**, *82*, 1375–1381. [CrossRef] [PubMed]

32. Brocker, C.N.; Vasiliou, V.; Nebert, D.W. Evolutionary divergence and functions of the ADAM and ADAMTS gene families. *Hum. Genom.* **2009**, *4*, 43–55. [CrossRef]

33. Wolfsberg, T.G.; Straight, P.D.; Gerena, R.L.; Huovila, A.P.; Primakoff, P.; Myles, D.G.; White, J.M. ADAM, a widely distributed and developmentally regulated gene family encoding membrane proteins with a disintegrin and metalloprotease domain. *Dev. Biol.* **1995**, *169*, 378–383. [CrossRef] [PubMed]

34. Apte, S.S. A disintegrin-like and metalloprotease (reprolysin type) with thrombospondin type 1 motifs: The ADAMTS family. *Int. J. Biochem. Cell. Biol.* **2004**, *36*, 981–985. [CrossRef] [PubMed]

35. Kelwick, R.; Desanlis, I.; Wheeler, G.N.; Edwards, D.R. The ADAMTS (A Disintegrin and Metalloproteinase with Thrombospondin motifs) family. *Genome Biol.* **2015**, *16*, 113. [CrossRef] [PubMed]

36. Rawlings, N.D.; Barrett, A.J. Evolutionary families of metallopeptidases. *Methods Enzymol.* **1995**, *248*, 183–228. [PubMed]

37. Manzetti, S.; McCulloch, D.R.; Herington, A.C.; van der Spoel, D. Modeling of enzyme-substrate complexes for the metalloproteases MMP-3, ADAM-9 and ADAM-10. *J. Comput. Aided Mol. Des.* **2003**, *17*, 551–565. [CrossRef] [PubMed]

38. Hunt, L.T.; Dayhoff, M.O. A surprising new protein superfamily containing ovalbumin, antithrombin-III, and alpha 1-proteinase inhibitor. *Biochem. Biophys. Res. Commun.* **1980**, *95*, 864–871. [CrossRef]

39. Rawlings, N.D.; Tolle, D.P.; Barrett, A.J. Evolutionary families of peptidase inhibitors. *Biochem. J.* **2004**, *378*, 705–716. [CrossRef] [PubMed]

40. Morjen, M.; Kallechziri, O.; Bazaa, A.; Othman, H.; Mabrouk, K.; Zouarikessentini, R.; Sanz, L.; Calvete, J.J.; Srairiabid, N.; El Ayeb, M. PIVL, a new serine protease inhibitor from Macrovipera lebetina transmediterranea venom, impairs motility of human glioblastoma cells. *Matrix Biol. J. Int. Soc. Matrix Biol.* **2013**, *32*, 52–62. [CrossRef] [PubMed]

41. Matsui, T.; Fujimura, Y.; Titani, K. Snake venom proteases affecting hemostasis and thrombosis. *Biochim. Biophys. Acta.* **2000**, *1477*, 146–156. [CrossRef]

42. Mourão, C.B.; Schwartz, E.F. Protease Inhibitors from Marine Venomous Animals and Their Counterparts in Terrestrial Venomous Animals. *Mar. Drugs* **2013**, *11*, 2069–2112. [CrossRef] [PubMed]

43. Zelko, I.N.; Mariani, T.J.; Folz, R.J. Superoxide dismutase multigene family: A comparison of the CuZn-SOD (SOD1), Mn-SOD (SOD2), and EC-SOD (SOD3) gene structures, evolution, and expression. *Free Radic Biol. Med.* **2002**, *33*, 337–349. [CrossRef]

44. Kim, K.Y.; Sang, Y.L.; Cho, Y.S.; Bang, I.C.; Kim, K.H.; Dong, S.K.; Nam, Y.K. Molecular characterization and mRNA expression during metal exposure and thermal stress of copper/zinc- and manganese-superoxide dismutases in disk abalone, Haliotis discus discus. *Fish. Shellfish Immunol.* **2007**, *23*, 1043–1059. [CrossRef] [PubMed]

45. Yang, J.; Dong, S.; Jiang, Q.; Si, Q.; Liu, X.; Yang, J. Characterization and expression of cytoplasmic copper/zinc superoxide dismutase (CuZn SOD) gene under temperature and hydrogen peroxide (H_2O_2) in rotifer Brachionus calyciflorus. *Gene* **2013**, *518*, 388–396. [CrossRef] [PubMed]

46. Umasuthan, N.; Bathige, S.D.; Revathy, K.S.; Lee, Y.; Whang, I.; Choi, C.Y.; Park, H.C.; Lee, J. A manganese superoxide dismutase (MnSOD) from Ruditapes philippinarum: Comparative structural- and expressional-analysis with copper/zinc superoxide dismutase (Cu/ZnSOD) and biochemical analysis of its antioxidant activities. *Fish. Shellfish Immunol.* **2012**, *33*, 753–765. [CrossRef] [PubMed]

47. Broxton, C.N.; Culotta, V.C. SOD Enzymes and Microbial Pathogens: Surviving the Oxidative Storm of Infection. *PLoS Pathog.* **2016**, *12*, e1005295. [CrossRef] [PubMed]

48. Lu, X.; Wang, C.; Liu, B. The role of Cu/Zn-SOD and Mn-SOD in the immune response to oxidative stress and pathogen challenge in the clam Meretrix meretrix. *Fish. Shellfish Immunol.* **2015**, *42*, 58–65. [CrossRef] [PubMed]

49. Bakshani, C.R.; Morales-Garcia, A.L.; Althaus, M.; Wilcox, M.D.; Pearson, J.P.; Bythell, J.C.; Burgess, J.G. Evolutionary conservation of the antimicrobial function of mucus: A first defence against infection. *NPJ Biofilms Microbiomes* **2018**, *4*, 14. [CrossRef] [PubMed]

50. Abrams, M.J.; Goentoro, L. Symmetrization in jellyfish: Reorganization to regain function, and not lost parts. *Zoology (Jena)* **2016**, *119*, 1–3. [CrossRef] [PubMed]

51. Johansson, M.E.; Sjovall, H.; Hansson, G.C. The gastrointestinal mucus system in health and disease. *Nat. Rev. Gastroenterol. Hepatol.* **2013**, *10*, 352–361. [CrossRef] [PubMed]

52. Yang, X.; Steukers, L.; Forier, K.; Xiong, R.; Braeckmans, K.; Reeth, K.V.; Nauwynck, H. A Beneficiary Role for Neuraminidase in Influenza Virus Penetration through the Respiratory Mucus. *PLoS ONE* **2014**, *9*, e110026. [CrossRef] [PubMed]

53. Hansen, L.K.; Becher, N.; Bastholm, S.; Glavind, J.; Ramsing, M.; Kim, C.J.; Romero, R.; Jensen, J.S.; Uldbjerg, N. The cervical mucus plug inhibits, but does not block, the passage of ascending bacteria from the vagina during pregnancy. *Acta Obstet. Gynecol Scand.* **2013**, *93*, 102–108. [CrossRef] [PubMed]

54. Casado, B.; Pannell, L.K.; Iadarola, P.; Baraniuk, J.N. Identification of human nasal mucous proteins using proteomics. *Proteomics* **2010**, *5*, 2949–2959. [CrossRef] [PubMed]

55. Panicker, G.; Ye, Y.; Wang, D.; Unger, E.R. Characterization of the Human Cervical Mucous Proteome. *Clin. Proteom.* **2010**, *6*, 18–28. [CrossRef] [PubMed]

56. De, D.A.; Idolo, A.; Bagordo, F.; Grassi, T.; Leomanni, A.; Serio, F.; Guido, M.; Canitano, M.; Zampardi, S.; Boero, F. Impact of stinging jellyfish proliferations along south Italian coasts: Human health hazards, treatment and social costs. *Int. J. Environ. Res. Public Health* **2014**, *11*, 2488–2503.

57. Wang, B.; Zhang, L.; Zheng, J.; Wang, Q.; Wang, T.; Lu, J.; Wen, X.; Zhang, B.; Liu, G.; Zhang, W.; et al. Multiple organ dysfunction: A delayed envenomation syndrome caused by tentacle extract from the jellyfish Cyanea capillata. *Toxicon* **2013**, *61*, 54–61. [CrossRef] [PubMed]

58. Nagai, H.; Takuwa, K.; Nakao, M.; Ito, E.; Miyake, M.; Noda, M.; Nakajima, T. Novel proteinaceous toxins from the box jellyfish (sea wasp) Carybdea rastoni. *Biochem. Biophys. Res. Commun.* **2000**, *275*, 582–588. [CrossRef] [PubMed]

59. Nagai, H.; Takuwa, K.; Nakao, M.; Sakamoto, B.; Crow, G.L.; Nakajima, T. Isolation and characterization of a novel protein toxin from the Hawaiian box jellyfish (sea wasp) Carybdea alata. *Biochem. Biophys. Res. Commun.* **2000**, *275*, 589–594. [CrossRef] [PubMed]

60. Brinkman, D.; Burnell, J. Identification, cloning and sequencing of two major venom proteins from the box jellyfish, *Chironex fleckeri. Toxicon* **2007**, *50*, 850–860. [CrossRef] [PubMed]

61. Brinkman, D.L.; Konstantakopoulos, N.; McInerney, B.V.; Mulvenna, J.; Seymour, J.E.; Isbister, G.K.; Hodgson, W.C. Chironex fleckeri (box jellyfish) venom proteins: Expansion of a cnidarian toxin family that elicits variable cytolytic and cardiovascular effects. *J. Biol. Chem.* **2014**, *289*, 4798–4812. [CrossRef] [PubMed]

62. Brinkman, D.L.; Burnell, J.N. Biochemical and molecular characterisation of cubozoan protein toxins. *Toxicon* **2009**, *54*, 1162–1173. [CrossRef] [PubMed]

63. Brinkman, D.L.; Jia, X.; Potriquet, J.; Kumar, D.; Dash, D.; Kvaskoff, D.; Mulvenna, J. Transcriptome and venom proteome of the box jellyfish *Chironex fleckeri. BMC Genom.* **2015**, *16*, 407. [CrossRef] [PubMed]

64. Li, R.; Yu, H.; Yue, Y.; Liu, S.; Xing, R.; Chen, X.; Li, P. Combined proteomics and transcriptomics identifies sting-related toxins of jellyfish *Cyanea nozakii. J. Proteom.* **2016**, *148*, 57–64. [CrossRef] [PubMed]

65. Ni, Y.; Xie, G.; Jia, W. Metabonomics of human colorectal cancer: New approaches for early diagnosis and biomarker discovery. *J. Proteome Res.* **2014**, *13*, 3857–3870. [CrossRef] [PubMed]

66. Peng, B.; Li, H.; Peng, X.X. Functional metabolomics: From biomarker discovery to metabolome reprogramming. *Protein Cell* **2015**, *6*, 628–637. [CrossRef] [PubMed]

67. Quinones, M.P.; Kaddurah-Daouk, R. Metabolomics tools for identifying biomarkers for neuropsychiatric diseases. *Neurobiol. Dis.* **2009**, *35*, 165–176. [CrossRef] [PubMed]

68. Lock, J.Y.; Carlson, T.L.; Carrier, R.L. Mucus models to evaluate the diffusion of drugs and particles. *Adv. Drug Deliv. Rev.* **2018**, *124*, 34–49. [CrossRef] [PubMed]

69. Murgia, X.; Loretz, B.; Hartwig, O.; Hittinger, M.; Lehr, C.M. The role of mucus on drug transport and its potential to affect therapeutic outcomes. *Adv. Drug Deliv. Rev.* **2018**, *124*, 82–97. [CrossRef] [PubMed]

70. Tittarelli, R.; Mannocchi, G.; Pantano, F.; Romolo, F.S. Recreational Use, Analysis and Toxicity of Tryptamines. *Curr. Neuropharmacol.* **2015**, *13*, 26–46. [CrossRef] [PubMed]

71. Goddijn, O.J.; Lohman, F.P.; de Kam, R.J.; Schilperoort, R.A.; Hoge, J.H. Nucleotide sequence of the tryptophan decarboxylase gene of *Catharanthus roseus* and expression of tdc-gusA gene fusions in *Nicotiana tabacum. Mol. Gen. Genet.* **1994**, *242*, 217–225. [CrossRef] [PubMed]

72. Guenter, J.; Lenartowski, R. Molecular characteristic and physiological role of DOPA-decarboxylase. *Postepy Higieny i Medycyny Doswiadczalnej* **2016**, *70*, 1424–1440. [CrossRef] [PubMed]

73. Brandt, S.D.; Freeman, S.; McGagh, P.; Abdul-Halim, N.; Alder, J.F. An analytical perspective on favoured synthetic routes to the psychoactive tryptamines. *J. Pharm. Biomed. Anal.* **2004**, *36*, 675–691. [CrossRef] [PubMed]

74. Mohammad-Zadeh, L.F.; Moses, L.; Gwaltney-Brant, S.M. Serotonin: A review. *J. Vet. Pharmacol. Ther.* **2008**, *31*, 187–199. [CrossRef] [PubMed]

75. Mayorova, T.D.; Kosevich, I.A.; Melekhova, O.P. On some features of embryonic development and metamorphosis of Aurelia aurita (Cnidaria, Scyphozoa). *Russ. J. Dev. Biol.* **2012**, *43*, 271–285. [CrossRef]

76. Fossat, P.; Bacque-Cazenave, J.; De Deurwaerdere, P.; Delbecque, J.P.; Cattaert, D. Comparative behavior. Anxiety-like behavior in crayfish is controlled by serotonin. *Science* **2014**, *344*, 1293–1297. [CrossRef] [PubMed]

77. Zhao, Y.T.; Valdivia, C.R.; Gurrola, G.B.; Powers, P.P.; Willis, B.C.; Moss, R.L.; Jalife, J.; Valdivia, H.H. Arrhythmogenesis in a catecholaminergic polymorphic ventricular tachycardia mutation that depresses ryanodine receptor function. *Proc. Natl. Acad. Sci. USA* **2015**, *112*, E1669–E1677. [CrossRef] [PubMed]

78. Da Silveira, R.B.; Wille, A.C.; Chaim, O.M.; Appel, M.H.; Silva, D.T.; Franco, C.R.; Toma, L.; Mangili, O.C.; Gremski, W.; Dietrich, C.P.; et al. Identification, cloning, expression and functional characterization of an astacin-like metalloprotease toxin from *Loxosceles intermedia* (brown spider) venom. *Biochem. J.* **2007**, *406*, 355–363. [CrossRef] [PubMed]

79. Molehin, A.J.; Gobert, G.N.; McManus, D.P. Serine protease inhibitors of parasitic helminths. *Parasitology* **2012**, *139*, 681–695. [CrossRef] [PubMed]

80. Zhao, Y.R.; Xu, Y.H.; Jiang, H.S.; Xu, S.; Zhao, X.F.; Wang, J.X. Antibacterial activity of serine protease inhibitor 1 from kuruma shrimp Marsupenaeus japonicus. *Dev. Comp. Immunol.* **2014**, *44*, 261–269. [CrossRef] [PubMed]

81. Alscher, R.G.; Erturk, N.; Heath, L.S. Role of superoxide dismutases (SODs) in controlling oxidative stress in plants. *J. Exp. Bot.* **2002**, *53*, 1331–1341. [CrossRef] [PubMed]

82. Wang, Y.; Branicky, R.; Noe, A.; Hekimi, S. Superoxide dismutases: Dual roles in controlling ROS damage and regulating ROS signaling. *J. Cell. Biol.* **2018**, *217*, 1915–1928. [CrossRef] [PubMed]

83. Poole, A.Z.; Kitchen, S.A.; Weis, V.M. The Role of Complement in Cnidarian-Dinoflagellate Symbiosis and Immune Challenge in the Sea Anemone Aiptasia pallida. *Front. Microbiol.* **2016**, *7*, 519. [CrossRef] [PubMed]

84. Kimura, A.; Sakaguchi, E.; Nonaka, M. Multi-component complement system of Cnidaria: C3, Bf, and MASP genes expressed in the endodermal tissues of a sea anemone, *Nematostella vectensis*. *Immunobiology* **2009**, *214*, 165–178. [CrossRef] [PubMed]

85. Zhou, Z.; Sun, D.; Yang, A.; Dong, Y.; Chen, Z.; Wang, X.; Guan, X.; Jiang, B.; Wang, B. Molecular characterization and expression analysis of a complement component 3 in the sea cucumber (*Apostichopus japonicus*). *Fish. Shellfish Immunol.* **2011**, *31*, 540–547. [CrossRef] [PubMed]

 *marine drugs* 

Article

The Large Jellyfish *Rhizostoma luteum* as Sustainable a Resource for Antioxidant Properties, Nutraceutical Value and Biomedical Applications

Laura Prieto [1,*], Angélica Enrique-Navarro [1], Rosalia Li Volsi [2] and María J. Ortega [3,*]

[1] Group Ecosystem Oceanography, Department of Ecology and Coastal Management, Instituto de Ciencias Marinas de Andalucía (CSIC), República Saharaui 2, 11519 Puerto Real, Cádiz, Spain; angelica.enrique@csic.es

[2] Department of Chemical, Biological, Pharmaceutical and Environmental Sciences, University of Messina, Viale F. Stagno D'Alcontres, 31, 98166 Messina, Italy; linda89lv@gmail.com

[3] Department of Organic Chemistry, University of Cádiz, Avda. República Saharaui s/n, 11510 Puerto Real, Cádiz, Spain

* Correspondence: laura.prieto@icman.csic.es (L.P.); mariajesus.ortega@uca.es (M.J.O.); Tel.: +34-956-832-612 (L.P.); +34-956-016-020 (M.J.O.); Fax: +34-956-834-701 (L.P.); +34-956-016-288 (M.J.O.)

Received: 26 September 2018; Accepted: 18 October 2018; Published: 21 October 2018

Abstract: Jellyfish is a compartment in the marine food web that often achieves high increases of biomass and that it is starting to be explored for several human potential uses. In this paper, a recently rediscovered large jellyfish, *Rhizostoma luteum*, is studied for the first time to describe its organic compounds for the isolation and production of bioactive compounds in several fields of food, cosmetics, or biomedical industries. The biogeochemical composition (Carbon, Nitrogen and Sulfur content), protein and phenols content, together with their antioxidant activity, and the analysis of lipid content (identifying each of the fatty acids presented) was analyzed. The results presented here suggested this jellyfish has the highest antioxidant activity ever measured in a jellyfish, but also with high content in polyunsaturated fatty acids (PUFAs), including the essential fatty acid linoleic. The large natural biomass of *Rhizostoma luteum* in nature, the wide geographical spread, the fact that already its life cycle has been completed in captivity, establishes a promising positive association of this giant jellyfish species and the isolation of bioactive compounds for future use in marine biotechnology.

Keywords: cnidarians; gelatinous zooplankton; bioprospecting; novel foods

---

1. Introduction

One of the challenges for coming decades in nutraceutical research is to find new potential resources that will provide an easy and low-cost approach to food necessities for the coming centuries. In this sense, jellyfish is a compartment in the marine food web (the greatest biome on earth) that often achieves high increases of biomass [1]. However, only recently has this wide group of organisms been starting to be seen from a more positive point of view [2], contrasting with the deleterious impact that it has on several economic human activities such as fisheries [3], clogging of power plants [4] and tourism [5]. This dark side of jellyfish blooms has been the only vision that has prevailed in the general human perception for many decades [6]. One example of the new positive point of view of the rapid increment of biomass in jellyfish is the cannonball, *Stomolophus meleagris*, in the Gulf of California, where a small Mexican community obtains benefits from switching for one month from fishing fish to fishing jellyfish, processes them, and sells them to the Asiatic market [7]. Another example of the explosive increment of jellyfish is the case of *Cotylorhiza tuberculata* in the Mar Menor lagoon in the

Mediterranean Sea, where during summer, fishermen are paid by regional authorities to catch jellyfish and bury them. Some summers they have caught more than five thousand tons of jellyfish [8,9].

The potential of jellyfish for human use is not only as a food resource [2,10–13], which is traditionally in Asian cuisine [14,15], but also recently as potential cosmeceutical and pharmacological applications [16]. Also, recently, jellyfish collagen has been studied for several biomedical applications [17], including antioxidant [18–20] and melanogenesis-inhibitory activity [18], skin photo-protection from Ultraviolet radiation [21,22], immunostimulatory effects [23,24], and antihypertensive effects [25].

The case we present here is unknown by the scientific community, and regards the potential use of the organic compounds and marine products from jellyfish. The jellyfish in the present work is *Rhizostoma luteum* Quoy and Gaimard, 1827 (Cnidaria: Scyphozoa: Rhizostomeae), which went scientifically unnoticed for more than 60 years until it was rediscovered and genetically characterized in 2012 [26]. Later, it was proved to be widely spread along the east coast of the Atlantic Ocean (from Portugal to South Africa) and in the Alboran Sea (Mediterranean Sea) and is much more frequent than previously thought, but was long misidentified for some of its congeners from the same geographical zones [27].

The purpose of the work is to describe the organic compounds for *Rhizostoma luteum* for the first time, with the hypothesis that it can be a candidate for the isolation and production of bioactive compounds in several fields of food, cosmetics, and biomedical industries. This hypothesis is based on three facts inherent to this jellyfish species: (1) it is from the family of the *Rhizostomaes*, which is the family of jellyfish preferable for consumption in Asiatic cuisine [28]; (2) The natural adult size of this jellyfish is very large, reaching more than 60 cm in umbrella size, weighing 13 kg and having 3 m tentacles [26,27,29], making one of the largest species of this family and therefore converting each individual into considerable amount of biomass; and (3) its life cycle has been recently described and closed entirely in captivity [29], which is a very difficult achievement since it is an open-ocean species. These three facts suggest that, in the case the potential use of *Rhizostoma luteum*, the exploitation of this species' biomass can be obtained both from the natural environment and from controlled culture in captivity. However, before using it, it is mandatory to characterize the different organic compounds and other properties inherent to this jellyfish.

The main aim of this work is the identification of the primary metabolites (fatty acids) with antioxidant activity of the jellyfish *Rhizostoma luteum*. This goal was achieved by combining the original data on biochemical composition, protein, and phenol contents with their antioxidant activity, and the analysis of lipid content (identifying each of the presented fatty acids). The results presented here suggest this jellyfish to be a promising sustainable source for the production of several natural products. *Rhizostoma luteum* appears all year around on the coasts of the east Atlantic Ocean and Alboran Sea [29], providing a highly unexploited biomass. Now a positive association of this giant jellyfish species and an isolation of its bioactive compounds can be established.

2. Results and Discussion

2.1. Jellyfish Biomass Characterization

Biometric and average biomass data of individuals of *Rhizostoma luteum* are shown in Table 1.

The specimens used are young medusae, with an age range from about 23–140 days (Table 1), with a proportionally increasing biomass with jellyfish age. The relation between fresh weight and diameter in *Rhizostoma luteum* has been described by [29], while the relationship between fresh weight (FW) and dry weight (DW) is documented for the first time for this species. The variability of the biometric measures, including the FW, the FW:diameter ratio and DW percentage values, is representative of the different growth stages of the jellyfish [30]. After lyophilization, the DW of *R. luteum* ranges from 3.9–31% of the FW, displaying a high variability in tissue consistency. The highest values correspond with the older jellyfish employed in this study.

Table 1. Biometric measures, fresh and dry weights of the different batches of *Rhizostoma luteum*.

Batch (*n* = Number of Jellyfish)	Days After Ephyra Release Range * Mean (days)	Umbrella Diameter Range * Mean (cm)	Fresh Weight Total (g)	Fresh Weight Per Individual Mean (g)	Ratio FW/Diameter	Dry Weight Total (g)	DW (% of FW)
1 (5)	109 ± 10	0.92 ± 0.03	2.9383	0.587 ± 0.106	0.6	0.912	31.0
2 (5)	92 ± 3	0.87 ± 0.01	3.0509	0.610 ± 0.121	0.7	0.808	26.5
3 (10)	89 ± 17	0.85 ± 0.06	9.6241	0.962 ± 0.142	1.1	0.350	3.6
4 (10)	85 ± 21	0.84 ± 0.07	5.6126	0.561 ± 0.144	0.7	0.2034	3.6
5 (10)	66 ± 10	0.77 ± 0.04	5.6181	0.560 ± 0.106	0.7	0.2747	4.9
6 (10)	72 ± 3	0.79 ± 0.01	4.5403	0.454 ± 0.086	0.6	0.1519	3.3
7 (41)	51 ± 22	0.69 ± 0.10	12.3213	0.300 ± 0.057	0.4	0.4609	3.7
8 (40)	42 ± 6	0.66 ± 0.04	5.9973	0.150 ± 0.028	0.2	0.1917	3.2
9 (61)	35 ± 6	0.62 ± 0.03	7.0676	0.116 ± 0.022	0.2	0.2384	3.4
10 (59)	51 ± 6	0.71 ± 0.03	9.9585	0.169 ± 0.032	0.2	0.388	3.9
11 (52)	53 ± 5	0.71 ± 0.03	8.7219	0.168 ± 0.032	0.2	0.332	3.8

FW, fresh weight; DW, dry weight; * data are expressed as range and/or means standard deviation ($5 < n < 61$).

The first parameter that should be assessed for evaluation in its potential use as biomass is the total energy or gross energy value, which is directly related to DW. The DW of *R. luteum* is similar to the measured range of the FW of *Cotylorhiza tuberculata* (3.9–32.4%), which is higher than *Aurelia* sp. 1 (2.2–3%) and *Rhizostoma pulmo* (4.1–6.8%) [16]. These values provide important information for the potential exploitation of *R. luteum* jellyfish biomass. It has recently been documented using different techniques that jellyfish may practically represent the total diet of several organisms, such as leatherback sea turtles and several fish [31–34].

2.2. Jellyfish Carbon, Nitrogen and Sulfur Content

The carbon, nitrogen and sulfur content of the jellyfish *Rhizostoma luteum* were 5.9 ± 0.3%, 1.6 ± 0.1% and 1.9 ± 0.2% of DW, respectively (mean ± standard deviation, *n* = 6 (Batch numbers: 3, 5–9)). The C:N molar ratio was 3.7 ± 0.1%. Similar values of carbon and nitrogen content have been reported in various of jellyfish such as *Aurelia aurita*, *Chrysaora fuscenses* and *Cyanea capillata* [35] and to the symbiotic jellyfish *Cassiopea xamachana* [36,37]. The C:N molar ratio of *R. luteum* is in the same range as overall medusae and ctenophores [38], and as the symbiotic adults of *Cotylorhiza tuberculata* [39] and *Cassiopea xamachana* [36,37].

2.3. Jellyfish Protein Content

The total freeze-dried tissue of 8 different batches were subjected to phosphate-buffered saline (PBS) solvent extraction treatment. The whole tissues of *Rhizostoma luteum* contained proteins soluble in polar solvents, specifically in aqueous solution with a mean value of 13.7 ± 3.4 mg of proteins of per g of dry weight (mean ± standard deviation, *n* = 8 (Batch numbers: 1, 2, 4–7, 9, 10)). The values ranged between 8.4–18.7 mg of proteins/g of DW. Compared to other species, such as *Aurelia* sp. 1, *Cotylorhiza tuberculata* and *Rhizostoma pulmo*, with 22, 35 and 37 mg of proteins/g of DW, respectively [16], *R. luteum* contained less soluble protein but in the same order of magnitude. The content of protein in jellyfish is consistently the most abundant organic fraction [38] and has been studied in other rhizostomae jellyfish such as *Rhopilema asamushi* [40] and from *Cyanea nozakii* [41] for collagen isolation and characterization, which has been suggested for multipurpose uses in cosmetics and nutraceutical sectors [42].

2.4. Phenolic Compound Content in Jellyfish Aqueous Soluble Extract

The total phenolic content of the jellyfish *Rhizostoma luteum* was 1964.9 ± 386.4 µg GAE (Gallic Acid Equivalent) per gram of DW in the PBS extract (mean ± standard deviation, *n* = 8 (Batch numbers: 1, 2, 4–7, 9, 10)). The values ranged between 1289.6–2597.1 µg GAE/g DW. The phenolic content in jellyfish is not well documented. Compared to *Aurelia* sp. 1 with 116 µg GAE/g DW in the PBS extract [16], *R. luteum* is in a position between *C. tuberculata* and *R. pulmo* reaching 1818 and 2079 µg

GAE/g DW, respectively [16]. Also, phenolic compounds were detected in the podocyst and adults of *Chrysaora quinquecirrha* [43,44] and in *Cyanea capillata* and *C. tuberculata* [30,44].

The biostability and biochemical properties of collagen-based tissues may be enhanced by the polyphenols through modulation of mechanisms of collagen fiber cross-linking at different levels (molecular, inter-molecular and inter-microfibrillar) [45,46]. The high phenolic content of *R. luteum* (this study), *C. tuberculata* and *R. pulmo* [16], all of them large rhizostomae jellyfish, may be the reason for the robust and hardened mesoglea, compared to the flexible and soft *Aurelia* spp.

2.5. Antioxidant Activity in Jellyfish Aqueous Soluble Extract

The total antioxidant activity of the jellyfish *Rhizostoma luteum* was 32,598 ± 9015 nmol of TE (Trolox Equivalent) per gram of DW in the PBS extract (means ± standard deviation, n = 8 (Batch numbers: 1, 2, 4–7, 9, 10)). The values ranged between 15,468–43,501 nmol of TE/g DW. These values of *R. luteum* are the highest values measured in a jellyfish so far, but in the same magnitude as other rhizostomae jellies such as *R. pulmo* and *C. tuberculata*, with antioxidant activity of 22,520 and 25,621 nmol of TE/g DW, respectively [16]. Meanwhile, *Aurelia* sp. 1 antioxidant activity was much lower, with measured values of 7651 nmol of TE/g DW [16].

The high values of *R. pulmo* and *C. tuberculata* compared to *Aurelia* sp. 1 were presumably attributed to the protein and phenols contents, although other unidentified compounds could be included [16]. Similarly, high values of antioxidant activity present in *R. luteum* in the present study could also be related to the protein and phenol content. The antioxidant activity referring to protein content, both in PBS extract, for *R. luteum* is 2379 nmol of TE/mg of proteins, showing evidence that this high antioxidant activity could be ascribed to the inherent protein properties of this jellyfish. Values of this ratio are double those reported by *R. pulmo* and *C. tuberculata* [16].

2.6. Lipid Content

Total lipid content of the jellyfish *Rhizostoma luteum* was around 0.94 g per 100 g of DW (standard deviation = 0.03, n = 6 (Batch numbers: 3, 5–8, 10)). This value was one order of magnitude lower than both *Aurelia* sp.1 and *R. pulmo* (4 g/100 g DW) and two orders of magnitude lower than *C. tuberculata* (12.3 g/100 g DW; [16]). The low values of *R. luteum* are probably due to the age of the specimens examined. In the case of symbiotic jellyfish such as *C. tuberculata*, the high lipid content could be related to the photosynthetic membranes of the zooxanthellae [30,47].

The fatty acid (FA) quantitative composition (as percentage values) of *R. luteum* showed that polyunsaturated fatty acids (PUFA) accounted for half of the total FA (49%), followed by saturated fatty acids (SFA), representing one-third of the total FA (about 30%), and finally monounsaturated fatty acids (MUFA), representing around 21% of the total FA (Table 2). In the case of *Aurelia* sp. 1, *C. tuberculata* and *R. pulmo*, the proportions were different to *R. luteum*, with SFA being the most abundant FA (two-thirds of the total FA), followed by PUFA (two-thirds of the total FA) and a small amount of MUFA (4–15%) [16].

Polyunsaturated fatty acids of *R. luteum* consisted mostly of ω-6 arachidonic ($C_{20:4}$), ω-3 linoleic ($C_{18:3}$) and the essential ω-6 linoleic ($C_{18:2}$) acids. Saturated fatty acids consisted mostly of stearic ($C_{18:0}$) and palmitic ($C_{16:0}$) acids. Among MUFA, oleic acid ($C_{18:1}$) was the prevalent FA (Table 2). This composition of fatty acids shows remarkable differences to the other three species studied previously (see Table 4 in [16]). The proportion of PUFA of both ω-6 arachidonic ($C_{20:4}$) and ω-3 linoleic ($C_{18:3}$) in *R. luteum* are the highest documented for jellyfish and the essential ω-6 linoleic ($C_{18:2}$) acid is only topped by the symbiotic jellyfish *C. tuberculata* [16], giving a peculiar composition of PUFA in *R. luteum*. In the case of MUFA oleic acid, ($C_{18:1}$) is in the same order of magnitude with *C. tuberculata* and higher than *Aurelia* sp.1 and *R. pulmo* [16]).

Table 2. Fatty acid composition of *Rhizostoma luteum* expressed as percentage of the total fatty acid ± standard deviation (SD).

Type of Fatty Acids (FA)	Name of FA	RT (min)	%	Total
Saturated FA (SFA)	Decanoic acid $C_{10:0}$	5.67	0.1 ± 0.0	
	Lauric acid $C_{12:0}$	8.65	0.3 ± 0.0	
	Tridecanoic acid $C_{13:0}$	10.20	0.1 ± 0.0	
	Myristic acid $C_{14:0}$	11.73	2.0 ± 0.1	
	Pentadecanoic acid $C_{15:0}$	13.21	0.4 ± 0.1	
	Palmitic acid $C_{16:0}$	14.65	11.0 ± 0.6	30.2
	Margaric acid $C_{17:0}$	16.03	0.7 ± 0.1	
	Stearic acid $C_{18:0}$	17.40	15.0 ± 1.5	
	Arachidic acid $C_{20:0}$	19.90	0.1 ± 0.0	
	Heneicosanoic acid $C_{21:0}$	21.13	0.1 ± 0.0	
	Docosanoic acid $C_{22:0}$	22.30	0.2 ± 0.0	
	Lignoceric acid $C_{24:0}$	24.90	0.2 ± 0.0	
Monounsaturated FA (MUFA)	Pentadec-10-enoic acid $C_{15:1}$ (ω5)	12.97	0.1 ± 0.1	
	Palmitoleic acid $C_{16:1}$	14.34	3.1 ± 0.2	
	Heptadec-10-enoic acid $C_{17:1}$ (ω7)	15.70	0.4 ± 0.0	
	Oleic acid $C_{18:1}$ (ω9)	17.02	11.3 ± 0.8	20.8
	Elaidic acid $C_{18:1}$ (ω9)	17.09	3.9 ± 0.1	
	Eicos-11-enoic acid $C_{20:1}$ (ω9)	19.58	1.0 ± 0.0	
	Erucic or cis-docos-13-enoic acid $C_{22:1}$ (ω9)	22.07	0.1 ± 0.0	
	Nervonic acid $C_{24:1}$ (ω9)	24.54	0.8 ± 0.0	
Polyunsaturated FA (PUFA)	Gamma-linolenic acid $C_{18:3}$ (ω6)	16.69	0.2 ± 0.0	
	Linoleic acid $C_{18:2}$ (ω6) *	16.92	4.6 ± 0.5	
	Linolelaidic acid $C_{18:2}$ (ω6, ω9)	17.02	1.0 ± 0.1	
	Linolenic acid $C_{18:3}$ (ω3)	17.02	9.8 ± 0.3	
	Arachidonic acid $C_{20:4}$ (ω6)	19.02	23.7 ± 1.3	49.0
	Eicosapentaenoic acid $C_{20:5}$ (ω3)	19.08	3.6 ± 0.2	
	Dihomo-gamma-linolenic acid $C_{20:3}$ (ω6)	19.25	3.2 ± 0.4	
	Eicosadien-11,14-oic acid $C_{20:2}$ (ω6)	19.50	0.7 ± 0.0	
	Eicosa-11,14,17-trienoic acid $C_{20:3}$ (ω3)	19.58	1.9 ± 0.1	
	Docosahexaenoic acid $C_{22:6}$ (ω3)	21.36	0.3 ± 0.1	

RT: retention time; SFA: saturated fatty acids; MUFA: monounsaturated fatty acids; PUFA: polyunsaturated fatty acids; *: essential fatty acids. $n = 6$.

Both ω-6 and ω-3 PUFAs were abundant in the jellyfish *Rhizostoma luteum*, with the ratio of ω-6 to ω-3 of 2.14. The Σ ω-6 was 33.4% while the Σ ω-3 was 15.6%. The total ω-3 PUFAs in *R. luteum* is similar to *Aurelia* sp. 1, *C. tuberculata* and *R. pulmo* [16], but the total ω-6 PUFAs in *R. luteum* is double the other three documented jellyfish. Therefore, the ratio ω-6 to ω-3 of the other jellyfish is lower (between 0.4 to 0.8) [16] than in *R. luteum*. The role of ω-3 PUFAs in a diverse biological process is well known, including development, growth, tissue, and cell homeostasis [48]. Regarding the numerous human health benefits of ω-3 PUFAs, these include antiarthritis, anti-inflammatory, antioxidant, antihypertensive, anticancer, hypo-trigleceridemic, antiaging and antidepressive effects [49]. A high ratio of ω-6 to ω-3 in the Western diet in humans appears to be related to pathogenesis of chronic disease due to its proinflammatory effects [50]. The fatty acid composition of *R. luteum* discovered in this study should be considered for new applications due to its nutraceutical value.

3. Materials and Methods

3.1. Materials and Chemicals

Methanol and ethanol were purchased from Merck (Darmstadt, Germany); potassium persulfate (dipotassium peroxodisulfate), 6-hydroxy-2,5,7,8-tetramethylchroman-2-carboxylic acid (Trolox), 2,2′-azinobis(3-ethylbenzothiazoline-6-sulfonic acid) diammonium salt (ABTS), gallic acid, PBS, Folin-Ciocalteu's phenol reagent, Bradford reagent, fatty acid methyl esters (FAME) Supelco 37 Component Mix were all purchased from Sigma-Aldrich Química (Madrid, Spain). All other reagents were of analytical grade.

3.2. Sample Collection and Preparation

Specimens of *Rhizostoma luteum* jellyfish were reared in the laboratory (ICMAN-CSIC) from polyps. The origin of this living collection is from planulae gathered directly from the gonad of a free-swimming female medusa offshore of the coast of the Alboran Sea (Spain) in October 2015 (details in Kienberger et al. [29]). For the present study, a total of 303 specimens were used. After the biometric measurements (weight and diameter) were taken, and due to their small size, specimens were grouped to reach enough biomass to be analyzed and lyophilized. Samples were freeze-dried for 4 days at $-55\,^{\circ}\mathrm{C}$ using a chamber pressure of 0.110 mbar in a freeze dryer (LyoAlfa 15, Telstar Life Science Solutions, Madrid, Spain). Lyophilized samples were weighed to annotate the dry weight and stored at $-20\,^{\circ}\mathrm{C}$ until use.

3.3. Elemental Analysis

Total carbon, nitrogen, and sulfur (C, N, S) content (in wt.%, approximately 100 mg per sample) were measured using an Elementary Chemical Analyzer LECO CHNS-932.

3.4. Polar Solvent Extraction

Lyophilized samples of total jellyfish were subjected to extraction in aqueous solvent PBS. Samples were stirred with 16 volumes of PBS (2 h at 4 $^{\circ}$C). Samples were then centrifuged at $9000\times g$ for 30 min at 4 $^{\circ}$C and the supernatants were essayed for protein and phenol contents and for antioxidant activity.

3.5. Protein Content

Total protein content was estimated by modified Bradford assay [51] using bovine serum albumin (BSA) as a standard.

3.6. Phenol Content

The total phenolic content was determined by using a modified Folin-Ciocalteau colorimetric method. Aliquots of extracts (100 µL) were mixed with 500 µL of Folin-Ciocalteu's phenol reagent and 500 µL of 7.5% sodium carbonate (Na_2CO_3). After 2 h of incubation at room temperature in the dark, the absorbance was spectrophotometrically measured at 760 nm. Gallic acid, ranging from 2.5 to 100 µg/mL, was used as a standard. The results were expressed as gallic acid equivalents (GAE) per gram of dry extract.

3.7. Antioxidant Activity

The total antioxidant activity was determined spectrophotometrically by the ABTS free radical decolorization assay developed by Re et al. [52], with some modifications. In brief, a solution of the radical cation $ABTS^+$ was prepared by mixing a solution of ABST (2,2′-azinobis(3-ethylbenzothiazoline-6-sulfonic acid) diammonium salt (7 mM) and a solution of potassium persulfate (2.45 mM) in H_2O. The mixture was kept in the dark overnight before use. Then, the $ABTS^+$ solution was diluted with EtOH to an absorbance of 0.70 ± 0.02 at 734 nm. Trolox was used as standard. The samples were prepared by adding 100 µL of extracts to 2 mL of $ABTS^+$ solution and the control by adding 100 µL of PBS solution to 2 mL of $ABTS^+$ solution. The measure at the absorbance at 734 nm was registered after 6 min of reaction. The percentage of inhibition of the absorbance was calculated by the following equation:

$$\%\text{inhibition} = ((A_0 - A_1)/A_0) \times 100 \tag{1}$$

where A_0 is the absorbance of the control and A_1 is the absorbance of the tested samples. Results were expressed as nmol of Trolox equivalents (TE) per gram of sample.

3.8. Total Lipid Extraction

Total lipids were extracted using the modified Bligh and Dyer method [53]. Lyophilized powder (100 mg) was mixed with a total of 10 mL solvent added in the sequence of chloroform, methanol, water, to achieve a final chloroform/methanol/water ratio of 1:2:0.8 (by volume). Samples were shaken for 15 s after addition of each solvent, and incubated overnight at 4 °C. After centrifugation at $6500 \times g$ for 10 min, the supernatant was transferred into a separating funnel, and phase separation of the biomass-solvent mixtures was achieved by adding chloroform and water to obtain a final chloroform/methanol/water ratio of 2:2:1.8 (by volume). After settling, the bottom phase was collected and evaporated under vacuum.

Fatty Acid Profiles Determination

Fatty acid methyl esters (FAMEs) were obtained using boron trifluoride (BF_3) according to [54] with some modifications. Total lipid extract was saponified at 90 °C for 20 min with 0.5 M KOH in methanol (3 mL). Forty-nine micrograms of the internal standard (methyl tricosanoate) were added before saponification. The fatty acids were methylated by adding 14% BF_3 in MeOH (2 mL) and heating at 90 °C for 10 min. After cooling, the mixture was extracted with hexane (1 mL × 2). After separation, the hexane layer was collected, taken to dryness under vacuum, dissolved in 1.0 mL of CH_2Cl_2 and analyzed by gas chromatography-mass spectrometry (GC-MS).

3.9. GC-MS Analysis

The analyses were performed on a GC-MS system consisting of high-resolution SYNAPT G2 (Waters, Milford, MA, USA) instrument equipped with a quadrupole-time-of-flight (QTOF) analyzer using atmospheric pressure ionization (API) in positive ionization mode. Compounds were separated on a DB-5 capillary column Agilent J&W DB-5ms column (250 µm × 30 m, 0.25 µm film) (Agilent, Santa Clara, CA, USA). The GC parameters were as follows: the column temperature was maintained at 90 °C for 1 min, then raised to 200 °C (20 °C/min) and to 300 °C (5 °C/min), and held at 300 °C for 2 min. Fatty acids were identified by comparison of retention time, molecular formula obtained by their high-resolution molecular ion $[M + H]^+$, and mass spectral data with FAME standards (Supelco-37) and a NIST library.

4. Conclusions

The potential use of the jellyfish *Rhizostoma luteum* as biomass, evaluating their nutraceutical value and antioxidant properties, has never been evaluated until the present study. The results presented here suggest this jellyfish to be an excellent candidate for the potentially sustainable production of nutraceutical, cosmeceutical and biomedical natural products due to the highest antioxidant activity ever measured in a jellyfish, but also for their high content in PUFAs, including the essential fatty acid linoleic. Because the present work has been performed with young medusae, further work should be carried out with adult specimens. The large natural biomass of *Rhizostoma luteum* in nature, the wide geographical spread, and the fact that already its life cycle has been completed in captivity establishes a promising positive potential of this giant jellyfish species and the isolation of bioactive compounds for future use in marine biotechnology.

Author Contributions: Conceptualization, Investigation, Resources, Funding Acquisition and Writing-Original Draft Preparation, L.P. and M.J.O.; Methodology and Validation, M.J.O.; Formal Analysis, A.E.-N., R.L.V. and M.J.O.; Data Curation, L.P., M.J.O. and A.E.-N.; Writing-Review and Editing, all authors; Supervision, L.P. and M.J.O.; Project Administration, L.P.

Funding: This research was funded by the Spanish Ministerio de Ciencia, Innovación y Universidades under grant number CTM2016-75487-R for the project MED2CA and the CSIC under grant number 2017301072. The research stay of R.L.V. was funded by Erasmus + Program.

Acknowledgments: We thank A. Moreno and P. Gento for their technical assistance in the laboratory to rear the jellyfish.

Conflicts of Interest: The authors declare no conflict of interest.

References

1. Brotz, L.; Cheung, W.W.L.; Kleisner, K.; Pakhomov, E.; Pauly, D. Increasing jellyfish populations: Trends in Large Marine Ecosystems. *Hydrobiologia* **2012**, *690*, 3–20. [CrossRef]
2. Doyle, T.K.; Hays, G.C.; Harrod, C.; Houghton, J.D.R. Ecological and societal benefits of jellyfish. In *Jellyfish Blooms*; Pitt, K.A., Lucas, C.H., Eds.; Springer Science + Business Media: Dordrecht, The Netherlands, 2014; pp. 105–127. [CrossRef]
3. Matsushita, Y.; Honda, N.; Kawamura, S. Design and tow trial of JET (Jellyfish Excluder for Towed fishing gear). *Nippon Suisan Gakk.* **2005**, *71*, 965–967. [CrossRef]
4. Purcell, J.E.; Uye, S.I.; Lo, W.T. Anthropogenic causes of jellyfish blooms and their direct consequences for humans: A review. *Mar. Ecol. Prog. Ser.* **2007**, *350*, 153–174. [CrossRef]
5. Kontogianni, A.D.; Emmanouilides, C.J. The cost of a gelatinous future and loss of critical habitats in the Mediterranean. *ICES J. Mar. Sci.* **2014**, *71*, 853–866. [CrossRef]
6. Condon, R.H.; Graham, W.M.; Duarte, C.M.; Pitt, K.A.; Lucas, C.H.; Haddock, S.H.; Sutherland, K.R.; Robinson, K.L.; Dawson, M.N.; Decker, M.B.; et al. Questioning the rise of gelatinous zooplankton in the world's oceans. *BioScience* **2012**, *62*, 160–169. [CrossRef]
7. Gibbons, M.J.; Boero, F.; Brotz, L. We should not assume that fishing jellyfish will solve our jellyfish problem. *ICES J. Mar. Sci.* **2015**, *73*, 1012–1018. [CrossRef]
8. Prieto, L.; Astorga, D.; Navarro, G.; Ruiz, J. Environmental control of phase transition and polyp survival of a massive-outbreaker jellyfish. *PLoS ONE* **2010**, *5*, e13793. [CrossRef] [PubMed]
9. Ruiz, J.; Prieto, L.; Astorga, D. A model for temperature control of jellyfish (*Cotylorhiza tuberculata*) outbreaks: A causal analysis in a Mediterranean coastal lagoon. *Ecol. Model.* **2012**, *233*, 59–69. [CrossRef]
10. Boero, F. Review of jellyfish blooms in the Mediterranean and Black Sea. In *General Fisheries Commission for the Mediterranean. Studies and Reviews*; No. 92; Food and Agriculture Organization of the United Nations (FAO), Ed.; FAO: Rome, Italy, 2013; p. 53, ISBN 978-92-5-107457-2.
11. Lucas, C.H.; Pitt, K.A.; Purcell, J.E.; Lebrato, M.; Condon, R.H. What's in a jellyfish? Proximate and elemental composition and biometric relationships for use in biogeochemical studies. *Ecology* **2011**, *92*, 1704. [CrossRef]
12. Hsieh, Y.H.P.; Rudloe, J. Potential of utilizing jellyfish as food in Western countries. *Trends Food Sci. Technol.* **1994**, *5*, 225–229. [CrossRef]
13. Hsieh, Y.H.P.; Leong, F.M.; Rudloe, J. Jellyfish as food. *Hydrobiologia* **2001**, *451*, 11–17. [CrossRef]
14. Li, J.; Hsieh, Y.H.P. Traditional Chinese food technology and cuisine. *Asia Pac. J. Clin. Nutr.* **2004**, *13*, 147–155. [PubMed]
15. Omori, M.; Nakano, E. Jellyfish fisheries in southeast Asia. *Hydrobiologia* **2001**, *451*, 19–26. [CrossRef]
16. Leone, A.; Lecci, R.M.; Durante, M.; Meli, F.; Piraino, S. The bright side of gelatinous blooms: Nutraceutical value and antioxidant properties of three Mediterranean jellyfish (Scyphozoa). *Mar. Drugs* **2015**, *13*, 4654–4681. [CrossRef] [PubMed]
17. Addad, S.; Exposito, J.Y.; Faye, C.; Ricard-Blum, S.; Lethias, C. Isolation, characterization and biological evaluation of jellyfish collagen for use in biomedical applications. *Mar. Drugs* **2011**, *9*, 967–983. [CrossRef] [PubMed]
18. Zhuang, Y.; Sun, L.; Zhao, X.; Wang, J.; Hou, H.; Li, B. Antioxidant and melanogenesis-inhibitory activities of collagen peptide from jellyfish (*Rhopilema esculentum*). *J. Sci. Food Agric.* **2009**, *89*, 1722–1727. [CrossRef]
19. Zhuang, Y.L.; Sun, L.P.; Zhao, X.; Hou, H.; Li, B.F. Investigation of gelatin polypeptides of jellyfish (*Rhopilema esculentum*) for their antioxidant activity in vitro. *Food Technol. Biotechnol.* **2010**, *48*, 222–228.
20. Barzideh, Z.; Latiff, A.A.; Gan, C.; Alias, A.K. ACE inhibitory and antioxidant activities of collagen hydrolysates from the ribbon jellyfish (*Chrysaora* sp.). *Food Technol. Biotechnol.* **2014**, *52*, 495–504. [CrossRef] [PubMed]
21. Zhuang, Y.; Hou, H.; Zhao, X.; Zhang, Z.; Li, B. Effects of collagen and collagen hydrolysate from jellyfish (*Rhopilema esculentum*) on mice skin photoaging induced by UV irradiation. *J. Food Sci.* **2009**, *74*, 183–188. [CrossRef] [PubMed]
22. Fan, J.; Zhuang, Y.; Li, B. Effects of collagen and collagen hydrolysate from jellyfish umbrella on histological and immunity changes of mice photoaging. *Nutrients* **2013**, *5*, 223–233. [CrossRef] [PubMed]

23. Nishimoto, S.; Goto, Y.; Morishige, H.; Shiraishi, R.; Doi, M.; Akiyama, K.; Yamauchi, S.; Sugahara, T. Mode of action of the immunostimulatory effect of collagen from jellyfish. *Biosci. Biotechnol. Biochem.* **2008**, *72*, 2806–2814. [CrossRef] [PubMed]

24. Morishige, H.; Sugahara, T.; Nishimoto, S.; Muranaka, A.; Ohno, F.; Shiraishi, R.; Doi, M. Immunostimulatory effects of collagen from jellyfish in vivo. *Cytotechnology* **2011**, *63*, 481–492. [CrossRef] [PubMed]

25. Zhuang, Y.; Sun, L.; Zhang, Y.; Liu, G. Antihypertensive effect of long-term oral administration of jellyfish (*Rhopilema esculentum*) collagen peptides on renovascular hypertension. *Mar. Drugs* **2012**, *10*, 417–426. [CrossRef] [PubMed]

26. Prieto, L.; Armani, A.; Macías, D. Recent strandings of the giant jellyfish *Rhizostoma luteum* Quoy and Gaimard, 1827 (Cnidaria: Scyphozoa: Rhizostomeae) on the Atlantic and Mediterranean coasts. *Mar. Biol.* **2013**, *160*, 3241–3247. [CrossRef]

27. Kienberger, K.; Prieto, L. The jellyfish *Rhizostoma luteum* (Quoy & Gaimard, 1827): Not such a rare species after all. *Mar. Biodivers.* **2017**, 1–8. [CrossRef]

28. Armani, A.; Tinacci, L.; Giusti, A.; Castigliego, L.; Gianfaldoni, D.; Guidi, A. What is inside the jar? Forensically informative nucleotide sequencing (FINS) of a short mitochondrial COI gene fragment reveals a high percentage of mislabeling in jellyfish food products. *Food Res. Int.* **2013**, *54*, 1383–1393. [CrossRef]

29. Kienberger, K.; Riera-Buch, M.; Schönemann, A.M.; Bartsch, V.; Halbauer, R.; Prieto, L. First description of the life cycle of the jellyfish *Rhizostoma luteum* (Scyphozoa: Rhizostomeae). *PLoS ONE* **2018**, *13*, e0202093. [CrossRef] [PubMed]

30. Leone, A.; Lecci, R.M.; Durante, M.; Piraino, S. Extract from the zooxanthellate jellyfish *Cotylorhiza tuberculata* modulates gap junction intercellular communication in human cell cultures. *Mar. Drugs* **2013**, *11*, 1728–1762. [CrossRef] [PubMed]

31. Heaslip, S.G.; Iverson, S.J.; Bowen, W.D.; James, M.C. Jellyfish support high energy intake of leatherback sea turtles (*Dermochelys coriacea*): Video evidence from animal-borne cameras. *PLoS ONE* **2012**, *7*. [CrossRef]

32. Milisenda, G.; Rosa, S.; Fuentes, V.L.; Boero, F.; Guglielmo, L.; Purcell, J.E.; Piraino, S. Jellyfish as prey: Frequency of predation and selective foraging of boops boops (vertebrata, actinopterygii) on the mauve stinger *Pelagia noctiluca* (Cnidaria, scyphozoa). *PLoS ONE* **2014**, *9*. [CrossRef] [PubMed]

33. D'Ambra, I.; Graham, W.M.; Carmichael, R.H.; Hernandez, F.J. Fish rely on scyphozoan hosts as a primary food source: Evidence from stable isotope analysis. *Mar. Biol.* **2014**, *162*, 247–252. [CrossRef]

34. Cardona, L.; De Quevedo, I.Á.; Borrell, A.; Aguilar, A. Massive consumption of gelatinous plankton by Mediterranean apex predators. *PLoS ONE* **2012**, *7*, e31329. [CrossRef] [PubMed]

35. Arai, M.N. *A Functional Biology of Scyphozoa*; Springer Science & Business Media: Dordrecht, The Netherlands, 1997. [CrossRef]

36. Bailey, T.G.; Youngbluth, M.J.; Owen, G.P. Chemical composition and metabolic rates of gelatinous zooplankton from midwater and benthic boundary layer environments off Cape Hatteras, North Carolina, USA. *Mar. Ecol. Prog. Ser.* **1995**, *122*, 121–134. [CrossRef]

37. Banaszak, A.T.; Lesser, M.P. Effects of solar ultraviolet radiation on coral reef organisms. *Photochem. Photobiol. Sci.* **2009**, *8*, 1276–1294. [CrossRef] [PubMed]

38. Pitt, K.A.; Welsh, D.T.; Condon, R.H. Influence of jellyfish blooms on carbon, nitrogen and phosphorus cycling and plankton production. *Hydrobiologia* **2009**, *616*, 133–149. [CrossRef]

39. Enrique-Navarro, A. Characterization of zooxanthellae living in symbiosis with the jellyfish *Cotylorhiza tuberculata*. Master Thesis, University of Cádiz, Cádiz, Spain, 2017.

40. Nagai, T.; Worawattanamateekul, W.; Suzuki, N.; Nakamura, T.; Ito, T.; Fujiki, K.; Nakao, M.; Yano, T. Isolation and characterization of collagen from rhizostomous jellyfish (*Rhopilema asamushi*). *Food Chem.* **2000**, *70*, 205–208. [CrossRef]

41. Zhang, J.; Duan, R.; Huang, L.; Song, Y.; Regenstein, J.M. Characterisation of acid-soluble and pepsin-solubilised collagen from jellyfish (*Cyanea nozakii* Kishinouye). *Food Chem.* **2014**, *150*, 22–26. [CrossRef] [PubMed]

42. Silva, T.; Moreira-Silva, J.; Marques, A.; Domingues, A.; Bayon, Y.; Reis, R. Marine origin collagens and its potential applications. *Mar. Drugs* **2014**, *12*, 5881–5901. [CrossRef] [PubMed]

43. Blanquet, R. Structural and chemical aspects of the podocyst cuticle of the scyphozoan medusa, *Chrysaora quinquecirrha*. *Biol. Bull.* **1972**, *142*, 1–10. [CrossRef]

44. Mitchell, K. Organic Compounds in *Cyanea capillata* and *Chrysaora quinquecirrha*. Available online: http://digitalcommons.iwu.edu/cgi/viewcontent.cgi?article=1012&context=bio_honproj (accessed on 26 March 2014).

45. Madhan, B.; Subramanian, V.; Rao, J.R.; Nair, B.U.; Ramasami, T. Stabilization of collagen using plant polyphenol: Role of catechin. *Int. J. Biol. Macromol.* **2005**, *37*, 47–53. [CrossRef] [PubMed]

46. Vidal, C.M.P.; Leme, A.A.; Aguiar, T.R.; Phansalkar, R.; Nam, J.W.; Bisson, J.; McAlpine, J.B.; Chen, S.N.; Pauli, G.F.; Bedran-Russo, A. Mimicking the hierarchical functions of dentin collagen cross-links with plant derived phenols and phenolic acids. *Langmuir* **2014**, *30*, 14887–14993. [CrossRef] [PubMed]

47. Leblond, J.D.; Chapman, P.J. Lipid class distribution of highly unsaturated long chain fatty acids in marine dinoflagellates. *J. Phycol.* **2000**, *36*, 1103–1108. [CrossRef]

48. Russo, G.L. Dietary n-6 and n-3 polyunsaturated fatty acids: From biochemistry to clinical implications in cardiovascular prevention. *Biochem. Pharmacol.* **2009**, *77*, 937–946. [CrossRef] [PubMed]

49. Siriwardhana, N.; Kalupahana, N.S.; Moustaid-Moussa, N. Health benefits of n-3 polyunsaturated fatty acids: Eicosapentaenoic acid and docosahexaenoic acid. *Adv. Food Nutr. Res.* **2012**, *65*, 211–222. [CrossRef] [PubMed]

50. Simopoulos, A.P. The importance of the ratio of omega-6/omega-3 essential fatty acids. *Biomed. Pharmacother.* **2002**, *56*, 365–379. [CrossRef]

51. Bradford, M.M. A rapid and sensitive method for the quantitation of microgram quantities of protein utilizing the principle of protein-dye binding. *Anal. Biochem.* **1976**, *72*, 248–254. [CrossRef]

52. Re, R.; Pellegrini, N.; Proteggente, A.; Pannala, A.; Yang, M.; Rice-Evans, C. Antioxidant activity applying and improved ABTS radical cation decolorization assay. *Free Radic. Biol. Med.* **1999**, *26*, 1231–1237. [CrossRef]

53. Bligh, E.G.; Dyer, W.J. A rapid method of total lipid extraction and purification. *Can. J. Biochem. Physiol.* **1959**, *37*, 911–917. [CrossRef] [PubMed]

54. Szczęsna-Antczak, M.; Antczak, T.; Piotrowicz-Wasiak, M.; Rzyska, M.; Binkowska, N.; Bielecki, S. Relationships between lipases and lipids in mycelia of two Mucor strains. *Enzyme Microb. Technol.* **2006**, *39*, 1214–1222. [CrossRef]

Review

The *Anemonia viridis* Venom: Coupling Biochemical Purification and RNA-Seq for Translational Research

Aldo Nicosia [1,*,†], Alexander Mikov [2,†], Matteo Cammarata [3], Paolo Colombo [4], Yaroslav Andreev [2,5], Sergey Kozlov [2] and Angela Cuttitta [1,*]

[1] National Research Council-Institute for the Study of Anthropogenic Impacts and Sustainability in the Marine Environment (IAS-CNR), Laboratory of Molecular Ecology and Biotechnology, Capo Granitola, Via del mare, Campobello di Mazara (TP), 91021 Sicily, Italy

[2] Shemyakin-Ovchinnikov Institute of Bioorganic Chemistry, RAS, GSP-7, ul. Miklukho-Maklaya, 16/10, 117997 Moscow, Russia; mikov.alexander@gmail.com (A.M.); ay@land.ru (Y.A.); serg@ibch.ru (S.K.)

[3] Department of Earth and Marine Sciences, University of Palermo, 90100 Palermo, Italy; matteo.cammarata@unipa.it

[4] Istituto di Biomedicina e di Immunologia Molecolare, Consiglio Nazionale delle Ricerche, Via Ugo La Malfa 153, 90146 Palermo, Italy; paolo.colombo@ibim.cnr.it

[5] Institute of Molecular Medicine, Ministry of Healthcare of the Russian Federation, Sechenov First Moscow State Medical University, 119991 Moscow, Russia

* Correspondence: aldo.nicosia@cnr.it (A.N.); angela.cuttitta@cnr.it (A.C.); Tel.: +39-0924-40600 (A.N. & A.C.)

† These authors have made equal contribution.

Received: 29 September 2018; Accepted: 24 October 2018; Published: 25 October 2018

Abstract: Blue biotechnologies implement marine bio-resources for addressing practical concerns. The isolation of biologically active molecules from marine animals is one of the main ways this field develops. Strikingly, cnidaria are considered as sustainable resources for this purpose, as they possess unique cells for attack and protection, producing an articulated cocktail of bioactive substances. The Mediterranean sea anemone *Anemonia viridis* has been studied extensively for years. In this short review, we summarize advances in bioprospecting of the *A. viridis* toxin arsenal. *A. viridis* RNA datasets and toxin data mining approaches are briefly described. Analysis reveals the major pool of neurotoxins of *A. viridis*, which are particularly active on sodium and potassium channels. This review therefore integrates progress in both RNA-Seq based and biochemical-based bioprospecting of *A. viridis* toxins for biotechnological exploitation.

Keywords: transcriptomics; bio-prospecting; computational biology; neurotoxins

---

1. Introduction

The exploitation of marine bio-resources aims to respond to critical needs of society supporting development in different fields of the blue biotechnologies which include drug discovery, bioremediation, biomaterials and aquafarming. Over the years, the screening of marine extracts for bioactive products has resulted in successful marine compounds screening and drug discovery, which has passed clinical trials [1,2]. Marine organisms are considered one of the largest reservoirs of natural molecules to be evaluated for drug activity, and among them, Cnidaria represent an ancient group of venomous animals specialized in toxins production and delivery [3].

Cnidaria are soft-bodied organisms which lack a traditional protection system such as as cuticle or exoskeleton, hemolymph and phagocytes, thus resulting in it being continuously exposed to diverse pathogens. Their defense systems are of growing interest, as they likely contain molecules able to counteract pathogen infection. Sea anemones (Actinaria, Cnidaria) are ancient sessile predators [4] whose survival mainly relies on their venom production [5] and nematocysts, specialized cells for stinging and delivery of venoms, are usually used to immobilize prey or dissuade predators.

Analyses of the venoms of many sea anemone species has uncovered a variegate arsenal of low molecular weight molecules [6]. Because the capture and killing of prey as well as mechanisms of defense and protection in sea anemones are closely related to toxin production, the presence of multiple toxin variants could provide some benefit [7].

The sea anemone *A. viridis*, previously known as *Anemonia sulcata*, is an extensively studied Mediterranean species [8–15]. More than 20 polypeptide toxins of different structures and functions have been isolated from crude extracts of this species. They include potassium channel blockers, such as kalicludines, kaliseptine, blood depressing substance (BDS), neurotoxins blocking sodium channels, and Kunitz-type inhibitors of proteolytic enzymes [16]. Rapid advances in DNA sequencing technologies resulted in growing bio-prospecting efforts for the screening and identification of novel sea anemone toxin sequences [17–19], thus enabling *A. viridis* to be considered as a reliable source of bioactive molecules.

2. *A. viridis* RNA Datasets and Data Mining

Large-scale datasets and transcriptome collections freely available on public databases represent priceless resources for the mining of gene products aiming at biotechnological exploitation.

In silico bioprospecting based on "big data" derived from RNA deep sequencing provided a boost in the screening of candidates for biomedicine and biotechnology applications, thus complementing canonical bio-guided fractionation procedures.

In short, projects of massively parallel sequencing have produced several millions of reads per run, and thus require extended computing resources for storage and data processing. Based on annotation procedures, homologs detection, motifs scan, and gene ontology analysis, different strategies are usually exploited for the identification of candidate gene products.

To date, 17 transcriptome datasets have been identified at NCBI using as query «*Anemonia viridis*» or «*Anemonia sulcata*» (Table 1). The majority of retrieved datasets were generated using Illumina HiSeq or SOLiD platforms; while the transcriptome of the symbiotic sea anemone *A. viridis*, at first was analyzed in depth using expressed sequence tags (EST) libraries combined with traditional Sanger sequencing [20], which retrieved 39,939 high quality ESTs, assembled into 14,504 unique sequences.

Table 1. Transcriptome datasets of *A. viridis* available at NCBI.

Accession	Experiment Title	Platform	Submitter	Amount
ERX1926108 ERX1926107 ERX1926106 ERX1926105	Study of mitogenome and corresponding transcriptome of sea anemones	Ion Torrent PGM and Sanger technology	The Arctic University of Norway (UiT)	unspecified
SRX3049371 SRX3049370 SRX3049369 SRX3049368 SRX3049367 SRX3049366 SRX3049365 SRX3049364 SRX3049363	Small RNA sequencing of *Anemonia viridis*	AB 5500Xl Genetic Analyzer	Urbarova et al., 2018	2.9×10^3 Mb
SRX699624	*Anemonia viridis* Transcriptome or Gene expression	Illumina HiSeq 2000	University of Haifa	5.4×10^3 Mb
Under different accession	Symbiotic sea anemone *A. viridis* cDNA library	ABI-3730 Genetic Analyze (Sanger Technology)	Sabourault et al., 2010	39,939 ESTs
SRX971460	Tissue specific transcriptomes of the emerging model organism *Anemonia sulcata*	Illumina HiSeq 1500	The Ohio State University	8.8×10^3 Mb
SRX971488	Tissue specific transcriptomes	Illumina HiSeq 2000	The Ohio State University	33.9×10^3 Mb

More recently, tissue-specific paired-end libraries were generated and sequenced on Illumina HiSeq platforms in order to characterize the venom composition among different anatomical districts of the sea anemone [19]. Additionally, in-depth small RNA libraries were produced and sequenced based on SOLiD technology enabling both the identification of novel miRNA and the analyses of differential miRNA expression under pH gradients [21].

For successful in silico bio-prospecting of toxin-like candidates, different approaches have been implemented. They exploit advances on homologues identification on the basis of hidden Markov models (HMMs) libraries built on a curated seed alignment of Pfam domains and motif recognition of specific key amino acid residues distribution which are characteristic of the different toxin types [17,18]. In particular, for candidate toxins of *A. viridis*, single residue distribution analysis (SRDA) was successfully used for the determination of a motif array based on alignment of curated cnidarian toxins, which takes into account the distribution pattern of Cys residues [17].

The RNA-Seq libraries were usually interrogated using these Cys-motifs (Cys-patterns), and translated sequences that satisfy each query were selected.

Thus, as a pipeline for in silico bio-prospecting, the workflow leading to candidate toxins identification is reported in Figure 1.

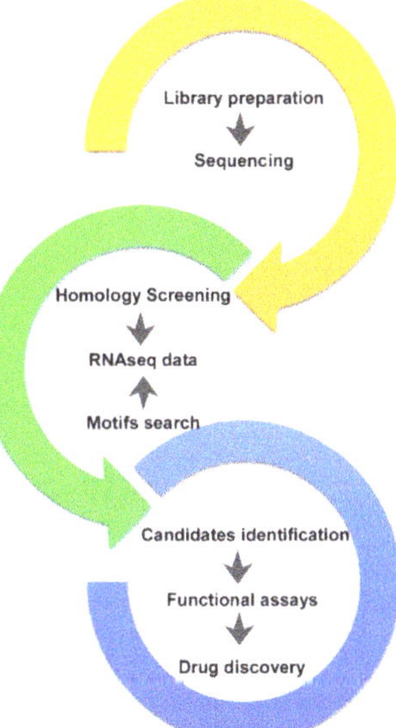

Figure 1. Pipeline for in silico bio-prospecting and candidate toxins identification. This includes library preparation, RNA deep sequencing, data analyses by motif and/or homology screening, and recovery of matching sequences, expression and subsequent functional testing.

Similar to NGS platforms, in recent years, the developments in mass spectrometry, public repository for proteomics data (such as PRoteomics IDEntifications –PRIDE– [22] PeptideAtlas-PASSEL– [23] ProteomicsDB [24], Mass Spectrometry Interactive Virtual Environment-MassIVE–) and algorithms for integrated searches have provided useful tools for proteogenomics applications [25,26] as venomics [27]. However, to date, no proteomics and mass spectrometry data have been found for *A. viridis*; while similar approaches have been successfully used in deep venomics of other organisms ([27] and references therein). The integration of RNA-seq data into the proteomics for validation of *A. viridis* candidate toxins against mass spectrometric data will establish a standardized OMICS approach for deep bio-discovery.

3. The Multifaceted Molecular Arsenal of Sea Anemone *A. viridis*

Sea anemones are known as a source of peptide toxins, acting mainly on ion channels as blockers or modulators, particularly in excitable cells. It has been reported that venomous animals are able to produce toxins that act on different molecular targets on the same physiological system. Toxins mainly act by altering the activity of neurons by either blocking/changing the kinetics of Na^+ channels [28,29] or by blocking of neuronal K^+ channels [30]. However, the chemical arsenal is not limited to modulators of excitability in prey; it also includes molecules altering membrane permeability or inducing the formation of pores in the membrane. Therefore, analyses of sea anemone venoms revealed that they contain a diversity of biologically active proteins and peptides which have not been fully explored until the exploiting of transcriptomic and/or proteomic procedures. Herein, data reporting canonical bioactive compounds in isolation were combined with in silico bio-prospecting on *A. viridis* RNA datasets to provide a comprehensive synopsis of a multifaceted molecular arsenal of the well-studied sea anemone *A. viridis*. To avoid confusion, toxins and other proteins herein collected were named accordingly to nomenclature provided in the databases or in previous reports.

3.1. Sodium Channel Peptide Toxins

Voltage-gated sodium channels (Na_Vs) display an important role in neuron signal transduction events, as they are responsible for the instauration of action potential and conduction of electric impulses. Reasonably, Na_Vs represent eligible targets for sea anemone toxins. Initially, sea anemone toxins were discovered in venoms composition utilizing classical biochemical and bio-guided fractionation procedures. This generally applies for all the sources of bioactive compound species, as well as for *A. viridis*, specifically. During the years, the screening of different cDNA and genome libraries [9], as well as a bio-prospecting approach on the EST and RNA-Seq collection of the Mediterranean sea anemone allowed for the identification of a novel class of candidate toxins and confirmed the existence of already known sodium channel toxins.

Historically, the sea anemone Na_V-toxins have been grouped into three types on the basis of amino acid sequences, length and S-S bonds arrangement (type I, II and III).

The *A. viridis* Type I-II toxins bind specifically to the Na_Vs of excitable tissue, prolonging the open state of the channels during the depolarization process. Thus, a delay in the inactivation process is usually achieved when applying these toxins on the soma membrane of crustacean neurons [31,32]. Because of this specialized activity on the Na_Vs channels, these toxins [33] may be used as reliable tools for physiological investigations focused on Na_Vs function.

Type I and II are represented by toxin molecules with 46–49 amino-acid residues. They were preferentially isolated from the organisms belonging to the families Actinidae and Stichodactylidae. The mature peptides inside each of these groups show similar structural features. Type I and II molecules share the same Cys-framework (I–V, II–IV, III–VI). More precisely, Type I disulphide connection pattern may be defined as (4–44, 6–34, 27–45), and for Type II as (3–17, 4–11, 6–22)—but the general framework is the same (I–V, II–IV, III–VI). Both molecules with confirmed experimental evidence at the protein level and those that were deduced only from transcriptomics data are included in Type I and Type II groups.

According to classification of Cys-distribution patterns in sea anemone venom toxins proposed by Mikov and Kozlov [34], type I *A. viridis* Na_V toxins belong to structural group 1a because they fall into the following pattern: $C1C\#\#C6C–X_1–CC\#$, where X_1 = 6–9 aa, ## = more than 9 aa, # = 1–9 aa. Moreover, type II *A. viridis* Na_V toxins belong to structural group 1a because they fall into the pattern: $CC1C–X_1–C5C4C\#$, where X_1 = 2 or 4 or 5 aa, # = 1–9 aa.

In the light of biotechnological exploitations, it is paramount to describe toxins with confirmed experimental evidence at the protein level (especially ones with established activity), because similar peptides are usually being found by inference from homology with them.

Type I Nav toxins. The following three toxins were found to be present in venom at the protein level:

(1)   neurotoxin 1 (δ-actitoxin-Avd1a; ATX Ia);

(2)   neurotoxin 2 (δ-actitoxin-Avd1c; ATX II);

(3)   neurotoxin 5 (δ-actitoxin-Avd1d; ATX-V).

Neurotoxin 1 (ATX Ia; UniProt ID: P01533) amino-acid sequence was published back in 1978 [35]. It was shown then that neurotoxin 1 has an effect on Na^+ currents inactivation, but it does not affect K^+ and Ca^{2+} currents in crustacean neurons with concentrations up to 5 μM [35]. In the presence of ATX Ia, inactivation of crustacean neuron Na^+ currents was incomplete. Moreover, some parameters of activation were also affected by ATX Ia, for example, the negative resistance branch of the peak Na^+ current-voltage relation had been shifted [35]. The secondary structure of neurotoxin 1 was determined using complete sequence-specific NMR-assignments [36]; it was shown that it comprises β-sheets consisting of four strands, and no evidence of helical structures was found. The three-dimensional structure of neurotoxin 1 in aqueous solution was determined by nuclear magnetic resonance [37]. Neurotoxin 1 has a well-defined molecular core formed by four-stranded β-sheet that is connected by two well-defined loops (defensin-like fold). The core is stabilized by three disulphide bridges (PDB ID: 1ATX). There is also an additional flexible loop made of 11 residues (8–18 aa).

Neurotoxin 2 (ATX II; UniProt ID: P0DL49) amino-acid sequence was established and published in 1976 [33], while disulphide bridges coordination pattern of this peptide was published later in 1978 [38]. However, the three-dimensional structure of ATX II has not been determined experimentally yet. First, physiological investigation of ATX II was described in the same paper as for ATX Ia [35]. In general, action of ATX II was reminiscent of that of ATX Ia—it inflicted selective Na^+ current inactivation slowdown with no effect on K^+ and Ca^{2+} currents and changed current-voltage relation for activation at the same time in crustacean neurons. Additionally, ATX II led to shortening of the action potential in crayfish neurons in experiments with repetitive stimulations [39].

Neurotoxin 1 was tested only on crayfish neurons. Neurotoxin 2, however, was subjected to more extensive investigations. ATX II was applied on α-subunits of hH1 (human heart subtype 1), rSkM1 (rat skeletal muscle subtype 1) and hSkM1 (human skeletal muscle) sodium channels expressed in the tsA101 line of human embryonic kidney cells to reveal broad pharmacological properties of this peptide [40]. Neurotoxin 2 revealed potent slowing of the inactivation phase of hH1 (IC_{50} 11 nM) as well as moderate slowing of the inactivation phase of rSkM1 (IC_{50} 51 nM) and hSkM1 (IC_{50} ~50 nM). There was no effect of ATX II on the activation of any of these sodium channels [40]. The example of studying pharmacological and physiological properties of neurotoxin 2 represents a good example of the urgency of testing novel toxins on a wide enlistment of channel types and subtypes as a possibility to reveal a real selectivity/range of actions of the particular toxin.

Neurotoxin 5 (ATX-V; UniProt ID: P01529) amino acid sequence was published in 1982 [40]; it was already known that this peptide is highly toxic for mammals. It was revealed that ATX-V delays inactivation of sodium currents and, therefore, strongly stimulates mammalian cardiac muscle contraction.

Type II Nav toxins. Concerning Type II molecules, only 1 peptide—neurotoxin 3 (δ-actitoxin-Avd2a; ATX III; Av3; UniProt ID: P01535) was present in venom at protein level. Neurotoxin 7 and neurotoxin 10 sequences were deduced from transcriptomes. Interestingly, ATX III, isolated earlier from the sea anemone venom [41], was not retrieved neither from the dbESTs, nor from any RNA-Seq datasets available to date. However, such peptides as Neurotoxin 7 and Neurotoxin 10 showed analogous features and exhibited a pattern of disulfide bonds connectivity (3–17, 4–11, 6–22; pattern: I–V, II–IV, III–VI) which is characteristic for ATX III (Figure 2).

Initially, is was shown that neurotoxin 3 had a similar activity mode towards crayfish neuron currents as toxins of Type I (slowing down inactivation of Na^+ channels, etc.) [30]. ATX III had no effect on K^+ and Ca^{2+} currents when tested on concentrations up to 5 μM. ATX III had selective effects on the Na^+ currents at the lowest tested concentration of 50 nM; the effect had the same nature as the action of ATX I and ATX II (influence on both inactivation and activation parameters). Moreover, it was shown that ATX III is toxic to crustaceans and is not toxic to mammals [35]. The structure-functional traits of neurotoxin 3 were studied in detail in the following comprehensive work by Moran et al. [42]. As previous data revealed, ATX III is highly toxic to crustaceans while being harmless to mice. Moran et al. suggested that ATX III might be selectively active on arthropods in general (not just crustaceans) [42]. The experiments demonstrated that Av3 is highly toxic to blowfly larvae (ED$_{50}$ 2.65 ± 0.46 pmol/100 mg) and that it effectively competes with classical site-3 scorpion toxin LqhαIT on binding to cockroach neuronal membranes [42]. Moreover, Av3 inhibited the inactivation of the fly *Drosophila melanogaster* channel DmNav1, but did not influence the functioning of mammalian Na$_V$ channels. Furthermore, action of ATX III was significantly intensified by receptor site-4 ligands, such as scorpion β-toxins [42]. Av3 presented itself as a quite unusual site-3 inhibitor of sodium channels: D1701R mutation in DmNav1 channel did not influence the action of Av3, while the actions of other site-3 ligands were abolished [42]; therefore, this toxin may possess a unique mode of interaction with DmNav1 and this needs to be studied in detail.

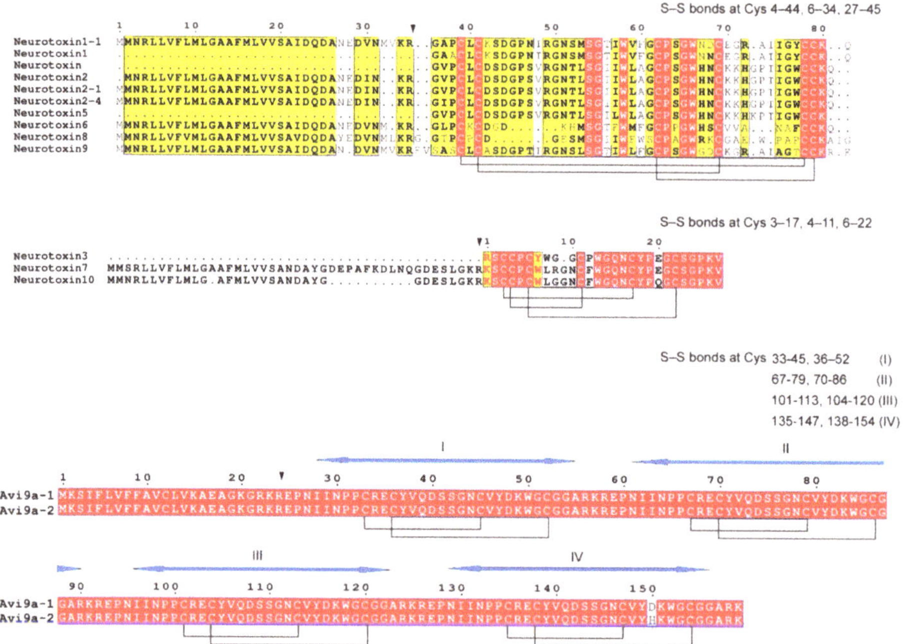

Figure 2. Multiple sequence alignment of the Na$_V$s toxins in *A. viridis*. Based on S-S bonds arrangement, Nav toxins are reported as Type I on the top, Type II on the middle and Type III on the bottom. Alignment was performed with T-coffee tool [43]. Similar residues are written in bold characters and boxed in yellow, whereas conserved residues are in white bold characters and boxed in red. The sequence numbering on the top refers to the alignment. For each alignment, the pattern of Cys residues forming disulfide bridges is shown. Pro-peptide processing sites are pointed out by an inverted black triangle. The four motifs of the active toxins for Avi 9a-1 and Avi 9a-2 are indicated by the blue arrows.

Type III Na$_V$ toxins. Type III includes shorter polypeptides with 27–31 aa residues. These toxins were also found in the arsenal of *A. viridis*. Type III toxins of A. viridis are encoded into complex 159 amino acid-long precursor proteins named AnmTX Avi 9a-1 and AnmTX Avi 9a-2 (short names are Avi 9a-1 and Avi 9a-2). Arbitrarily, mature toxins coded by these long precursors are also named Avi 9a-1 and Avi 9a-2. As mature toxins, Avi 9a-1 and Avi 9a-2 are short peptides almost identical in their primary structure (the only difference is His-22 in Avi 9a-2, instead of Asp-22 in Avi 9a-1). Interestingly, precursor protein AnmTX Avi 9a-1 includes four copies of Avi 9a-1 toxin, while precursor protein AnmTX Avi 9a-2 includes three copies of Avi 9a-1 toxin and one copy of Avi 9a-2 toxin (see Figure 2). To the best of our knowledge, there are no works confirming evidence of short toxins

Avi 9a-1 and Avi 9a-2 on protein level and there are no works on the activity of these toxins; there is only evidence on transcript level [17]. Mature Avi 9a-1 and Avi 9a-2 toxins are homologous to toxin π-AnmTX Ugr 9a-1 which is known as an inhibitor of human type 3 asid-sensing ion channels (hASIC3) [44]. Moreover, they are similar to Am I toxin from the sea anemone *Antheopsis maculata* in their primary structure [45]. The homology does not allow making any conclusions on Avi 9a-1 and Avi 9a-2 actual neuropharmacology, so solid experimental evidence is needed. Concerning Cys-distribution pattern, Type III *A. viridis* Na_V toxins belong to structural group 9a because they comply with the following pattern: C2C#C#C#, where # = 1–9 aa (according to Reference [33]).

3.2. Potassium Channel Peptide Toxins

Sea anemone K^+ channel-blocking toxins have been grouped into different classes (type 1–4) based on the number of amino acid residues, disulfide bridge patterns and molecular structure [46,47]. The mature form of Type 1 and 2 toxins consists of about 40 and 58 amino acid residues, respectively; whereas three disulphide bridges stabilize the structures of these molecules. Type 1 peptides include ShK toxins from the sea anemone *Stichodactyla helianthus*. The ShK represents one of first K^+ channel-blocking toxins from marine organisms which was structurally and functionally characterised because of specific activity on $K_V1.3$, $K_V1.1$ and $K_V3.2$ [48,49] channels; while type 2 peptides, which include the kalicludines, are similar to Kunitz-type protease inhibitors.

Type 1 K_V blockers. *A. viridis* members of Type 1 family include kaliseptine (AsKS; κ-actitoxin-Avd6a; UniProt ID: Q9TWG1)—36 aa peptide, which has no sequence homology with kalicludines from *A. viridis* or with dendrotoxins from mamba snake venoms. Kaliseptine was isolated from *A. viridis* venom during the separation combined with tests of fractions on their ability to compete with ^{125}I-dendrotoxin I (^{125}I-DTX$_I$) for binding its receptors, K^+ channels, in rat brain membranes [50]. IC_{50} for inhibition of ^{125}I-DTX$_I$ binding by AsKS is 27 nM, while DTX$_I$ itself competes with ^{125}I-DTX$_I$ much more actively than with IC_{50} 0.14 nM. Interestingly, AsKS does not compete with ^{125}I-calcicludine for its binding to Ca^{2+} channels of the rat brain up to a concentration of 5 μM [50]. Electrophysiological experiments using different K_V channels expressed in Xenopus oocytes revealed that kaliseptine inhibits $K_V1.2$ currents with IC_{50} 140 nM, while DTX$_I$ had IC_{50} 2.1 nM [50]. Unlike kalicludines, kaliseptine has no homology with bovine pancreatic trypsin inhibitor and, therefore, does not inhibit trypsin even at very high concentrations.

Based on polypeptide length and exhibition of a specific pattern of S-S bonds (2–36, 11–29, 20–33; I–VI, II–IV, III–V), biochemical isolation and motif scan on RNA datasets retrieved the existence of a Type 1 K^+ toxin subset including 12 peptide toxins with substantial homology to kaliseptine (AvTx1–11 [17] and KTX-1 [19] (Figure 3). It may be easily predicted by the SignalP algorithm [51] that after the proteolytic releases mature toxins, AvTx1–11 and KTX-1 consist of about 40 amino-acid residues. None of these molecules were tested for pharmacology properties, so whether these toxins possess K_V-blocking or proteinase-inhibiting activity is an open issue for future investigations. Type 1 A. viridis K_V blockers belong to structural group 2a, because they conform with the following Cys-pattern: C8C–X$_1$–C*C3C2C, where X$_1$ = 8 or 4, and *—is for highly variable intervals (according to Reference [33]).

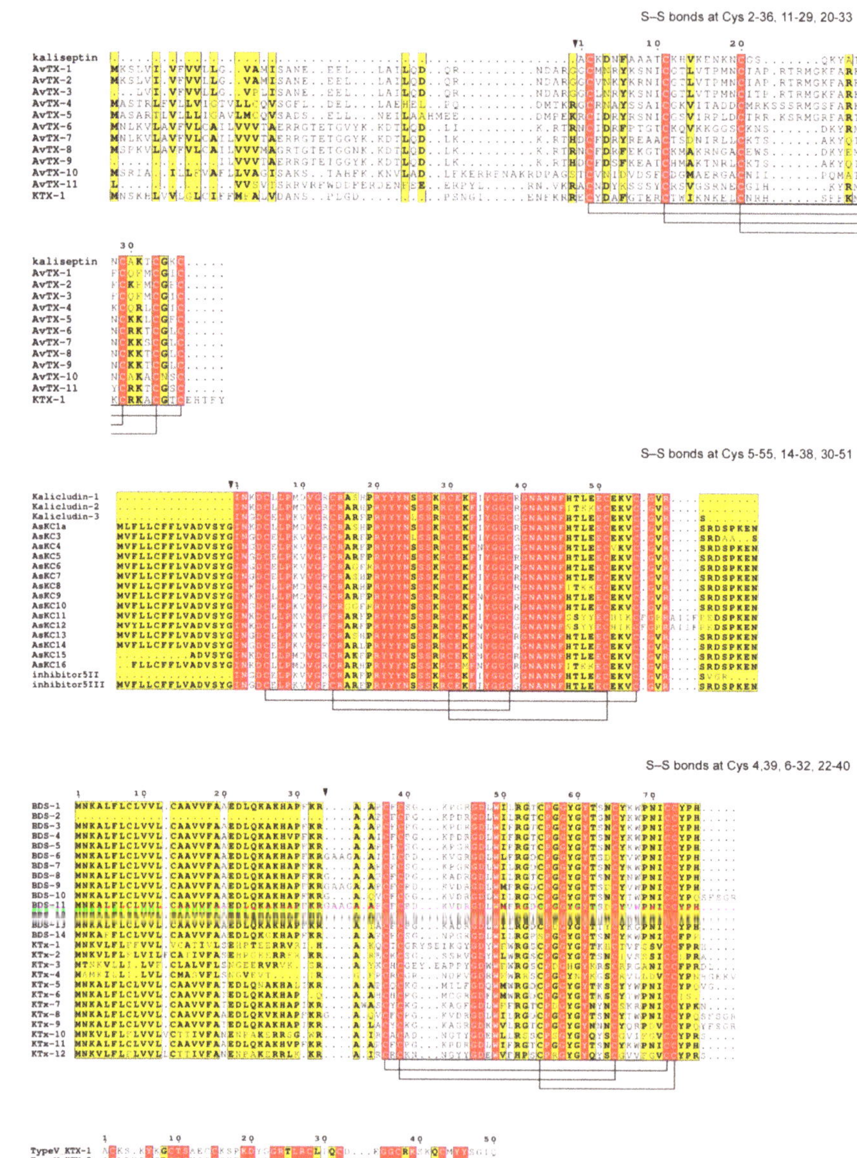

Figure 3. Multiple sequence alignment of the K$_V$s toxins in *A. viridis*. Alignment was performed with the T-coffee tool [43]. Similar residues are written in bold characters and boxed in yellow, whereas conserved residues are in white bold characters and boxed in red. The sequence numbering on the top refers to the alignment. For each alignment, the pattern of Cys residues forming disulfide bridges is shown. Type 1, 2, 3 and 5 K$_V$ blockers are reported; while no member of Type 4 has been identified to date. No S-S bonds and Cys pattern are defined for type V KTx because of the absence of any 3D structure experimentally determined to date.

Type 2 K$_V$ blockers. Several members of type 2 K$^+$ channel-blocking toxins were also found in *A. viridis*. The complex array included 22 kalicludine variants (AsKCn), 2 proteinase inhibitors (5 II and 5 III) and 24 additional type 2 peptides (KTx2n) showing a molecular architecture similar to

Kunitz-type protease inhibitors. Similar to kaliseptines, kalicludines are able to block K_V channels containing $K_V1.2$ subunits [50]. Interestingly, in silico bio-prospecting also revealed the production of two non-canonical type V KTx toxins. The founder members of K_V type V toxins were recently isolated from the venom of the sea anemone *Bunodosoma caissarum* and showed a high affinity for different channels including $K_V1.1$, $K_V1.2$, $K_V1.3$, $K_V1.6$ and Shaker IR [52].

The initial pipeline for identification of three kalicludines (AsKC1–3) was the same as for kaliseptine [50]. Similar to kaliseptine, kalicludines are able to block rat brain K_V channels containing $K_V1.2$ subunits [42]. The IC_{50} for inhibition of the $K_V1.2$ current was 1.1 mM for AsKC2, 1.3 mM for AsKC3, and 2.8 mM for AsKC1 (compared to IC_{50} 2.1 nM for DTX_I); so kalicludines are quite weak blockers of $K_V1.2$ channels [50]. Similar to AsKS, kalicludines do not compete with ^{125}I-calcicludine for its binding to Ca^{2+} channels of the rat brain up to a concentration of 5 µM. However, in contrast to AsKS, AsKCs exhibited proteinase-inhibiting properties, and inhibited trypsin with K_d below 30 nM; therefore, kalicludines are molecules with dual type of activity. Combination of trypsin inhibition and K_V-blocking properties is very peculiar to toxins with Kunitz-type fold. So, despite the fact that 3D structures of kalicludines have not been determined yet, a combination of activity and Cys-distribution patterns may be proposed so that AsKCs possess a 3D structure similar to classical Kunitz-type molecules. Disulphide connection framework is CysI–CysVI, CysII–CysIV, CysIII–CysV and the structure consists of a combination of N-terminal 3_{10}-helix, C-terminal α-helix and 3-stranded anti-parallel β-structure in the core [53–56]. Sequences of kalicludines AsKC1-3 have been extracted from EST databases, so their real presence in the venom needs to be validated by rigorous proteomics investigations and their biological activity is to be established in the future. Type 2 *A. viridis* K_V blockers belong to structural group 3a because they comply with the following pattern: C8C15C7C12C3C (classification [1]).

Type 3 K_V blockers. The Mediterranean sea anemone *A. viridis* also produces Type 3 toxins which are represented by BDS peptides (1–14) and 12 Type 3 KTx. Members of this class include peptides of various origins. For example, APETx1 (inhibits $K_V11.1$, $K_V11.3$ and blocks $Na_V1.2$, $Na_V1.3$, $Na_V1.4$, $Na_V1.5$, $Na_V1.6$, $Na_V1.8$ of mammals) and APETx2 (inhibits mammalian $K_V3.4$, $K_V11.1$, some Na_V and blocks ASIC3-containing trimers) were isolated from *Anthopleura elegantissima* venom [57–59]. Another example is Am-II from *Antheopsis maculata* which is of unknown pharmacology but is toxic to crabs [45]. As might be seen, Type 3 toxins may exhibit a lot of diversified properties besides being K_V-blockers.

The initial members of this class from *A. viridis* venom—BDS-1 (UniProtID: P11494) and BDS-2 (UniProtID: P59084)—were originally described as blood pressure-reducing substances with antiviral activities [60]. These are 43 amino acid long, cysteine-rich polypeptides stabilized by three disulphide bonds. The peptides voltage-dependently inhibit K^+ channels containing K_V3 subunits such as $K_V3.1$, $K_V3.2$, and fast inactivating channel $K_V3.4$ [61–63]. Additionally, it has been recently reported that BDS-1 also slows down inactivation of the human $Na_V1.7$/SCN9A channels (and $Na_V1.7$ in rat SCG neurons as well), and weakly inhibits $Na_V1.3$ channels [64].

Type 3 K_V blockers belong to a Cys-distribution pattern named "1b" (classification [1]) which implies the following distribution scheme: C1C##C9C–X_1–CC#, where X_1 = 6–8, # = 1–9 aa, ## = more than 9 aa. Concerning *A. viridis* molecules, the spatial structure was resolved for BDS-1 by NMR [65] (concerning other species, 3D-structures were also determined for APETx1 [66] and APETx2 [67]). These 3D-structures are quite similar for BDS-1, APETx1 and APETx2—each of them may be described as a combination of the compact disulphide-stabilized nucleus consisting of a four-stranded β-sheet with N- and C-termini protruding from it (the so-called defensin-like fold, the same as for structural pattern 1a).

Type 4 K_V blockers. Toxins of this Type were not identified in the venom of *A. viridis* as well as in its transcriptomes. This may be both due to technical limitations and difficulties or simply because *A. viridis* does not produce Type 4 K_V blockers. This issue is to be resolved in future investigations.

Non-canonical Type 5 KTx-like peptides. Interestingly, in silico bioprospecting also revealed the production of two non-canonical Type V KTx-like peptides. The founder members of K_V Type 5 toxins were recently isolated from the venom of the sea anemone *B. caissarum* and showed a high affinity for different channels including $K_V1.1$, $K_V1.2$, $K_V1.3$, $K_V1.6$ and Shaker IR [52].

3.3. Other Candidate Toxins of A. viridis

In addition to the peptide toxins whose functions have been well characterized, several putative polypeptides were retrieved only by means of bio-prospecting on transcriptome datasets.

Gigantoxin homologs. This group includes gigantoxin 4 (U-AITX-Avd12a; UniProtID: P0DMY9) and gigantoxin 5 (U-AITX-Avd12b; UniProtID: P0DMZ0). They show significant similarity with gigantoxin I from the giant carpet sea anemone *Stichodactyla gigantea* [68]. Gigantoxin I (ω-stichotoxin-Sgt1a; UniProtID: Q76CA1) is known to exert a potent paralytic and weak lethal effect on crabs (PD_{50} is 215 µg/kg; LD_{50} is > 1000 µg/kg) [68]. Moreover, gigantoxin I is capable of binding to epidermal growth factor receptor (EGFR) and, therefore, induces morphological changes (rounding of the cells) and tyrosine phosphorylation of the EGFR in cells (was shown using epidermoid carcinoma A431 cell line) [68]. Despite the fact that gigantoxin 4 and gigantoxin 5 are homological to gigantoxin I, no conclusion on their potential biological activity might be deduced from this fact.

Acrorhagin homologs. Although the presence of specialized aggressive organs named acrorhagi remains controversial in *A. viridis*, analyses of RNA-Seq data provides evidence for the production of four candidate toxins belonging to acrorhagin 1 and acrorhagin 2 subtypes [19]. Founder members of this class have been isolated from the acrorhagi of the sea anemone *Actinia equina* and have been reported to induce mortality when injected on crabs with LD_{50} values corresponding to 520 and 80 µg/kg for acrorhagin 1 and acrorhagin 2, respectively [69]. Computational tools for sequence and relation analysis fail to identify homologues superfamily and consistently they do not show significant sequence similarity (lower than 30% identity) with any toxins from other sources. Despite the fact that biological activity and functions have not yet been determined for this group of proteins, they may contribute to the articulated sea anemone venom assemblage.

All candidate toxins were found only on transcript level and their presence in venom needs to be validated in the future. Because the toxin expression profiles may vary greatly even inside single species, massive proteomics studies of venoms from different *A. viridis* specimens need to be carried out.

4. From Bioprospecting to Translational Research

Cnidarians possess exciting strategies for survival, which includes an articulated and finely defined toxin arsenal. Because of neurotoxicity and cytotoxicity of the sea anemone peptides, they represent promising putative pharmacological agents for translational research and biomedical applications. Moreover, different lines of evidences have defined cnidarian venom as an experimental tool for cell physiology [70]. The two *A. viridis* non-canonical Type 5 KTx represent interesting candidates for immune system targeting. Because the $K_V1.3$ channel is responsible for the activity human effector memory T cells, the pharmacological application of recombinant or sinthetic Type 5 KTx may take place in the treatment of autoimmune diseases mediated by T cells as already described for the ShK toxin from the sea anemone *S. helianthus* [71,72]. Thus, blockers of $K_V1.3$ channels show interesting application for the treatment of type 1 diabetes mellitus, rheumatoid arthritis and multiple sclerosis. BDS-1 has been pointed out as a promising potential tool for targeting Kv3.4 subunit, which has been implicated in CNS disorders as Parkinson's and Alzheimer's disease [73,74]. Several studies have reported the use of such an antagonist as a pharmacological tool to assign functional roles to channels in CNS neurons [74–76]. Effects were reported on several channels ($K_V3.4$, $K_V3.1$, $K_V3.2$, and $Na_V1.7$), thus instilling doubts on its future practical applications. It is noteworthy to specify that these reports take the advances of the use of purified sea anemone extracts, consisting of the BDS variants (1–14) which may display different specificity. Therefore, the development of new peptides with restricted activity by means of recombinant BDSs may allow the selective targeting of these channels.

Although several *A. viridis* toxins represent interesting candidates for biomedical applications, pharmacological investigations are still mandatory to reveal potency and selectivity of other *A. viridis* toxins, including those that have been only detected on the transcriptional level. High-throughput activity-testing systems are required to be developed to accelerate this process, thus allowing them to reach a different phase of clinical trials.

Author Contributions: A.N. and A.M. collected data and prepared the manuscript, M.C., P.C., Y.A., S.K., A.C. revised the manuscript.

Funding: This research was funded by [Russian Science Foundation] grant number [16-15-00167] to A.M. and Y.A.

Conflicts of Interest: The authors declare no conflict of interest.

References

1. Molinski, T.F.; Dalisay, D.S.; Lievens, S.L.; Saludes, J.P. Drug development from marine natural products. *Nat. Rev. Drug Discov.* **2009**, *8*, 69–85. [CrossRef] [PubMed]

2. Glaser, K.B.; Mayer, A.M.S. A renaissance in marine pharmacology: from preclinical curiosity to clinical reality. *Biochem. Pharmacol.* **2009**, *78*, 440–448. [CrossRef] [PubMed]

3. Rocha, J.; Peixe, L.; Gomes, N.C.M.; Calado, R. Cnidarians as a source of new marine bioactive compounds—An overview of the last decade and future steps for bioprospecting. *Mar. Drugs* **2011**, *9*, 1860–1886. [CrossRef] [PubMed]

4. Chen, J.-Y.; Oliveri, P.; Gao, F.; Dornbos, S.Q.; Li, C.-W.; Bottjer, D.J.; Davidson, E.H. Precambrian animal life: probable developmental and adult cnidarian forms from Southwest China. *Dev. Biol.* **2002**, *248*, 182–196. [CrossRef] [PubMed]

5. Ruppert, E.E.; Barnes, R.D. *Invertebrate Zoology*, 6th ed.; Saunders College Publishing: Fort Worth, TX, USA, 1994.

6. Beress, L. Biologically active compounds from coelenterates. *Pure Appl. Chem.* **1982**, *54*, 1981–1994. [CrossRef]

7. Nicosia, A.; Maggio, T.; Mazzola, S.; Gianguzza, F.; Cuttitta, A.; Costa, S. Characterization of small HSPs from *Anemonia viridis* reveals insights into molecular evolution of alpha crystallin genes among cnidarians. *PLoS ONE* **2014**, *9*, e105908. [CrossRef] [PubMed]

8. Nicosia, A.; Maggio, T.; Mazzola, S.; Cuttitta, A. Evidence of accelerated evolution and ectodermal-specific expression of presumptive BDS toxin cDNAs from *Anemonia viridis*. *Mar. Drugs* **2013**, *11*, 4213–4231. [CrossRef] [PubMed]

9. Moran, Y.; Weinberger, H.; Sullivan, J.C.; Reitzel, A.M.; Finnerty, J.R.; Gurevitz, M. Concerted evolution of sea anemone neurotoxin genes is revealed through analysis of the *Nematostella vectensis* genome. *Mol. Biol. Evol.* **2008**, *25*, 737–747. [CrossRef] [PubMed]

10. Moran, Y.; Weinberger, H.; Lazarus, N.; Gur, M.; Kahn, R.; Gordon, D.; Gurevitz, M. Fusion and retrotransposition events in the evolution of the sea anemone *Anemonia viridis* neurotoxin genes. *J. Mol. Evol.* **2009**, *69*, 115–124. [CrossRef] [PubMed]

11. Castañeda, O.; Harvey, A.L. Discovery and characterization of cnidarian peptide toxins that affect neuronal potassium ion channels. *Toxicon* **2009**, *54*, 1119–1124. [CrossRef] [PubMed]

12. Nicosia, A.; Maggio, T.; Costa, S.; Salamone, M.; Tagliavia, M.; Mazzola, S.; Gianguzza, F.; Cuttitta, A. Maintenance of a Protein Structure in the Dynamic Evolution of TIMPs over 600 Million Years. *Genome Biol. Evol.* **2016**, *8*, 1056–1071. [CrossRef] [PubMed]

13. Bulati, M.; Longo, A.; Masullo, T.; Vlah, S.; Bennici, C.; Bonura, A.; Salamone, M.; Tagliavia, M.; Nicosia, A.; Mazzola, S.; et al. Partially Purified Extracts of Sea Anemone *Anemonia viridis* Affect the Growth and Viability of Selected Tumour Cell Lines. *Biomed Res. Int.* **2016**, *2016*, 3849897. [CrossRef] [PubMed]

14. Cuttitta, A.; Ragusa, M.A.; Costa, S.; Bennici, C.; Colombo, P.; Mazzola, S.; Gianguzza, F.; Nicosia, A. Evolutionary conserved mechanisms pervade structure and transcriptional modulation of allograft inflammatory factor-1 from sea anemone *Anemonia viridis*. *Fish Shellfish Immunol.* **2017**, *67*, 86–94. [CrossRef] [PubMed]

15. Nicosia, A.; Bennici, C.; Biondo, G.; Costa, S.; Di Natale, M.; Masullo, T.; Monastero, C.; Ragusa, M.A.; Tagliavia, M.; Cuttitta, A. Characterization of Translationally Controlled Tumour Protein from the Sea Anemone *Anemonia viridis* and Transcriptome Wide Identification of Cnidarian Homologues. *Genes* **2018**, *9*, 30. [CrossRef] [PubMed]

16. Frazão, B.; Vasconcelos, V.; Antunes, A. Sea anemone (Cnidaria, Anthozoa, Actiniaria) toxins: An overview. *Mar. Drugs* **2012**, *10*, 1812–1851. [CrossRef] [PubMed]

17. Kozlov, S.; Grishin, E. The mining of toxin-like polypeptides from EST database by single residue distribution analysis. *BMC Genom.* **2011**, *12*, 88. [CrossRef] [PubMed]

18. Kozlov, S.; Grishin, E. Convenient nomenclature of cysteine-rich polypeptide toxins from sea anemones. *Peptides* **2012**, *33*, 240–244. [CrossRef] [PubMed]

19. Macrander, J.; Broe, M.; Daly, M. Tissue-Specific Venom Composition and Differential Gene Expression in Sea Anemones. *Genome Biol. Evol.* **2016**, *8*, 2358–2375. [CrossRef] [PubMed]

20. Sabourault, C.; Ganot, P.; Deleury, E.; Allemand, D.; Furla, P. Comprehensive EST analysis of the symbiotic sea anemone, *Anemonia viridis*. *BMC Genom.* **2009**, *10*, 333. [CrossRef] [PubMed]

21. Urbarova, I.; Patel, H.; Forêt, S.; Karlsen, B.O.; Jørgensen, T.E.; Hall-Spencer, J.M.; Johansen, S.D. Elucidating the Small Regulatory RNA Repertoire of the Sea Anemone *Anemonia viridis* Based on Whole Genome and Small RNA Sequencing. *Genome Biol. Evol.* **2018**, *10*, 410–426. [CrossRef] [PubMed]

22. Martens, L.; Hermjakob, H.; Jones, P.; Adamski, M.; Taylor, C.; States, D.; Gevaert, K.; Vandekerckhove, J.; Apweiler, R. PRIDE: The proteomics identifications database. *Proteomics* **2005**, *5*, 3537–3545. [CrossRef] [PubMed]

23. Farrah, T.; Deutsch, E.W.; Kreisberg, R.; Sun, Z.; Campbell, D.S.; Mendoza, L.; Kusebauch, U.; Brusniak, M.-Y.; Hüttenhain, R.; Schiess, R.; et al. PASSEL: The PeptideAtlas SRMexperiment library. *Proteomics* **2012**, *12*, 1170–1175. [CrossRef] [PubMed]

24. Wilhelm, M.; Schlegl, J.; Hahne, H.; Gholami, A.M.; Lieberenz, M.; Savitski, M.M.; Ziegler, E.; Butzmann, L.; Gessulat, S.; Marx, H.; et al. Mass-spectrometry-based draft of the human proteome. *Nature* **2014**, *509*, 582–587. [CrossRef] [PubMed]

25. Martens, L.; Vizcaíno, J.A. A Golden Age for Working with Public Proteomics Data. *Trends Biochem. Sci.* **2017**, *42*, 333–341. [CrossRef] [PubMed]

26. Salamone, M.; Nicosia, A.; Bennici, C.; Quatrini, P.; Catania, V.; Mazzola, S.; Ghersi, G.; Cuttitta, A. Comprehensive Analysis of a *Vibrio parahaemolyticus* Strain Extracellular Serine Protease VpSP37. *PLoS ONE* **2015**, *10*, e0126349. [CrossRef] [PubMed]

27. Himaya, S.W.A.; Lewis, R.J. Venomics-Accelerated Cone Snail Venom Peptide Discovery. *Int. J. Mol. Sci.* **2018**, *19*, 788. [CrossRef] [PubMed]

28. Salgado, V.L.; Kem, W.R. Actions of three structurally distinct sea anemone toxins on crustacean and insect sodium channels. *Toxicon* **1992**, *30*, 1365–1381. [CrossRef]

29. Norton, R.S. Structure and structure-function relationships of sea anemone proteins that interact with the sodium channel. *Toxicon* **1991**, *29*, 1051–1084. [CrossRef]

30. Molgó, J.; Mallart, A. Effects of *Anemonia sulcata* toxin II on presynaptic currents and evoked transmitter release at neuromuscular junctions of the mouse. *Pflugers Arch.* **1985**, *405*, 349–353. [CrossRef] [PubMed]

31. Schweitz, H.; Vincent, J.P.; Barhanin, J.; Frelin, C.; Linden, G.; Hugues, M.; Lazdunski, M. Purification and pharmacological properties of eight sea anemone toxins from *Anemonia sulcata*, *Anthopleura xanthogrammica*, *Stoichactis giganteus*, and *Actinodendron plumosum*. *Biochemistry* **1981**, *20*, 5245–5252. [CrossRef] [PubMed]

32. Warashina, A.; Ogura, T.; Fujita, S. Binding properties of sea anemone toxins to sodium channels in the crayfish giant axon. *Comp. Biochem. Physiol. C, Comp. Pharmacol. Toxicol.* **1988**, *90*, 351–359. [PubMed]

33. Wunderer, G.; Fritz, H.; Wachter, E.; Machleidt, W. Amino-acid sequence of a coelenterate toxin: Toxin II from *Anemonia sulcata*. *Eur. J. Biochem.* **1976**, *68*, 193–198. [CrossRef] [PubMed]

34. Mikov, A.N.; Kozlov, S.A. Structural Features of Cysteine-Stabilized Polypeptides from Sea Anemones Venoms. *Bioorg. Khim.* **2015**, *41*, 511–523. [PubMed]

35. Hartung, K.; Rathmayer, W. *Anemonia sulcata* toxins modify activation and inactivation of Na+ currents in a crayfish neurone. *Pflugers Arch.* **1985**, *404*, 119–125. [CrossRef] [PubMed]

36. Widmer, H.; Wagner, G.; Schweitz, H.; Lazdunski, M.; Wüthrich, K. The secondary structure of the toxin ATX Ia from *Anemonia sulcata* in aqueous solution determined on the basis of complete sequence-specific 1H-NMR assignments. *Eur. J. Biochem.* **1988**, *171*, 177–192. [CrossRef] [PubMed]

37. Widmer, H.; Billeter, M.; Wüthrich, K. Three-dimensional structure of the neurotoxin ATX Ia from *Anemonia sulcata* in aqueous solution determined by nuclear magnetic resonance spectroscopy. *Proteins* **1989**, *6*, 357–371. [CrossRef] [PubMed]

38. Wunderer, G. The disulfide bridges of toxin II from *Anemonia sulcata*. *Hoppe-Seyler's Z. Physiol. Chem.* **1978**, *359*, 1193–1201. [CrossRef] [PubMed]

39. Chahine, M.; Plante, E.; Kallen, R.G. Sea anemone toxin (ATX II) modulation of heart and skeletal muscle sodium channel alpha-subunits expressed in tsA201 cells. *J. Membr. Biol.* **1996**, *152*, 39–48. [CrossRef] [PubMed]

40. Scheffler, J.J.; Tsugita, A.; Linden, G.; Schweitz, H.; Lazdunski, M. The amino acid sequence of toxin V from *Anemonia sulcata*. *Biochem. Biophys. Res. Commun.* **1982**, *107*, 272–278. [CrossRef]

41. Martinez, G.; Kopeyan, C. Toxin III from *Anemonia sulcata*: Primary structure. *FEBS Lett.* **1977**, *84*, 247–252. [CrossRef]

42. Moran, Y.; Kahn, R.; Cohen, L.; Gur, M.; Karbat, I.; Gordon, D.; Gurevitz, M. Molecular analysis of the sea anemone toxin Av3 reveals selectivity to insects and demonstrates the heterogeneity of receptor site-3 on voltage-gated Na+ channels. *Biochem. J.* **2007**, *406*, 41–48. [CrossRef] [PubMed]

43. Notredame, C.; Higgins, D.G.; Heringa, J. T-Coffee: A novel method for fast and accurate multiple sequence alignment. *J. Mol. Biol.* **2000**, *302*, 205–217. [CrossRef] [PubMed]

44. Osmakov, D.I.; Kozlov, S.A.; Andreev, Y.A.; Koshelev, S.G.; Sanamyan, N.P.; Sanamyan, K.E.; Dyachenko, I.A.; Bondarenko, D.A.; Murashev, A.N.; Mineev, K.S.; et al. Sea Anemone Peptide with Uncommon β-Hairpin Structure Inhibits Acid-sensing Ion Channel 3 (ASIC3) and Reveals Analgesic Activity. *J. Biol. Chem.* **2013**, *288*, 23116–23127. [CrossRef] [PubMed]

45. Honma, T.; Hasegawa, Y.; Ishida, M.; Nagai, H.; Nagashima, Y.; Shiomi, K. Isolation and molecular cloning of novel peptide toxins from the sea anemone *Antheopsis maculata*. *Toxicon* **2005**, *45*, 33–41. [CrossRef] [PubMed]

46. Honma, T.; Shiomi, K. Peptide toxins in sea anemones: Structural and functional aspects. *Mar. Biotechnol.* **2006**, *8*, 1–10. [CrossRef] [PubMed]

47. Honma, T.; Kawahata, S.; Ishida, M.; Nagai, H.; Nagashima, Y.; Shiomi, K. Novel peptide toxins from the sea anemone *Stichodactyla haddoni*. *Peptides* **2008**, *29*, 536–544. [CrossRef] [PubMed]

48. Tudor, J.E.; Pallaghy, P.K.; Pennington, M.W.; Norton, R.S. Solution structure of ShK toxin, a novel potassium channel inhibitor from a sea anemone. *Nat. Struct. Biol.* **1996**, *3*, 317–320. [CrossRef] [PubMed]

49. Pennington, M.W.; Kem, W.R.; Mahnir, V.M.; Byrnes, M.E.; Zaydenberg, I.; Khaytin, I.; Krafte, D.S.; Hill, R. Identification of essential residues in the potassium channel inhibitor ShK toxin: Analysis of monosubstituted analogs. In *Peptides: Chemistry, Structure and Biology*; Kaumaya, P.T.P., Hodges, R.S., Eds.; Escom: Leiden, The Netherlands, 1995; pp. 14–16.

50. Schweitz, H.; Bruhn, T.; Guillemare, E.; Moinier, D.; Lancelin, J.-M.; Béress, L.; Lazdunski, M. Kalicludines and Kaliseptine Two different classes of sea anemone toxins for voltage sensitive K+ channels. *J. Biol. Chem.* **1995**, *270*, 25121–25126. [CrossRef] [PubMed]

51. Petersen, T.N.; Brunak, S.; von Heijne, G.; Nielsen, H. SignalP 4.0: Discriminating signal peptides from transmembrane regions. *Nat. Methods* **2011**, *8*, 785–786. [CrossRef] [PubMed]

52. Orts, D.J.B.; Peigneur, S.; Madio, B.; Cassoli, J.S.; Montandon, G.G.; Pimenta, A.M.C.; Bicudo, J.E.P.W.; Freitas, J.C.; Zaharenko, A.J.; Tytgat, J. Biochemical and electrophysiological characterization of two sea anemone type 1 potassium toxins from a geographically distant population of *Bunodosoma caissarum*. *Mar. Drugs* **2013**, *11*, 655–679. [CrossRef] [PubMed]

53. Scheidig, A.J.; Hynes, T.R.; Pelletier, L.A.; Wells, J.A.; Kossiakoff, A.A. Crystal structures of bovine chymotrypsin and trypsin complexed to the inhibitor domain of Alzheimer's amyloid beta-protein precursor (APPI) and basic pancreatic trypsin inhibitor (BPTI): Engineering of inhibitors with altered specificities. *Protein Sci.* **1997**, *6*, 1806–1824. [CrossRef] [PubMed]

54. Zweckstetter, M.; Czisch, M.; Mayer, U.; Chu, M.L.; Zinth, W.; Timpl, R.; Holak, T.A. Structure and multiple conformations of the kunitz-type domain from human type VI collagen alpha3(VI) chain in solution. *Structure* **1996**, *4*, 195–209. [CrossRef]

55. Chen, C.; Hsu, C.H.; Su, N.Y.; Lin, Y.C.; Chiou, S.H.; Wu, S.H. Solution structure of a Kunitz-type chymotrypsin inhibitor isolated from the elapid snake *Bungarus fasciatus*. *J. Biol. Chem.* **2001**, *276*, 45079–45087. [CrossRef] [PubMed]

56. Antuch, W.; Berndt, K.D.; Chávez, M.A.; Delfín, J.; Wüthrich, K. The NMR solution structure of a Kunitz-type proteinase inhibitor from the sea anemone *Stichodactyla helianthus*. *Eur. J. Biochem.* **1993**, *212*, 675–684. [CrossRef] [PubMed]

57. Diochot, S.; Loret, E.; Bruhn, T.; Béress, L.; Lazdunski, M. APETx1, a new toxin from the sea anemone *Anthopleura elegantissima*, blocks voltage-gated human ether-a-go-go-related gene potassium channels. *Mol. Pharmacol.* **2003**, *64*, 59–69. [CrossRef] [PubMed]

58. Diochot, S.; Baron, A.; Rash, L.D.; Deval, E.; Escoubas, P.; Scarzello, S.; Salinas, M.; Lazdunski, M. A new sea anemone peptide, APETx2, inhibits ASIC3, a major acid-sensitive channel in sensory neurons. *EMBO J.* **2004**, *23*, 1516–1525. [CrossRef] [PubMed]

59. Blanchard, M.G.; Rash, L.D.; Kellenberger, S. Inhibition of voltage-gated Na(+) currents in sensory neurones by the sea anemone toxin APETx2. *Br. J. Pharmacol.* **2012**, *165*, 2167–2177. [CrossRef] [PubMed]

60. Béress, L.; Doppelfeld, I.S.; Etschenberg, E.; Graf, E.; Henschen, A.; Zwick, J. Polypeptides, Methods of Production and Their Uses as Antihypertensives. German Patent DE3324689A1, 17 January 1985.

61. Diochot, S.; Schweitz, H.; Béress, L.; Lazdunski, M. Sea anemone peptides with a specific blocking activity against the fast inactivating potassium channel Kv3.4. *J. Biol. Chem.* **1998**, *273*, 6744–6749. [CrossRef] [PubMed]

62. Yeung, S.Y.M.; Thompson, D.; Wang, Z.; Fedida, D.; Robertson, B. Modulation of Kv3 subfamily potassium currents by the sea anemone toxin BDS: Significance for CNS and biophysical studies. *J. Neurosci.* **2005**, *25*, 8735–8745. [CrossRef] [PubMed]

63. Abbott, G.W.; Butler, M.H.; Bendahhou, S.; Dalakas, M.C.; Ptacek, L.J.; Goldstein, S.A. MiRP2 forms potassium channels in skeletal muscle with Kv3.4 and is associated with periodic paralysis. *Cell* **2001**, *104*, 217–231. [CrossRef]

64. Liu, P.; Jo, S.; Bean, B.P. Modulation of neuronal sodium channels by the sea anemone peptide BDS-I. *J. Neurophysiol.* **2012**, *107*, 3155–3167. [CrossRef] [PubMed]

65. Driscoll, P.C.; Gronenborn, A.M.; Beress, L.; Clore, G.M. Determination of the three-dimensional solution structure of the antihypertensive and antiviral protein BDS-I from the sea anemone *Anemonia sulcata*: A study using nuclear magnetic resonance and hybrid distance geometry-dynamical simulated annealing. *Biochemistry* **1989**, *28*, 2188–2198. [CrossRef] [PubMed]

66. Chagot, B.; Diochot, S.; Pimentel, C.; Lazdunski, M.; Darbon, H. Solution structure of APETx1 from the sea anemone Anthopleura elegantissima: A new fold for an HERG toxin. *Proteins* **2005**, *59*, 380–386. [CrossRef] [PubMed]

67. Chagot, B.; Escoubas, P.; Diochot, S.; Bernard, C.; Lazdunski, M.; Darbon, H. Solution structure of APETx2, a specific peptide inhibitor of ASIC3 proton-gated channels. *Protein Sci.* **2005**, *14*, 2003–2010. [CrossRef] [PubMed]

68. Shiomi, K.; Honma, T.; Ide, M.; Nagashima, Y.; Ishida, M.; Chino, M. An epidermal growth factor-like toxin and two sodium channel toxins from the sea anemone *Stichodactyla gigantea*. *Toxicon* **2003**, *41*, 229–236. [CrossRef]

69. Honma, T.; Minagawa, S.; Nagai, H.; Ishida, M.; Nagashima, Y.; Shiomi, K. Novel peptide toxins from acrorhagi, aggressive organs of the sea anemone *Actinia equina*. *Toxicon* **2005**, *46*, 768–774. [CrossRef] [PubMed]

70. Morabito, R.; La Spada, G.; Crupi, R.; Esposito, E.; Marino, A. Crude Venom from Nematocysts of the Jellyfish *Pelagia noctiluca* as a Tool to Study Cell Physiology. *Cent. Nerv. Syst. Agents Med. Chem.* **2015**, *15*, 68–73. [CrossRef] [PubMed]

71. Prentis, P.J.; Pavasovic, A.; Norton, R.S. Sea Anemones: Quiet Achievers in the Field of Peptide Toxins. *Toxins* **2018**, *10*, 36. [CrossRef] [PubMed]

72. Chi, V.; Pennington, M.W.; Norton, R.S.; Tarcha, E.J.; Londono, L.M.; Sims-Fahey, B.; Upadhyay, S.K.; Lakey, J.T.; Iadonato, S.; Wulff, H.; et al. Development of a sea anemone toxin as an immunomodulator for therapy of autoimmune diseases. *Toxicon* **2012**, *59*, 529–546. [CrossRef] [PubMed]

73. Baranauskas, G.; Tkatch, T.; Nagata, K.; Yeh, J.Z.; Surmeier, D.J. Kv3.4 subunits enhance the repolarizing efficiency of Kv3.1 channels in fast-spiking neurons. *Nat. Neurosci.* **2003**, *6*, 258–266. [CrossRef] [PubMed]

74. Angulo, E.; Noe, V.; Casado, V.; Mallol, J.; Gomez-Isla, T.; Lluis, C.; Ferrer, I.; Ciudad, C.J.; Franco, R. Up-regulation of the Kv3.4 potassium channel subunit in early stages of Alzheimer's disease. *J. Neurochem.* **2004**, *91*, 547–557. [CrossRef] [PubMed]

75. Chabbert, C.; Chambard, J.M.; Sans, A.; Desmadryl, G. Three types of depolarization-activated potassium currents in acutely isolated mouse vestibular neurons. *J. Neurophysiol.* **2001**, *85*, 1017–1026. [CrossRef] [PubMed]

76. Shevchenko, T.; Teruyama, R.; Armstrong, W.E. High-threshold, Kv3-like potassium currents in magnocellular neurosecretory neurons and their role in spike repolarization. *J. Neurophysiol.* **2004**, *92*, 3043–3055. [CrossRef] [PubMed]

 *marine drugs*

Article

The Jellyfish *Rhizostoma pulmo* (Cnidaria): Biochemical Composition of Ovaries and Antibacterial Lysozyme-like Activity of the Oocyte Lysate

Loredana Stabili [1,2,*], Lucia Rizzo [3,4], Francesco Paolo Fanizzi [1], Federica Angilè [1], Laura Del Coco [1], Chiara Roberta Girelli [1], Silvia Lomartire [1], Stefano Piraino [1,3,*] and Lorena Basso [1,3,*]

[1] Department of Biological and Environmental Sciences and Technologies, University of Salento, Via Prov.le Lecce Monteroni, 73100 Lecce, Italy; fp.fanizzi@unisalento.it (F.P.F.); federica.angile@unisalento.it (F.A.); laura.delcoco@unisalento.it (L.D.C.); chiara.girelli@unisalento.it (C.R.G.); silvia_03_@hotmail.it (S.L.)

[2] Water Research Institute (IRSA) of the National Research Council, S.S. Talassografico of Taranto, Via Roma 3, 74122 Taranto, Italy

[3] Consorzio Nazionale Interuniversitario per le Scienze del Mare, CoNISMa, Piazzale Flaminio 00196, 9- Roma, Italy; lucia.rizzo@szn.it

[4] Integrative Marine Ecology, Stazione Zoologica Anton Dohrn, Villa Comunale, 80121 Napoli, Italy

* Correspondence: loredana.stabili@irsa.cnr.it (L.S.); stefano.piraino@unisalento.it (S.P.); lorena_basso@libero.it (L.B.); Tel.: +39-0832298971 (L.S.)

Received: 18 October 2018; Accepted: 23 December 2018; Published: 29 December 2018

Abstract: Jellyfish outbreaks in marine coastal areas represent an emergent problem worldwide, with negative consequences on human activities and ecosystem functioning. However, potential positive effects of jellyfish biomass proliferation may be envisaged as a natural source of bioactive compounds of pharmaceutical interest. We investigated the biochemical composition of mature female gonads and lysozyme antibacterial activity of oocytes in the Mediterranean barrel jellyfish *Rhizostoma pulmo*. Chemical characterization was performed by means of multinuclear and multidimensional NMR spectroscopy. The ovaries of *R. pulmo* were mainly composed of water (93.7 $\pm$ 1.9% of wet weight), with organic matter (OM) and dry weight made respectively of proteins (761.76 $\pm$ 25.11 µg mg^{-1} and 45.7 $\pm$ 1.5%), lipids (192.17 $\pm$ 10.56 µg mg^{-1} and 9.6 $\pm$ 0.6%), and carbohydrates (59.66 $\pm$ 2.72 µg mg^{-1} and 3.7 $\pm$ 0.3%). The aqueous extract of *R. pulmo* gonads contained free amino acids, organic acids, and derivatives; the lipid extract was composed of triglycerides (TG), polyunsaturated fatty acids (PUFAs), diunsaturated fatty acids (DUFAs), monounsaturated fatty acids (MUFAs), saturated fatty acids (SFAs), and minor components such as sterols and phospholipids. The *R. pulmo* oocyte lysate exhibited an antibacterial lysozyme-like activity (mean diameter of lysis of 9.33 $\pm$ 0.32 mm corresponding to 1.21 mg/mL of hen egg-white lysozyme). The occurrence of defense molecules is a crucial mechanism to grant healthy development of mature eggs and fertilized embryos (and the reproductive success of the species) by preventing marine bacterial overgrowth. As a corollary, these results call for future investigations for an exploitation of *R. pulmo* biomasses as a resource of bioactive metabolites of biotechnological importance including pharmaceuticals and nutrition.

Keywords: antibacterial activity; NMR spectroscopy; biochemical characterization; jellyfish blooms

---

1. Introduction

Anthozoans and medusozoans (commonly known as polyps and jellyfish, respectively) belong to Cnidaria, a group of approximately 10,000 marine invertebrates known to produce complex proteinaceous venomous mixtures used for defense and prey capture and delivered through highly specialized, epithelial mechano-sensor cells (the cnidocytes). A relatively small number of jellyfish (including scyphozoans, cubozoans, and hydrozoans) exhibit life history traits promoting reproductive success and formation of large aggregated populations [1]. Their regular outbreaks represent an emergent problem in worldwide coastal areas, favored by the rise of sea surface temperatures [2–5].

Routinely exposed to marine microbes, including viruses, bacteria, protists, and parasites, cnidarians seem unaffected by attacks of pathogens [6] even in the absence of typical protection systems of other metazoans (e.g., cuticle, hemolymph, phagocytic cells). Nevertheless, cnidarians possess a repertory of defense mechanisms, including the production of bioactive compounds involved in the recognition and neutralization of invaders [7]. In many cnidarians, sexual reproduction requires external fertilization, and the presence of defense molecules is crucial in the environment to secure protection of eggs and embryos, which contain energy-rich materials, against eukaryotic predators and bacteria [8]. Mortality rates of marine invertebrate eggs and larvae are high due to predation and diseases caused by marine microorganisms [9]. However, gametes of several invertebrates are equipped with several defensive substances, including IgM-like molecules [10], lectins [11], and antifungal and antibacterial proteins [12]. In *Hydra*, for instance, embryos are defended by a maternally produced antimicrobial peptide (AMP) of the periculin peptide family, which controls microbial colonizers during embryogenesis [8]. The endogenous AMPs are among the most important effectors of invertebrate innate immunity. A wide variety of antimicrobial peptides has been extracted from sponges, annelids, mollusks, crustaceans, tunicates, and cnidarians, however, AMPs still remain a largely unexplored resource, representing the starting point for the development of new antibiotics with a natural broad spectrum of action [13]. Some of these compounds do not easily allow bacteria to develop resistance towards them [14]. Over the recent decades, the metabolomics profiling and biochemical evaluation of a number of cnidarian species led to the discovery of more than 2000 natural compounds with antimicrobial/antibiotic properties [15].

Among the antimicrobial enzymes, lysozyme is the best characterized lytic agent, capable of breaking the peptidoglycan-based bacterial cell wall, causing high-pressure osmotic cytolysis and burst [16]. Several methods are available to measure lysozyme activity, including the spectrophotometric method and the standard assay on inoculated Petri dishes [17,18]. Lysozyme widely occurs in marine protostomes and deuterostomes, including polychaetes and echinoderms [19], particularly in secretions such as mucus. Recently, an antibacterial lysozyme-like activity was found in the anthozoan *Actinia equina* mucus [20], presumably with a defensive role against potential predation by the surrounding microorganisms. In this framework, the so-called "white barrel" or "sea lung" *Rhizostoma pulmo* (Scyphozoa) is one of the largest and most abundant jellyfishes along the Mediterranean coasts. The medusa stage of this species is known to produce considerable amounts of sticky mucus used to either entrap food particles or as a deterrent against predators. The large biomass reached by its populations along the Mediterranean coasts has recently suggested this species as a top candidate for isolation and sustainable production of bioactive compounds for pharmaceutical applications or for nutritional purposes [21]. In this framework, the biochemical composition and the metabolic profile (by ^{1}H NMR spectroscopy) of *R. pulmo* ovaries as well as the antimicrobial properties of oocytes were investigated.

2. Results

2.1. Biochemical Composition of Ovaries

The water content of *R. pulmo* ovaries was 93.7 $\pm$ 1.9% (Figure 1A). After dehydration, organic matter was the main part (60.43 $\pm$ 10.14%) of ovary dry weight (Figure 1B). The organic residue of the ovaries was mainly composed of proteins (761.76 $\pm$ 25.11 µg mg^{-1} OM), lipids (192.17 $\pm$ 10.56 µg mg^{-1} OM), and carbohydrates (59.66 $\pm$ 2.72 µg mg^{-1} OM); Figure 1B).

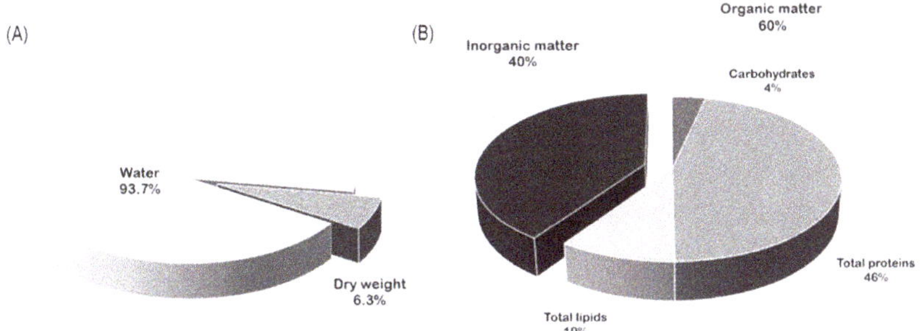

Figure 1. *Rhizostoma pulmo* ovaries composition: (**A**) water content and dried weight; (**B**) inorganic and organic residues.

Among lipids, the mean triglyceride and cholesterol concentrations were 0.12 ± 0.008 mg/mL and 0.24 ± 0.0023 mg/mL, respectively.

2.2. NMR Spectroscopy

2.2.1. NMR Analysis of Female Gonads Aqueous Extracts

The ^{1}H NMR spectrum of the aqueous extract of *R. pulmo* ovaries was characterized by free amino acids, organic acids, and derivatives (Figure 2). Many signals due to different compounds, such as betaine (δ 3.27 and 3.90), taurine (δ 3.27 and 3.41), homarine (δ 4.37, 7.97, 8.04, 8.55, 8.71), lactate (δ 1.33 and 4.16), succinate (δ 2.41), acetate (δ 1.92), and formate (δ 8.46), were identified. High signals at δ 3.57 and the doublet at δ 1.48 were also assigned to glycine and alanine, respectively. The multiplet of proline appeared at δ 2.1–2.0, 2.32–2.36, 3.30–3.40, 4.10–4.14, while the multiplet at δ 2.07 and 2.36 were assigned to glutamate. Other amino acids such as leucine (δ 0.96, 1.70), isoleucine (δ 1.01, 1.97), valine (δ 1.05, 2.29), threonine (overlapping doublets at δ 1.33 and multiplets at δ 3.68 and 4.29) were detected. Using 2D NMR experiments (Figure 3), and by comparison with literature data [22–24], two other osmolytes were identified: β-alanine (triplets at δ 2.56 and 3.20) and hypotaurine (triplets at δ 2.65 and 3.37). In the aromatic region, low-intensity signals at δ 6.90 and 7.19 were assigned to tyrosine, while δ 8.08, 8.84, and 9.13 signals were also identified for trigonelline (*N*-methylpicolinic acid). Quantitative analysis [25] showed that taurine, betaine, and glycine (concentrations >10 mM) were the most abundant free metabolites. Homarine, β-alanine, and alanine contents ranged from 5 to 3 mM. Finally, trigonelline, acetate, valine, formate, succinate, and hypotaurine were present in very low concentrations (≤1 mM).

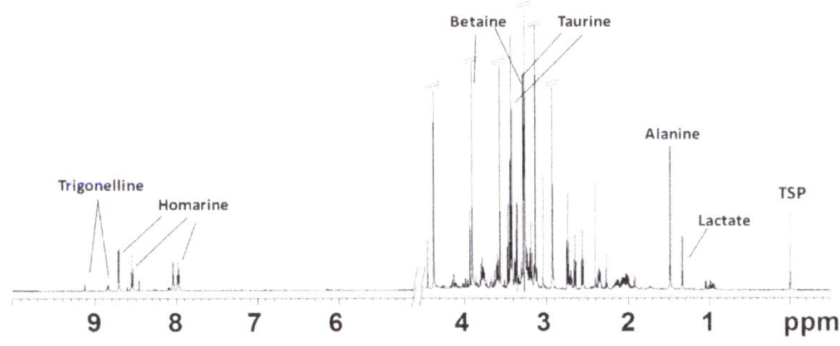

Figure 2. Typical ^{1}H NMR spectrum obtained at 600 MHz of *R. pulmo* ovaries aqueous extract.

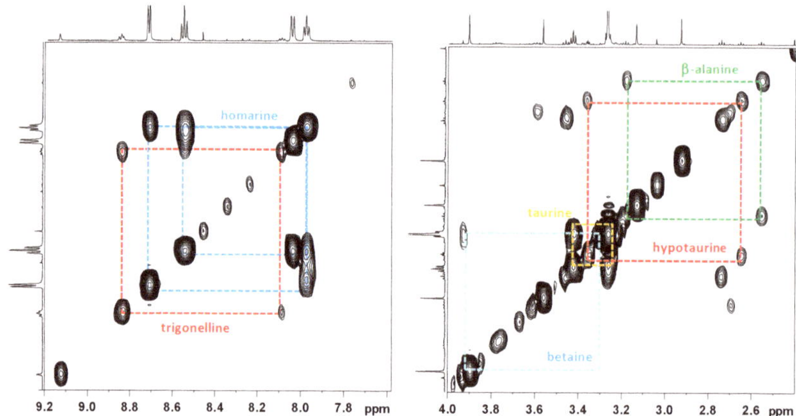

Figure 3. Expansions of COSY spectrum of *Rhizostoma pulmo* aqueous extract. Colored boxed regions correlate with the various resonances of homarine, trigonelline, betaine, taurine, hypotaurine, and glycine.

2.2.2. NMR Analysis of Female Gonads Lipid Extracts

The lipid extracts of the examined jellyfish female gonads were characterized by the presence of triglycerides (TG), polyunsaturated fatty acids (PUFAs), diunsaturated fatty acids (DUFAs), monounsaturated fatty acids (MUFAs), saturated fatty acids (SFAs), and minor components such as sterols (cholesterol) and phospholipids. The main signals, marked in the spectrum (Figure 4), corresponded to $-CH_2$ in α and β-position to the carboxylic acid esters ($COOCH_2CH_2$), unsaturations ($CH=CH-CH_2-CH=CH$), and monounsaturated fatty acids (docosahexaenoic, DHA C22:6, and eicosapentaenoic acids, EPA C20:5, ω-3) or other PUFAs (two and more than two double bonds) of long fatty acids alkyl chain, including PUFA CH_3s and terminal CH_3s of phospholipids.

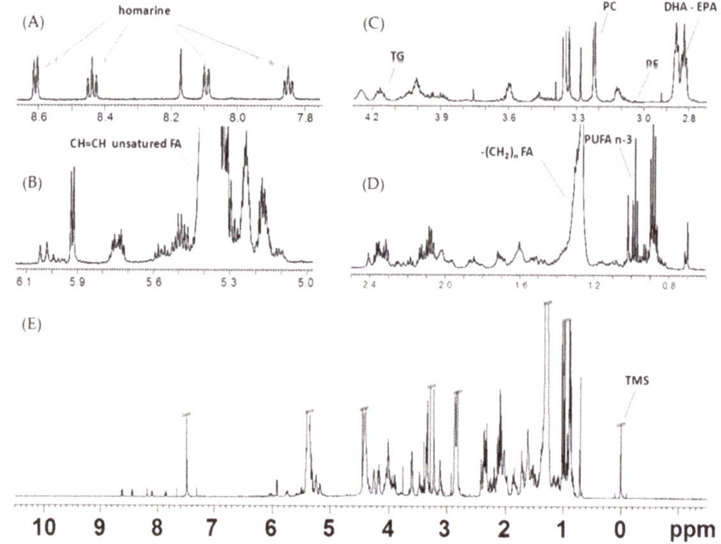

Figure 4. Typical ^{1}H NMR spectrum obtained at 600MHz of $CD_3OD/CDCl_3$ *R. pulmo* female gonad lipid extract (**a**) high-, (**b**,**c**) middle-, and (**d**) low-frequency regions, (**e**) full spectrum.

The COSY cross peaks correlated with the multiplets from the glycerol moiety of TG appeared at δ 4.14 and 4.11 (sn 1,3) and δ 5.24 (sn 2), with very low intensities. The signals in the range of δ 2.32–2.27 and δ 1.66–1.57 were assigned to protons of $COOCH_2$ and $COOCH_2CH_2$, respectively, for all the fatty acids chains, except for DHA (signal at δ 2.38, $COOCH_2CH_2$,) and EPA (signal at δ 1.70 $COOCH_2CH_2$).

The presence of ω-3 PUFAs is confirmed by the appearance of a triplet at δ 0.98 related to the terminal methyl group. This terminal methyl group is clearly separated from other methyl groups at δ 0.80 and 0.91, ascribable to all other non-ω-3 fatty acids, such as DUFAs, MUFAs, and SFAs. The spectra also indicated intense signals in the range δ 2.88–2.75 for the presence of bis-allylic ($CH=CH$-CH_2-$CH=CH$) protons of long alkyl chain fatty acids components. In particular, the multiplet at δH 2.85–2.80 were assigned to bis-allylic protons ($CH=CH$-CH_2-$CH=CH$) of PUFAs (such as DHA and EPA), while bis-allylic protons of other PUFAs, such as α-linolenic fatty acid and DUFAs, appeared at δH 2.77. The presence of partially overlapping singlets at δH 3.22 are due to the polar head group ($N(CH_3)_3$) of phosphatidylcholine (PC), while the signal at δH 3.03 is attributed to the CH_2N group of phosphatidylethanolamine (PE). The presence of phospholipids in the extracts was confirmed by ^{31}P NMR analysis of few samples (data not shown). Furthermore, signals at δH 0.68–0.69, 0.92, and 1.01, due to characteristic resonances of cholesterol (CHO) and multiplets in ranges of δH 5.26–5.13 and 4.28–4.12, assigned to 1,2-diacylglycerols (DAGs), were also observed. Finally, homarine signals appeared at δH 8.61, 8.44, 8.09, 7.85. The assignments were confirmed by 2D experiments (Figure 5) and literature data [26–29].

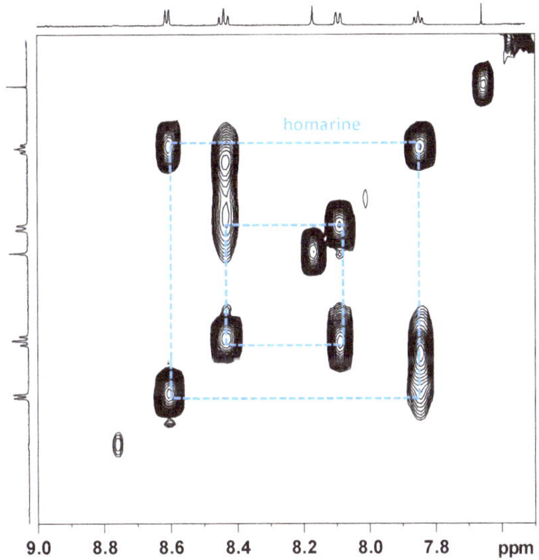

Figure 5. Expansion of the COSY spectrum of *R. pulmo* lipid extract. Colored boxed regions correlate with the various resonances of homarine.

2.3. Lysozyme-like Activity in R. pulmo Oocyte Lysate

Oocyte lysate of *R. pulmo* showed a natural lysozyme-like activity. By the standard assay on Petri dishes, a diameter of lysis of 9.33 ± 0.32 mm corresponding to 1.21 mg/mL of hen egg-white lysozyme was observed (Figure 6A). The lysozyme activity of the egg lysate was significantly affected by temperature ($p = 0.0002$), ionic strength ($p = 0.0028$), and pH ($p = 0.0002$) of the incubation (Figure 6B–D; Table 1). Post hoc analyses clarified better responses at different experimental conditions tested (Table 2). Increasing the temperature improved proportionally lysozyme-like activity (Figure 6B), showing significant differences among different conditions. The lytic activity increased significantly after dialysis against PB at I = 0.175 (Figure 6C, Table 2). Among all experiments, the maximum diameter of lysis was reported at pH 4.0, although there were no significant differences among the measured diameters at pH 4 or 6 (Figure 6D, Table 2). A dose-response correlation was obtained when increasing amounts of oocyte lysate were plotted against the respective lysis area diameters (Figure 7). The diameter of the lysis area was positively correlated with the sample volume.

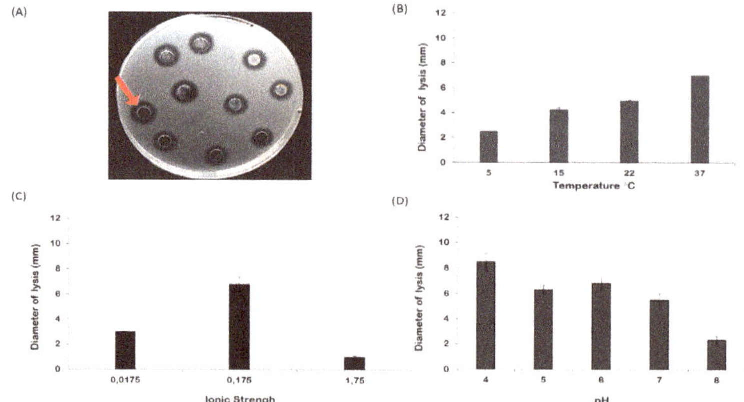

Figure 6. (**A**) Lysozyme-like activity of *R. pulmo* oocyte lysate measured on Petri dish. The arrow indicates the diameter of lysis around each well (6.3 mm in diameter) in which the oocyte lysate (30 μL) was loaded. All the wells were loaded with 30 μL of oocyte lysate and represent replicates; (**B**) effect of the temperature (5, 15, 22, and 37 °C) on lysozyme-like activity measured at ionic strength (I) = 0.175 and pH 6.0; (**C**) the effect of ionic strength (I = 0.0175, 0.175, 1.75) on lysozyme-like activity measured at temperature 37 °C and pH 6.0; (**D**) the effect of the pH (4, 5, 6, 7, 8) on lysozyme-like activity measured at 37 °C and I = 0.175. Data are reported as mean value ± standard error.

Table 1. Results from the multivariate permutational analysis (PERMANOVA) showing differences in lysozyme activity among different tested conditions.

	df	MS	F	*p*
Temperature	3	10.416	640.72	***
Residual	8	0.0163		
Total	11			
Ionic Strength	2	27.417	81.117	**
Residual	6	0.33799		
Total	8			
pH	4	21.35	36.6	***
Residual	10	0.58333		
Total	14			

df = degree of freedom; MS = mean sum of squares; F = F value by permutation; p = p-value by permutation. ** = $p < 0.01$; *** = $p < 0.001$.

Table 2. Results of the pairwise tests showing differences in lysozyme activity among various levels in different laboratory conditions (temperature, ionic strength, pH).

Temperature	T	P(MC)	Ionic Strength	T	P(MC)	pH	T	P(MC)
5 vs. 15	12.12	***	0.0175 vs. 0.175	6.88	0.0029	4 vs. 5	4.025	*
5 vs. 22	84.87	***	0.0175 vs. 1.75	28.73	0.0001	4 vs. 6	2.646	ns
5 vs. 37	1561	***	0.175 vs. 1.75	10.21	0.0006	4 vs. 7	5.277	**
15 vs. 22	4.756	**				4 vs. 8	11	***
15 vs. 37	19.08	***				5 vs. 6	1	ns
22 vs. 37	70.83	***				5 vs. 7	1.89	ns
						5 vs. 8	13	***
						6 vs. 7	2.324	ns
						6 vs. 8	8.66	**
						7 vs. 8	12.12	***

T = T value; P(MC) = probability level after Monte Carlo simulations. * = $p < 0.05$; ** = $p < 0.01$; *** = $p < 0.001$; ns = not significant.

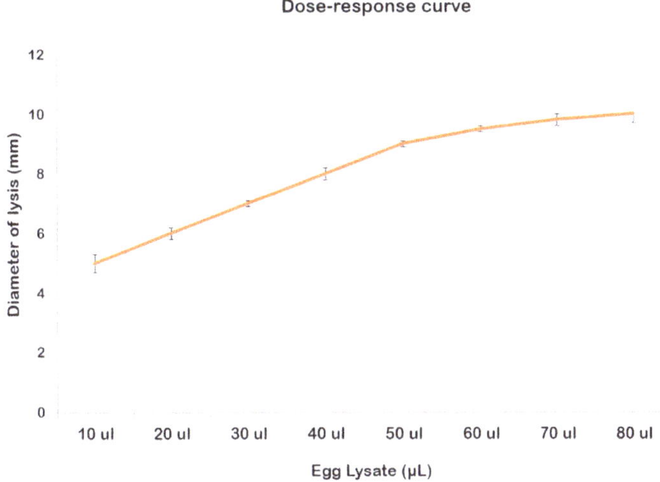

Figure 7. Dose-response curve of lysozyme-like activity of *R. pulmo* oocyte lysate.

3. Discussion

The apparent increase of global jellyfish abundance in coastal marine ecosystems has recently attracted scientific interest for the potential impacts on human activities and ecosystem functioning. Also, the possible use of jellyfish biomass as a source of energy and bioactive compounds useful for pharmaceutical and nutritional applications has been suggested [30,31]. In this context, understanding the biological mechanisms underlying jellyfish outbreaks is crucial to predict and/or mitigate impacts of recurrent bloom events. The occurrence of jellyfish outbreaks is also directly linked to the reproductive success. The present paper represents the first insight into the biochemical composition of ovaries and the lysozyme antibacterial activity associated with oocytes of the barrel jellyfish *R. pulmo* in order to investigate aspects related to the mechanisms boosting the success of sexual reproduction.

In *R. pulmo*, more than 90% of the ovary volume is composed of water in accordance with previous studies on *R. octopus* [32] and other scyphozoan jellyfish gonads (e.g., *Cyanea capillata*, *Chrysaora hysoscella*, and *Pelagia noctiluca* [5,32]). The organic matter in *R. pulmo* ovaries represented 60% of dry weight and was composed mainly of proteins, lipids, and low content of carbohydrates, corroborating the general trend in gonadal composition recorded in other scyphozoans, including *P. noctiluca* [5,32]. In particular, proteins were twofold concentrated in *R. pulmo* female gonads compared to *C. capillata* and threefold compared to *R. octopus* gonads [32]. The biochemical composition of *R. pulmo* ovaries reflects the composition of the entire jellyfish in which the organic content is mainly represented by protein, followed by lipid and carbohydrate fractions [21].

The ¹H NMR characterization of both lipid and aqueous extracts of *R. pulmo* female gonads provided advanced information on the chemistry of this jellyfish compartment. As known [22,33], the untargeted ¹H NMR-based metabolomic approach could be used to provide simultaneous determination of the end products of metabolism, such as small-molecular-weight molecules in solution [24]. In the present study, many ¹H NMR signals of the aqueous extract were due to free amino acids, such as leucine, isoleucine, valine, threonine (essential amino acids) and alanine, glycine, proline, and glutamate, representing also the dominant amino acids in the gonads of the edible Asiatic jellyfish *Rhopilema esculentum* [34] and other edible jellyfish species [35]. In the aromatic region, ¹H NMR analysis revealed also the presence of tyrosine, which is considered an essential amino acid for humans, useful during stages of human prenatal development [36]. Alanine, glycine, and glutamate are also implicated in other physiological processes and on account of these features are considered useful as antioxidant constituents in foodstuffs and beverages [37]. Thus, the *R. pulmo* ovaries could represent a source of amino acids exploitable for nutraceutical and pharmaceutical applications as well as a source of proteins for the development of innovative dietary supplements for fish nourishment. In this framework indeed, considering the growing cost of fish feed worldwide, each innovative natural resource as a potential ingredient in their preparation must be considered. Furthermore, the aqueous extract was characterized by the presence of compounds with important metabolic roles in marine invertebrates [23,38,39], such as the osmolytes trimethylamine *N*-oxide (TMAO), betaine, taurine, and homarine. Interestingly, in the colonial hydropolyp *Hydractinia echinata* (Hydrozoa), it was observed that betaine, homarine, and trigonelline, when present simultaneously, play a regulatory role during the development and prevent the onset of metamorphosis [40,41]. On account of the high amount evidenced in *R. pulmo* gonads, it is likely that, as in the case of *H. echinata*, also in the here-investigated jellyfish, these compounds could be involved in maintaining the larval state until an appropriate signal allows metamorphosis. Moreover, TMAO is an osmolyte that commonly occurs in several marine animals and it has been found to neutralize the effects of hydrostatic pressure on cnidarian, fish, and mammalian [42]. The densities differ widely among habitats, species, and season and ontogeny within species A relationship exists between the concentration of TMAO (and betaine) in muscle tissue and lipid. At present, we do not rule out the possibility that, in *R. pulmo* ovaries, TMAO plays a role for its protein-stabilizing attributes as already hypothesized for several marine organisms [43].

In the female gonads of *R. pulmo*, a higher content of total lipids was also recorded in comparison with the total jellyfish [21]. As reported in literature [21,44], the lipid composition of jellyfish can be considerably influenced by several external factors such as diet, size, and age of organism. The lipid NMR analysis of *R. pulmo* gonads showed the presence of different lipid classes. As already observed in the whole jellyfish [21], the analysis of *R. pulmo* ovaries confirmed the presence of a high content of ω-3 PUFAs. In the gonads of marine invertebrates, there is a notable richness of PUFAs, particularly 20:5 ω-3 and 22:6 ω-3 [45]. PUFAs actively participate in gonad maturation, egg quality [46], and larval growth of fish [47]. In the common octopus, *Octopus vulgaris*, PUFAs can improve membrane fluidity and flexibility of spermatozoa membrane, and are actively implicated in the regulation of cellular movement, gonadal metabolism of lipids, and fusion capacity [48]. In crustaceans and mollusks, PUFAs not only determine hatching and growth [49] but also play an important role in metabolism processes, like production of prostaglandins and hormones, and regulate ionic fluxes [50]. Among the categories of fatty acids (FAs) of *R. pulmo* ovaries, the two signals of EPA and DHA were also revealed. Noteworthy, these categories of fatty acids were already recorded in other jellyfish, including *Aurelia* sp., whose fatty acid profiles were broadly similar to 16:0, EPA, 18:0, AA, and DHA as the five main components accounting for around 66% of the total FAs. Furthermore, EPA and DHA have been detected in different species of *Aurelia* jellyfish [44,51]. The presence of ω-3 PUFAs, mainly DHA and EPA, in the gonads of *R. pulmo* suggests their potential exploitation as sources of these compounds for the application in the pharmaceutical field. It is well known indeed that ω-3 PUFAs, DHA and EPA, possess antioxidant and anti-inflammatory properties useful for potential treatment strategies for mental health and neuro-inflammation-induced memory deficits [52,53]. Moreover, taking into account that diets for fish are usually enriched with additional supplements of EPA and DHA, the gonads of *R. pulmo* could furnish these essential FAs to be added in the production of fish feed.

An antibacterial lysozyme-like activity was detected in the oocyte lysate of *R. pulmo*. As in the majority of scyphozoans, also in *R. pulmo*, reproduction is external; thus, an antibacterial activity may prevent eggs and embryos from being overgrown and killed by pathogenic bacteria. To our knowledge, this is the first record of an antimicrobial activity in jellyfish eggs. Many marine taxa synthesize specific metabolites to protect themselves against the settlement and growth of microbial agents. For example, surface attachment and growth of several marine bacteria are inhibited by the extracts from the eggs of several coral species [54]; potentially pathogenic bacteria, including a *Vibrio* sp., are subjected to the toxic action induced by the extracts of various developmental stages of the soft coral, *Parerythropodum fulvum* [55]. Gunthorpe and Cameron [56,57] found that in some species of soft corals, the extracts exerted an antibiotic activity negatively related to the presence of immature gonads, suggesting that reproductive status represented a significant predictor of antimicrobial activity [56]. A similar phenomenon was found in the octocoral *Lobophytum compactum*; indeed, antimicrobial diterpenes were selectively included into egg lipid material [58] detectable in adults before spawning and absent afterwards. Moreover, the extracts of the damselfish *Pomacentrus mollucensis* eggs is defended chemically [59]. In addition, egg extracts of the coral species *Montipora digitata* are able to produce growth inhibition of *Escherischia coli* [60]. Among marine invertebrates, also in echinoderms, characterized by external fertilization, the eggs and larvae from the regular echinoid *Paracentrotus lividus* exert an antibacterial lysozyme-like activity [16]. Lysozyme-like proteins have also already been evidenced in other cnidarians [61,62]. Regarded as the best and most active enzyme involved in the innate immunity [63], lysozyme is a glycoside hydrolase whose constitutive levels defend the organism from pathogenic bacteria present in the surrounding environment and regulate natural symbiotic microflora. Besides antimicrobial activity, lysozymes play a role in digestion, antiviral, anti-inflammatory, and antitumor activities, taking part in the innate immunity as first defensive line [64]. It is well known that lysozyme activity is affected by various factors such as temperature, pH, and salts [65,66]. In the case of *R. pulmo* egg lysate, the highest lysozyme-like activity is detected at pH 4 and the ionic strength 0.175. A similar result was already obtained from egg lysate of the sea star *Marhasterias glacialis* (maximum of activity at pH 4.2 and the ionic strength 0.175) [16]. Further studies will be undertaken to assess whether the category of lysozyme involved in egg protection is the same in marine invertebrates as well as to estimate the antibacterial activity against other living microorganisms besides *Micrococcus luteus*. At this stage, we only focused on the occurrence of a lysozyme-like activity in the whole oocyte lysate. However, to further elucidate the mechanisms related to *R. pulmo* egg defenses, it is necessary to perform the isolation, purification, and quantification of the effectors of such antimicrobial activity which will be carried out in the near future.

The evidence of a lysozyme-like activity in *R. pulmo* oocyte lysate suggests that this species may also represent a new and exciting resource for the extraction of potent antibacterial agents and encourages the potential use of the jellyfish for lysozyme-based preparations in pharmaceutical research. Currently, lysozyme is used for pharmaceutical preparations due to the therapeutic effectiveness of lysozyme based not only on its ability to control the growth of bacteria but also to modulate the immune responses of the host. Moreover, the treatment with lysozyme leads to a regression in the growth of some tumor cells [67]. Lysozyme can be also used in the treatment of a wide range of infections in humans since it has no toxic effect on humans and thus it is a good candidate for the use of epidermal and cosmetic formulations. Finally, considering that controlling bacterial infections is currently one of the main problems of aquaculture on an industrial scale, lysozyme is attracting the interest of researchers for its potential applications in aquaculture.

4. Materials and Methods

4.1. Sample Collection and Preparation

Sixty specimens of *R. pulmo* adult medusae (umbrella diameter > 25 cm at sexual maturity; Basso, personal observations) were sampled at the Ginosa Marina (Ionian Sea 40°25.7′ N, 16°53.1′ E; Italy) throughout 2017 with a 1 cm mesh hand net from a boat. Immediately after sampling, jellyfish were transported into the laboratory and washed with filter-sterilized seawater (0.2 μm, Millipore) to remove the mucus layer produced during transport. Ovaries appeared from pink to orange, with easily distinguishable eggs. When gender determination was uncertain visually, a small piece of gonad tissue was removed and examined under the stereomicroscope. The ovaries were carefully dissected with microscissors at the stereomicroscope to avoid loss of gonadic tissue or accidental inclusion of subumbrellar or exumbrellar tissues and mature eggs were collected from a number of mature gonads. Each gonad was then divided in two aliquots. The first aliquot was frozen at −80 °C in liquid nitrogen to be lyophilized and then employed for the biochemical and NMR analyses; the second aliquot was employed to obtain the mature eggs. In particular, the eggs were obtained in pasteurized seawater (PSW) by placing ovaries on four layers of gauze. Eggs were allowed to settle and, after removal of the supernatant, were resuspended in sufficient PSW to obtain a 10% (*v*/*v*) suspension. After that, the eggs were gently swirled and then centrifuged at 12,000 g for 30 min [68]. The resulting supernatant (oocyte lysate) was dialyzed against distilled water, then lyophilized, and then concentrated 10-fold in PSW and used to evaluate the lysozyme-like antibacterial activity.

4.2. Biochemical Analysis

The ovary organic matter content of *R. pulmo* was evaluated using approximately 10 mg (±0.01 mg) of dry tissue reduced to ash for 4 h at 500 °C in a muffle furnace (BICASA B.E. 34). This content was expressed as the percentage of organic matter of total tissue dry weight (DW)). The weight of organic matter (OM) was determined as the difference between the gonad DW and the ash weight [69]. Female gonads biochemical composition was performed in order to detect contents in protein, carbohydrate, and total lipid (*n* = 15). Ovary tissue was frozen in liquid nitrogen, temporarily stored at −20 °C, and briefly transferred one hour before lyophylisation to −80 °C to facilitate freeze-drying (48 h) for the biochemical and NMR analyses.

Carbohydrate, protein, and lipid quantification was performed by colorimetric determination at 480 nm, 750 nm, and 520 nm, respectively. In order to calculate the carbohydrate content in the ovarian tissue, approximately 10 mg (±0.1 mg) of each lyophilized sample was homogenized in 3 mL of double distilled water [70] with glucose as a standard. The content of proteins was estimated by employing approximately 10 mg (±0.1 mg) of each lyophilized tissue sample homogenized in 2 mL of 1N NaOH [71] with albumin as a standard. Finally, total lipids were determined by homogenizing approximately 10 mg (±0.1 mg) of each lyophilized tissue sample in 3 mL of chloroform–methanol (2:1) with cholesterol as a standard [72]. Quantities were expressed as μg mg^{-1} of OM.

Cholesterol content was evaluated by homogenizing approximately 150 mg (±0.1 mg) of each lyophilized sample in 4 mL of distilled water and was calculated by the colorimetric enzymatic method using the commercial kit (10028 Cholesterol, SGM, Rome, Italy) based on Jacobs et al. [73] with known amounts of cholesterol standard. Finally, triglycerides were estimated by homogenizing approximately 150 mg (±0.1 mg) of each lyophilized tissue sample in 4 mL of distilled water and were measured by the colorimetric enzymatic method using the commercial kit (10160 Triglycerides, SGM, Rome, Italy) based on Bucolo and David [74].

4.3. NMR Analysis

Samples were prepared according to a modified Bligh and Dyer extraction method [75,76]. Lyophilized gonads (~100 mg) were added to 400 μL methanol, 400 μL deionized filtered water, and 400 μL chloroform. The solution was mixed and placed on ice for 10 min before centrifugation at 10000 rpm for 20 min at 4 °C. The polar and lipophilic phases were separated and dried by a SpeedVac concentrator (SC 100, Savant, Ramsey MN, USA). The lipid extracts were dissolved in 700 μL of $CD_3OD/CDCl_3$ (1:2 mix) containing 0.03% *v*/*v* tetramethylsilane (TMS, δ = 0.00) as internal standard. The aqueous extracts were dissolved in 160 μL 0.2 M Na_2HPO_4/NaH_2PO_4 buffer (pH 7.4) and 540 μL D_2O containing 3-(trimethylsilyl)-propionic-2,2,3,3-d4 acid (TSP δ = 0.00) as internal standard). Both extracts were transferred into 5 mm NMR tubes for NMR analyses.

NMR Spectroscopy and Data Processing

NMR spectra were obtained using a Bruker Avance III 600 Ascend NMR spectrometer (Bruker, Ettliger, Germany) operating at 606.13 MHz for ^{1}H observation with a z axis gradient coil and automatic tuning-matching (ATM). Experiments were performed at 300 K in automation mode after loading individual samples on a Bruker Automatic Sample Changer, interfaced with the software IconNMR (Bruker). For aqueous extracts, a one-dimensional zgcppr Bruker standard pulse sequence was applied to suppress the residual water signal. A number of 32 repetitions of scans (with 4 dummy scans) were collected into 64 k data points with relaxation delay set to 5 s, a spectral width of δ 20.0276 (12,019.230 Hz), 90° pulse of 11.11 μs. For lipid extracts, a one-dimensional zg Bruker standard pulse sequence was acquired, with repetitions of 64 scans (with 4 dummy scans), a spectral width of δ 20.0276 (12,019.23 Hz) and 90° pulse of 10 μs. For both aqueous and lipid extracts, the FIDs were multiplied by an exponential weighting function corresponding to a line broadening of 0.3 Hz before Fourier transformation, phasing and baseline correction. Metabolites identified by the ^{1}H NMR spectra were assigned on the basis of analysis of 2D NMR spectral analysis (2D ^{1}H J *res*, COSY, HSQC, and HMBC) and by comparison with literature data [23,28,40]. Whenever possible, also quantitative analysis of free metabolites was determined by integrating selected unbiased NMR signals, using TSP for chemical shift calibration and quantification [25]. In particular, signals corresponding to valine (δ 1.05), alanine (δ 1.48), acetate (δ 1.92), succinate (δ 2.41), β-alanine, (δ 2.56), hypotaurine (δ 2.65), betaine (δ 3.90), taurine (δ 3.41), glycine (δ 3.57), homarine (δ 8.04), formate (δ 8.46), and trigonelline (δ 9.13) were integrated. All chemical reagents for analysis were of analytical grade. D_2O, $CDCl_3$, CD_3OD (99.8 atom%D), TSP, 3-(trimethylsilyl)-propionic-2,2,3,3,d4 acid, and tetramethylsilane, TMS (0.03 $v/v\%$) were purchased from Armar Chemicals (Döttingen, Switzerland).

4.4. Lysozyme-Like Activity

Normally, a spectrophotometric method is used to demonstrate the occurrence of lysozyme activity [17]; however, the standard assay on inoculated Petri dishes can be used as an alternative method to demonstrate the occurrence of lysozyme activity [18]. Here, the presence of lysozyme activity was detected by the standard assay on Petri dishes, which resulted as a quick, sensitive, low-cost, and therefore very versatile method [16,18,20,77–79]. Briefly, 700 μL of 5 mg/mL of dried *Micrococcus luteus* cell walls (Sigma, Saint Louis, MO, USA) were suspended in 7 mL of 0.05 M PB-agarose (1.2%, pH 5.0) then spread on a Petri dish. Four wells of 6.3 mm diameters were sunk in agarose gel and each filled with 30 μL of sample (oocyte lysate). After overnight incubation at 37 °C, the diameter of the cleared zone of at least four replicates was measured. Diameters of lysis were compared with those of reference obtained with known amounts of standard hen egg-white lysozyme (Merck, Darmstadt, Germany). The effects of pH, ionic strength (I), and temperature were assessed for each sample. The pH effect was tested by dialyzing (7000-MW cut-off), the samples in PB 0.05 M, ionic strength, I = 0.175, adjusted at pH 4, 5, 6, 7, 8, and by dissolving agarose in PB at the same I and pH values. The ionic strength effect was tested in PB 0.05 M (pH 6.0), adjusted at I = 0.0175, 0.175, 1.75. Agarose was dissolved in PB at the same I values. The temperature effect was evaluated by performing the Petri dish assays (in PB, at pH 6.0, and I = 0.175) and incubating the plates at 5, 15, 22, and 37 °C. The dose-response curve of lysozyme-like activity was constructed by using Petri dish assays (in PB, at pH 6.0, and I = 0.175) with different amounts of sample (10, 20, 30, 40, 50, 60, or 80 μL of sample in each well in triplicate).

4.5. Statistical Analysis

To test the effects of temperature, pH, and ionic strength on antibacterial activity of oocyte lysate, one-way permutational multivariate analyses of variance (PERMANOVA) [80] were performed based on Euclidean distances on untransformed data, using 9999 random permutations of the appropriate units [81], following three different designs with one factor, separately: temperature (Te, as fixed factor with four levels); pH (pH, as fixed factor with five levels); ionic strength (IS, as fixed factor with three levels). When significant differences were found ($p < 0.05$), post hoc pairwise tests were carried out to ascertain the consistency of the differences among several different conditions tested. Because of the restricted number of unique permutations in the pairwise tests, p values were obtained from Monte Carlo samplings. The analyses were performed using the software PRIMER v. 6 [82].

5. Conclusions

In conclusion, *R. pulmo* ovaries and oocytes could represent a promising source of bioactive compounds for different applications mainly in the pharmaceutical field or as specialty feed.

In particular, the antibacterial lysozyme-like activity suggests that this jellyfish species may represent a new and renewable resource for drugs discovery. Moreover, the presence of ω-3 PUFAs encourages their potential exploitation as sources of these compounds in the production of fish feed. Further studies will help to standardize a sustainable exploitation pilot system to use different jellyfish fractions for different purposes (e.g., food, feed) and beneficial services for human wellbeing.

Author Contributions: Conceptualization, L.S., F.P.F., S.P., and L.B.; Methodology, L.S., F.A., L.D.C., C.R.G., L.R., and L.B.; Validation, L.S., L.R., F.P.F., S.P., and L.B.; Investigation, L.S., L.R., F.A., L.D.C., C.R.G., S.L., and L.B.; Writing-Review & Editing, L.S., L.R., F.P.F., S.P., and L.B.; Supervision, L.S., F.P.F., S.P.; Funding Acquisition, S.P. and L.B.

Funding: This research was funded by the European Community's Horizon 2020 Programme—project PULMO, Marie Sklodowska-Curie Individual Fellowships, grant number 708698, and project CERES, Research and Innovation Programme, grant number 678193.

Conflicts of Interest: The authors declare no conflict of interest.

References

1. Hamner, W.M.; Dawson, M.N. A review and synthesis on the systematics and evolution of jellyfish blooms: Advantageous aggregations and adaptive assemblages. *Hydrobiologia* **2009**, *616*, 161–191. [CrossRef]

2. Falkenhaug, T. Review of Jellyfish Blooms in the Mediterranean and Black Sea. *Mar. Biol. Res.* **2014**, *10*, 1038–1039. [CrossRef]

3. Benedetti-Cecchi, L.; Canepa, A.; Fuentes, V.; Tamburello, L.; Purcell, J.E.; Piraino, S.; Roberts, J.; Boero, F.; Halpin, P. Deterministic factors overwhelm stochastic environmental fluctuations as drivers of jellyfish outbreaks. *PLoS ONE* **2015**, *10*, e0141060. [CrossRef] [PubMed]

4. Boero, F.; Brotz, L.; Gibbons, M.J.; Piraino, S.; Zampardi, S. Impacts and effects of ocean warming on jellyfish. In *Explaining Ocean Warming: Causes, Scale, Effects and Consequences*; Full Report; IUCN: Gland, Switzerland, 2016; pp. 213–237.

5. Milisenda, G.; Martinez-Quintana, A.; Fuentes, V.L.; Bosch-Belmar, M.; Aglieri, G.; Boero, R.R.; Piraino, S. Reproductive and bloom patterns of Pelagia noctiluca in the Strait of Messina, Italy. *Estuar. Coast. Shelf Sci.* **2018**, *201*, 29–39. [CrossRef]

6. Bosch, T.C.G. The Path Less Explored: Innate Immune Reactions in Cnidarians. In *Innate Immunity of Plants, Animals, and Humans*; Springer: Berlin/Heidelberg, Germany, 2008; pp. 27–42.

7. Mariottini, G.L.; Grice, I.D. Antimicrobials from cnidarians. A new perspective for anti-infective therapy? *Mar. Drugs* **2016**, *14*. [CrossRef] [PubMed]

8. Jung, S.; Dingley, A.J.; Augustin, R.; Anton-Erxleben, F.; Stanisak, M.; Gelhaus, C.; Gutsmann, T.; Hammer, M.U.; Podschun, R.; Bonvin, A.M.J.J.; et al. Hydramacin-1, structure and antibacterial activity of a protein from the basal metazoan hydra. *J. Biol. Chem.* **2009**, *284*, 1896–1905. [CrossRef] [PubMed]

9. Rumrill, S.S. Natural mortality of marine invertebrate larvae. *Ophelia* **1990**, *32*, 163–198. [CrossRef]

10. Fuda, H.; Hara, A.; Yamazaki, F.; Kobayashi, K. A peculiar immunoglobulin M (IgM) identified in eggs of chum salmon (*Oncorhynchus keta*). *Dev. Comp. Immunol.* **1992**, *16*, 415–423. [CrossRef]

11. Ozaki, H.; Ohwaki, M.; Fukada, T. Studies on lectins of amago (*Oncorhyncus rhodurus*) I. Amago ova lectin and its receptor on homologous macrophages. *Dev. Comp. Immunol.* **1983**, *7*, 77–87. [CrossRef]

12. Kudo, S.; Teshima, C. Enzyme activities and antifungal action of fertilization envelope extract from fish eggs. *J. Exp. Zool.* **1991**, *259*, 392–398. [CrossRef]

13. Ovchinnikova, T.V.; Balandin, S.V.; Aleshina, G.M.; Tagaev, A.A.; Leonova, Y.F.; Krasnodembsky, E.D.; Men'shenin, A.V.; Kokryakov, V.N. Aurelin, a novel antimicrobial peptide from jellyfish *Aurelia aurita* with structural features of defensins and channel-blocking toxins. *Biochem. Biophys. Res. Commun.* **2006**, *348*, 514–523. [CrossRef] [PubMed]

14. Hancock, R.E.W.; Scott, M.G. The role of antimicrobial peptides in animal defenses. *Proc. Natl. Acad. Sci. USA* **2000**, *97*, 8856–8861. [CrossRef] [PubMed]

15. Rocha, J.; Peixe, L.; Gomes, N.C.M.; Calado, R. Cnidarians as a source of new marine bioactive compounds—An overview of the last decade and future steps for bioprospecting. *Mar. Drugs* **2011**, *9*, 1860–1886. [CrossRef] [PubMed]

16. Stabili, L.; Licciano, M.; Pagliara, P. Evidence of antibacterial and lysozyme-like activity in different planktonic larval stages of *Paracentrotus lividus*. *Mar. Biol.* **1994**, *119*. [CrossRef]

17. Shugar, D. The measurement of lysozyme activity and the ultra-violet inactivation of lysozyme. *Biochim. Biophys. Acta* **1952**, *8*, 302–309. [CrossRef]

18. Canicatti, C.; Roch, P. Studies on *Holothuria polii* (Echinodermata) antibacterial proteins. I. Evidence for and activity of a coelomocyte lysozyme. *Experientia* **1989**, *45*, 756–759. [CrossRef]

19. Canicatti, C. Protease activity in holothuria polh coelomic fluid and coelomocyte lysate. *Comp. Biochem. Physiol.* **1990**, *95B*, 145–148. [CrossRef]

20. Stabili, L.; Schirosi, R.; Parisi, M.G.; Piraino, S.; Cammarata, M. The mucus of *Actinia equina* (Anthozoa, Cnidaria): An unexplored resource for potential applicative purposes. *Mar. Drugs* **2015**, *13*, 5276–5296. [CrossRef]

21. Leone, A.; Lecci, R.M.; Durante, M.; Meli, F.; Piraino, S. The bright side of gelatinous blooms: Nutraceutical value and antioxidant properties of three Mediterranean jellyfish (Scyphozoa). *Mar. Drugs* **2015**, *13*, 4654–4681. [CrossRef]

22. Zotti, M.; De Pascali, S.A.; Del Coco, L.; Migoni, D.; Carrozzo, L.; Mancinelli, G.; Fanizzi, F.P. ^{1}H NMR metabolomic profiling of the blue crab (*Callinectes sapidus*) from the Adriatic Sea (SE Italy): A comparison with warty crab (*Eriphia verrucosa*), and edible crab (*Cancer pagurus*). *Food Chem.* **2016**, *196*, 601–609. [CrossRef]

23. Tikunov, A.P.; Johnson, C.B.; Lee, H.; Stoskopf, M.K.; Macdonald, J.M. Metabolomic investigations of American oysters using ^{1}H-NMR spectroscopy. *Mar. Drugs* **2010**, *8*, 2578–2596. [CrossRef] [PubMed]

24. De Pascali, S.A.; Del Coco, L.; Felline, S.; Mollo, E.; Terlizzi, A.; Fanizzi, F.P. ^{1}H NMR spectroscopy and MVA analysis of *Diplodus sargus* eating the exotic pest *Caulerpa cylindracea*. *Mar. Drugs* **2015**, *13*, 3550–3566. [CrossRef] [PubMed]

25. Girelli, C.R.; De Pascali, S.A.; Del Coco, L.; Fanizzi, F.P. Metabolic profile comparison of fruit juice from certified sweet cherry trees (*Prunus avium* L.) of Ferrovia and Giorgia cultivars: A preliminary study. *Food Res. Int.* **2016**, *90*, 281–287. [CrossRef] [PubMed]

26. Igarashi, T.; Aursand, M.; Hirata, Y.; Gribbestad, I.S.; Wada, S.; Nanaka, M. Nondestructive quantitative determination of docosahexaenoic acid and n-3 fatty acids in fish oils by high-resolution ^{1}H Nuclear Magnetic Resonance Spectroscopy. *J. Am. Oil Chem. Soc.* **2000**, *77*, 737–748. [CrossRef]

27. Sacchi, R.; Savarese, M.; Falcigno, L.; Giudicianni, I.; Paolillo, L. Proton NMR of fish oils and lipids. In *Modern Magnetic Resonance*; Springer International Publishing: Cham, Switzerland, 2008; pp. 919–923.

28. Shumilina, E.; Ciampa, A.; Capozzi, F.; Rustad, T.; Dikiy, A. NMR approach for monitoring post-mortem changes in Atlantic salmon fillets stored at 0 and 4 °C. *Food Chem.* **2015**, *184*, 12–22. [CrossRef] [PubMed]

29. Nestor, G.; Bankefors, J.; Schlechtriem, C.; BräNnäS, E.; Pickova, J.; Sandström, C. High-resolution ^{1}H magic angle spinning nmr spectroscopy of intact arctic char (*Salvelinus alpinus*) muscle. quantitative analysis of n-3 fatty acids, EPA and DHA. *J. Agric. Food Chem.* **2010**, *58*, 10799–10803. [CrossRef] [PubMed]

30. Brotz, L.; Cheung, W.W.L.; Kleisner, K.; Pakhomov, E.; Pauly, D. Increasing jellyfish populations: Trends in Large Marine Ecosystems. *Hydrobiologia* **2012**, *690*, 3–20. [CrossRef]

31. Leone, A.; Lecci, R.M.; Durante, M.; Piraino, S. Extract from the zooxanthellate jellyfish *Cotylorhiza tuberculata* modulates gap junction intercellular communication in human cell cultures. *Mar. Drugs* **2013**, *11*, 1728–1762. [CrossRef]

32. Doyle, T.K.; Houghton, J.D.R.; McDevitt, R.; Davenport, J.; Hays, G.C. The energy density of jellyfish: Estimates from bomb-calorimetry and proximate-composition. *J. Exp. Mar. Biol. Ecol.* **2007**, *343*, 239–252. [CrossRef]

33. Nicholson, J.K.; Foxall, P.J.D.; Spraul, M.; Farrant, R.D.; Lindon, J.C. 750 MHz 1H and 1H-13C NMR Spectroscopy of Human Blood Plasma. *Anal. Chem.* **1995**, *67*, 793–811. [CrossRef]

34. Yu, H.; Li, R.; Liu, S.; Xing, R.E.; Chen, X.; Li, P. Amino acid composition and nutritional quality of gonad from jellyfish *Rhopilema esculentum*. *Biomed. Prev. Nutr.* **2014**, *4*, 399–402. [CrossRef]

35. Khong, N.M.; Yusoff, F.M.; Jamilah, B.; Basri, M.; Maznah, I.; Chan, K.W.; Nishikawa, J. Nutritional composition and total collagen content of three commercially important edible jellyfish. *Food Chem.* **2016**, *196*, 953–960. [CrossRef] [PubMed]

36. Imura, K.; Okada, A. Amino acid metabolism in pediatric patients. *Nutrition* **1998**, *14*, 143–148. [CrossRef]

37. Gülçin, I. Antioxidant activity of food constituents: An overview. *Arch. Toxicol.* **2012**, *86*, 345–391. [CrossRef] [PubMed]

38. Zotti, G. Open-source virtual archaeoastronomy. *Mediterr. Archaeol. Archaeom.* **2016**, *16*, 17–24. [CrossRef]
39. Mannina, L.; Sobolev, A.P.; Capitani, D.; Iaffaldano, N.; Rosato, M.P.; Ragni, P.; Reale, A.; Sorrentino, E.; D'Amico, I.; Coppola, R. NMR metabolic profiling of organic and aqueous sea bass extracts: Implications in the discrimination of wild and cultured sea bass. *Talanta* **2008**, *77*, 433–444. [CrossRef] [PubMed]
40. Berking, S. 3 Hydrozoa Metamorphosis and Pattern Formation. *Curr. Top. Dev. Biol.* **1997**, *38*, 81–131. [CrossRef]
41. Siefker, B.; Kroiler, M.; Berking, S. Induction of metamorphosis from the larval to the polyp stage is similar in Hydrozoa and a subgroup of Scyphozoa (Cnidaria, Semaeostomeae). *Helgol. Mar. Res.* **2000**, *54*, 230–236. [CrossRef]
42. Yancey, P.H. Organic osmolytes as compatible, metabolic and counteracting cytoprotectants in high osmolarity and other stresses. *J. Exp. Biol.* **2005**, *208*, 2819–2830. [CrossRef]
43. Seibel, B.A.; Walsh, P.J. Trimethylamine oxide accumulation in marine animals: Relationship to acylglycerol storage. *J. Exp. Biol.* **2002**, *205*, 297–306. [CrossRef]
44. Abdullah, A.; Nurjanah, N.; Hidayat, T.; Aji, D.U. Fatty Acid Profile of Jellyfish (*Aurelia aurita*) as a Source Raw Material of Aquatic Result Rich Beneft. *Int. J. Chem. Biol. Sci.* **2015**, *1*, 12–16.
45. Besnard, J.Y. Etude des Constituents Lipidiques Dans la Gonade Femelle et les Larves de *Pecten maximus* L. (Mollusque Lamellibranche). Ph.D. Dissertation, Université de Caen, Caen, France, 1988; p. 154.
46. Izquierdo, M.S.; Fernández-Palacios, H.; Tacon, A.G.J. Effect of broodstock nutrition on reproductive performance of fish. *Aquaculture* **2001**, *197*, 25–42. [CrossRef]
47. Tulli, F.; Tibaldi, E. Changes in amino acids and essential fatty acids during early larval rearing of dentex. *Aquac. Int.* **1997**, *5*, 229–236. [CrossRef]
48. Miliou, H.; Fintikaki, M.; Tzitzinakis, M.; Kountouris, T.; Verriopoulos, G. Fatty acid composition of the common octopus, *Octopus vulgaris*, in relation to rearing temperature and body weight. *Aquaculture* **2006**, *256*, 311–322. [CrossRef]
49. Xu, Y.; Cleary, L.J.; Byrne, J.H. Identification and characterization of pleural neurons that inhibit tail sensory neurons and motor neurons in Aplysia: Correlation with FMRFamide immunoreactivity. *J. Neurosci.* **1994**, *14*, 3565–3577. [CrossRef] [PubMed]
50. Stanley-Samuelson, D.W. Physiological roles of prostaglandins and other eicosanoids in invertebrates. *Biol. Bull.* **1987**, *173*, 92–109. [CrossRef] [PubMed]
51. Nichols, P.D.; Danaher, K.T.; Koslow, J.A. Occurrence of High Levels of Tetracosahexaenoic Acid in the Jellyfish *Aurelia* sp. *Lipids* **2003**, *38*, 1207–1210. [CrossRef] [PubMed]
52. Song, C.; Zhao, S. Omega-3 fatty acid eicosapentaenoic acid. A new treatment for psychiatric and neurodegenerative diseases: A review of clinical investigations. *Expert Opin. Investig. Drugs* **2007**, *16*, 1627–1638. [CrossRef]
53. Song, C.; Phillips, A.G.; Leonard, B.E. Interleukin-1β enhances conditioned fear memory but impairs spatial learning in rats: Possible involvement of glucocorticoids and monoamine function in the amygdala and hippocampus. *Eur. J. Neurosci.* **2003**, *18*, 1739–1743. [CrossRef]
54. Marquis, C.P.; Baird, A.H.; De Nys, R.; Holmström, C.; Koziumi, N. An evaluation of the antimicrobial properties of the eggs of 11 species of scleractinian corals. *Coral Reefs* **2005**, *24*, 248–253. [CrossRef]
55. Kelman, D.; Kushmaro, A.; Loya, Y.; Kashman, Y.; Benayahu, Y. Antimicrobial activity of a Red Sea soft coral, *Parerythropodium fulvum fulvum*: Reproductive and developmental considerations. *Mar. Ecol. Prog. Ser.* **1998**, *169*, 87–95. [CrossRef]
56. Gunthorpe, L.; Cameron, A.M. Widespread but variable toxicity in scleractinian corals. *Toxicon* **1990**, *28*, 1199–1219. [CrossRef]
57. Gunthorpe, L.; Cameron, A.M. Intracolonial variation in toxicity in scleractinian corals. *Toxicon* **1990**, *28*, 1221–1227. [CrossRef]
58. Bowden, B.; Coll, J.; Willis, R. Some chemical aspects of spawning in alcyonacean corals. In Proceedings of the Fifth International Coral Reef Congress, Tahiti, French Polynesia, 27 May–1 June 1985; Volume 4, pp. 325–329.
59. Baird, A.H.; Pratchett, M.S.; Gibson, D.J.; Koziumi, N.; Marquis, C.P. Variable palatability of coral eggs to a planktivorous fish. *Mar. Freshw. Res.* **2001**, *52*, 865–868. [CrossRef]
60. Fusetani, N.; Toyoda, T.; Asai, N.; Matsunaga, S.; Maruyama, T. Montiporic acids A and B, cytotoxic and antimicrobial polyacetylene carboxylic acids from eggs of the scleractinian coral *Montipora digitata*. *J. Nat. Prod.* **1996**, *59*, 796–797. [CrossRef] [PubMed]

61. Lesser, M.P.; Stochaj, W.R.; Tapley, D.W.; Shick, J.M. Bleaching in coral reef anthozoans: Effects of irradiance, ultraviolet radiation, and temperature on the activities of protective enzymes against active oxygen. *Coral Reefs* **1995**, *8*, 225–232. [CrossRef]

62. Leclerc, M. Humoral factors in marine invertebrates. In *Molecular and Subcellular Biology: Invertebrate Immunology*; Rinkevich, B., Muller, W.E.G., Eds.; Springer: Berlin, Germany, 1996; pp. 1–9.

63. Stryer, L. *Biochemie*; Spektrum Akademischer Verlag: Heidelberg/Berlin, Germany, 2013; New York, korr. Nachdruck 1991 der völlig neubearb. Aufl. 1990; pp. 1–1024.

64. Jielian, W.; Baoqing, H.; Chungen, W.; Peipei, Y. Characterization and roles of lysozyme in molluscs. *ISJ* **2017**, *14*, 432–442.

65. Chang, K.Y.; Carr, C.W. Studies on the structure and function of lysozyme: I. The effect of pH and cation concentration on lysozyme activity. *Biochim. Biophys. Acta* **1917**, *229*, 496–503. [CrossRef]

66. Davies, R.C.; Neuberger, A.; Wilson, B.M. The dependence of lysozyme activity on pH and ionic strength. *Biochim. Biophys. Acta* **1969**, *178*, 294–305. [CrossRef]

67. Labro, M.T. The prohost effect of antimicrobial agents as a predictor of clinical outcome. *J. Chemother.* **1997**, *9*, 100–108.

68. SeGall, G.K.; Lennarz, W.J. Chemical characterization of the component of the jelly coat from sea urchin eggs responsible for induction of the acrosome reaction. *Dev. Biol.* **1979**, *71*, 33–48. [CrossRef]

69. Slattery, M.; McClintock, J.B. Population structure and feeding deterrence in three shallow-water antarctic soft corals. *Mar. Biol.* **1995**, *122*, 461–470. [CrossRef]

70. Dubois, M.; Gille, K.A.; Hamilton, J.K.; Rebers, P.A.; Smith, F. Colorimetric method for determination of sugars and related substances. *Anal. Chem.* **1956**, *28*, 350–356. [CrossRef]

71. Lowry, O.H.; Rosebrough, N.J.; Farr, A.L.; Randall, R.J. Protein measurement with the Folin phenol reagent [Folin-Ciocalteu reagent]. *J. Biol. Chem.* **1951**, *193*, 265–275. [CrossRef] [PubMed]

72. Barnes, H.; Blackstock, J. Estimation of lipids in marine animals and tissues: Detailed investigation of the sulphophosphovanilun method for "total" lipids. *J. Exp. Mar. Bio. Ecol.* **1973**, *12*, 103–118. [CrossRef]

73. Jacobs, D.S.; Kasten, B.L.; Demott, W.R. Laboratory Test Handbook, 2nd ed. *J. Intraven. Nurs.* **1991**, *14*, 354.

74. Bucolo, G.; David, H. Quantitative determination of serum triglycerides by the use of enzymes. *Clin. Chem.* **1973**, *19*, 476–482.

75. Bligh, E.G.; Dyer, W.J. A rapid method of total lipid extraction and purification. *Can. J. Biochem. Physiol.* **1959**, *37*, 911–917. [CrossRef]

76. Wu, H.; Southam, A.D.; Hines, A.; Viant, M.R. High-throughput tissue extraction protocol for NMR- and MS-based metabolomics. *Anal. Biochem.* **2008**, *372*, 204–212. [CrossRef]

77. Moreira-Ferro, C.K.; Daffre, S.; James, A.A.; Marinotti, O. A lysozyme in the salivary glands of the malaria vector *Anopheles darlingi. Insect Mol. Biol.* **1998**, *7*, 257–264. [CrossRef]

78. Daffre, S.; Kylsten, P.; Samakovlis, C.; Hultmark, D. The lysozyme locus in *Drosophila melanogaster*: An expanded gene family adapted for expression in the digestive tract. *Mol. Gen. Genet.* **1994**, *242*, 152–162. [CrossRef] [PubMed]

79. Smith, L.C.; Arizza, V.; Barela Hudgell, M.A.; Barone, G.; Bodnar, A.G.; Buckley, K.M.; Cunsolo, V.; Dheilly, N.M.; Franchi, N.; Fugmann, S.D.; et al. Echinodermata: The Complex Immune System in Echinoderms. In *Advances in Comparative Immunology*; Cooper, E.L., Ed.; Springer International Publishing: Cham, Switzerland, 2018; pp. 409–501. ISBN 978-3-319-76768-0.

80. Anderson, M.J. A new method for non-parametric multivariate analysis of variance. *Austral Ecol.* **2001**, *26*, 32–46. [CrossRef]

81. Anderson, M.; Braak, C. Ter Permutation tests for multi-factorial analysis of variance. *J. Stat. Comput. Simul.* **2003**, *73*, 85–113. [CrossRef]

82. Clarke, K.R. Non-parametric multivariate analyses of changes in community structure. *Aust. J. Ecol.* **1993**, *18*, 117–143. [CrossRef]

 *marine drugs*

Review

Jellyfish-Associated Microbiome in the Marine Environment: Exploring Its Biotechnological Potential

Tinkara Tinta [1,2,*], Tjaša Kogovšek [2], Katja Klun [2], Alenka Malej [2], Gerhard J. Herndl [1,3] and Valentina Turk [2]

[1] Department of Limnology and Bio-Oceanography, University of Vienna, Althanstrasse 14, A-1090 Vienna, Austria; gerhard.herndl@univie.ac.at
[2] Marine Biology Station Piran, National Institute of Biology, Fornače 41, 6330 Piran, Slovenia; tjasa.kogovsek@nib.si (T.K.); katja.klun@nib.si (K.K.); alenka.malej@nib.si (A.M.); valentina.turk@nib.si (V.T.)
[3] NIOZ, Department of Marine Microbiology and Biogeochemistry, Royal Netherlands Institute for Sea Research, Utrecht University, 1790 AB Den Burg, The Netherlands
* Correspondence: tinkara.tinta@univie.ac.at; Tel.: +43-1-4277-764-40

Received: 9 January 2019; Accepted: 29 January 2019; Published: 1 February 2019

Abstract: Despite accumulating evidence of the importance of the jellyfish-associated microbiome to jellyfish, its potential relevance to blue biotechnology has only recently been recognized. In this review, we emphasize the biotechnological potential of host–microorganism systems and focus on gelatinous zooplankton as a host for the microbiome with biotechnological potential. The basic characteristics of jellyfish-associated microbial communities, the mechanisms underlying the jellyfish-microbe relationship, and the role/function of the jellyfish-associated microbiome and its biotechnological potential are reviewed. It appears that the jellyfish-associated microbiome is discrete from the microbial community in the ambient seawater, exhibiting a certain degree of specialization with some preferences for specific jellyfish taxa and for specific jellyfish populations, life stages, and body parts. In addition, different sampling approaches and methodologies to study the phylogenetic diversity of the jellyfish-associated microbiome are described and discussed. Finally, some general conclusions are drawn from the existing literature and future research directions are highlighted on the jellyfish-associated microbiome.

Keywords: Cnidaria; Ctenophora; biodiversity; bioactive compounds; microbial communities; blue biotechnology

---

1. Introduction

1.1. Biotechnological Potential of Host–Microorganism Systems in the Ocean

Our fascination with the hidden treasures of the ocean is evident since Jules Verne's classic masterpiece Twenty Thousand Leagues Under the Seas, but only recently have we started to understand and unravel the incredible biotechnological potential of the ocean, in particular the potential locked in the vast diversity of marine microorganisms. Oceans represent the largest biosphere on the planet, and their smallest but most abundant, productive, and diverse residents, marine microorganisms, inhabit every marine habitat; with their diverse metabolic pathways, they play a central role in biogeochemical cycles in the oceans [1–3]. Consequently, marine microorganisms are potentially also a source of biotechnologically important enzymes, other compounds, and molecules [4,5]. For example, polymer-degrading enzymes and robust enzymes isolated from extremophilic microorganisms are being successfully applied in several branches of industry, from laundry detergents and food processing to sophisticated molecular biology reagents.

Biosurfactants and (extracellular) polymeric substances from marine bacteria are also finding increasing applications in bioremediation, industrial processes, manufacturing, and food processing and are used as underwater surface coatings, bio-adhesives, drag-reducing coatings on ship hulls, dyes, sunscreens, biodegradable plastics, etc. (reviewed in Reference [6]). In addition, a large set of bioactive compounds from marine microbes has been tested for their biomedical potential, such as antibacterial, antifungal and antiviral agents, anticancer and anti-inflammatory drugs, drug delivery agents, and others (reviewed in References [7–10]). In addition, it seems that the more peculiar, rich, or extreme their habitat is, the more biotechnologically interesting molecules microorganisms produce.

In the ocean, where microorganisms are constantly facing changing environmental conditions on a microscale level, one of the adaptation/survival strategies of some microbes is to establish long-term relationships with other organisms. At the same time, these host-microorganism systems are production hotspots of chemical compounds and/or secondary metabolites that can serve as the system's own defense mechanism against predators, colonization, and/or disease [11–14]. It is becoming evident that many of the bioactive compounds isolated from these systems are actually a result of the microorganisms' rather than the host's biosynthesis/metabolism or the interaction of both. Therefore, explorations of the taxonomic and metabolic diversity of the host-associated microbiome and investigations into mechanisms underlying these associations provide answers to questions regarding the evolution and ecology of these systems. This research is also generating datasets that can be screened for novel microbial strains, genes, secondary metabolites, byproducts, and other compounds of microbial (and host) origin that could be exploited by the fast-growing blue biotechnology sector, in a process known as bioprospecting [1].

1.2. Gelatinous Zooplankton as Host for Specific Microbiome

Marine invertebrates have been extensively studied as hosts of microorganisms producing compounds with biotechnological potential (e.g., [11,12,14–16]). Recently, among the marine invertebrates, Porifera (sponges), Annelida (within which are Polychaeta, marine worms), and Cnidaria (within which are corals, mostly octocorals) have been investigated, focusing on their associated microbiome and potential biotechnological applications [17]. The biotechnological potential of the cnidarians and their interactions with their microbiome was recently reviewed ([18] this issue), with the main focus on the coral holobiont. In addition, cnidarian–microbe interactions were investigated in detail in *Hydra* as a model host and its holobiont [19]. Medusozoans, characterized by the presence of a pelagic phase in their life cycle, are less well studied. To the best of our knowledge, medusozoans have never been comprehensively reviewed in terms of their interactions with their microbiome or as hosts of a microbiome with biotechnological potential.

In this review, we use the term "jellyfish" to describe gelatinous marine plankton belonging to the cnidarian subphylum Medusozoa (Scyphozoa, Cubozoa, and Hydrozoa) and phylum Ctenophora. Their convergent features are transparency and fragility, their body surface is coated with mucus, and they lack a hard skeleton. In addition, the proteinaceous body has a high water content (>95%) accompanied by a low content of organic matter on a wet mass basis. Ctenophora spend all their life in the pelagic environment, while the life cycle of a large majority of Medusozoa is characterized by a shift of planktonic medusa and a benthic polyp phase. Therefore, the Medusozoa-associated microbiome experiences changes in terms of morphology and biochemical structure of their hosts and a complete shift of the hosts' lifestyle from benthic/attached to pelagic/swimming/free-living. Another important fact to consider is that as the medusa stage drifts with ocean currents over long distances, its associated microbiome could be subjected to changing environmental conditions as could represent pressure on the new environment (e.g., jellyfish as vectors of allochthonous microbes' transmission in the marine environment). Taking into account their unique simple anatomy, evolutionary age, alteration between different life stages, wide distribution, and important role in diverse marine ecosystems worldwide, Medusozoa and Ctenophora could potentially harbor taxonomically and metabolically diverse microorganisms with great biotechnological potential. This seems even more important (and justified) in light of the recently reported increase of jellyfish blooms in several marine environments [20], with potentially serious ecological and socio-economic consequences. Due to high reproductive output and fast growth, jellyfish form blooms when conditions are favorable, reaching high biomass within a short period of time ([21] and references therein), and represent a major source of organic matter for the marine ecosystem. It was shown that jellyfish blooms influence the diversity of marine food webs and may affect biogeochemical cycles in the ocean [21–24]. Especially at the end of their lifespan, jellyfish debris represents hotspots for growth and development of specific, even pathogenic, microbial phylotypes, potentially affecting human health and well-being [25–29].

The aim of this paper is to critically review the existing literature on the microbiome associated with jellyfish in terms of: (i) screening which jellyfish taxa have already been investigated for their associated microbiome and how many remain to be explored; (ii) different methodological approaches applied to study the jellyfish-associated microbiome; (iii) understanding the characteristics of the jellyfish-associated microbiome in terms of (a) degree of microbiome specialization (e.g., generalist versus specialist), (b) preference of the microbiome for specific jellyfish taxa, and (c) specificity of the microbiome at the jellyfish population, (d) life stage, and (e) body part level; (iv) gaining insights into the composition and function of the jellyfish-associated microbiome; and (v) determining the biotechnological potential of the jellyfish-associated microbiome.

2. Jellyfish-Associated Microbiome

Thus far, only a limited number of studies have focused on the microbiome associated with jellyfish during their life span (Table 1). Early reports on microbes associated with jellyfish were simple observations, where jellyfish was a primary object of research [30–32]. Later, studies specifically focused on selected pathogenic bacteria associated with jellyfish, e.g., investigating them as vectors of fish pathogens [33,34]. Already these studies addressed questions on the ecology, composition, and role of microbial communities associated with jellyfish, the mechanisms underlying these interactions, and the nature of the relationships between jellyfish and their associated microbiome. Based on this, researchers have started to focus on the microbial counterpart of jellyfish–microbe associations, aiming at assessing the diversity of microbial communities associated with different jellyfish species from various ecosystems and with their different life stages and body compartments.

Table 1. Overview of publications on jellyfish-associated microbiome in terms of species studied (and their taxonomy) and jellyfish life stage, body compartment, sampling location, and methodology applied to analyze the composition and/or structure of the associated microbiome. FISH, fluorescence in situ hybridization; DGGE, denaturing gradient gel electrophoresis; ARISA, automated ribosomal intergenic spacer analysis; ITS, internal transcribed spacer; NGS, next-generation sequencing; T-RFLP, terminal restriction fragment length polymorphism.

Jellyfish Taxonomy					Study Design				
Phylum/ Subphylum	Class	Order	Family	Species	Life Stage	Body Part (Adult Medusae)	Sampling Location	Methodology to Study Associated Microbiota	Publication
Medusozoa	Scyphozoa	Semaeostomeae	Ulmaridae	*Aurelia aurita*	Adult medusae	Whole body	North Atlantic coastal waters	16S rRNA gene clone libraries	[36]
					Polyps, strobila, ephyra, juvenile adult medusae	Mucus, gastric cavity	Kiel Bight, Baltic Sea, Southern English Channel, North Sea	Confocal laser scanning microscopy FISH NGS 454 technology V1–V2 16S rRNA region	[35]
					Adult medusae	Oral arms umbrella gastric cavity	Northern Adriatic	Culturing, DGGE, 16S rRNA gene clone libraries	[37]
			Cyaneidae	*Cyanea capillata*	Adult medusae	Tentacles	Scottish waters (Orkney)	Culturing DGGE, sequencing bands	[38]
				Cyanea lamarckii	Adult medusae	Tentacles	Scottish waters (Orkney)	Culturing DGGE, sequencing bands	[38]
					Larvae polyps adult medusae	Tentacles, umbrella, mouth arm, gonads	German Bight	ARISA of ITS region	[39]
			Pelagiidae	*Pelagia noctiluca*	Adult medusae	Mouth	Ireland	Sequencing of specific bacterial 16S rRNA gene	[33]
				Chrysaora plocamia	Polyps podocyst excyst	Whole body	Northern Chile	NGS Illumina MiSeq platform 2 × 300 bp paired end, V1–V2 16S rRNA region	[40]
				Chrysaora hysoscella	Larvae polyps adult medusae	Tentacles, umbrella, mouth arm, gonads	German Bight	ARISA of ITS region	[39]
		Rhizostomeae	Mastigiidae	*Mastigias cf. papua*	Adult medusae	Dome, tentacles	Indonesian marine lakes	NGS 454 technology, V3–V4 region	[41]
			Cepheidae	*Cotylorhiza tuberculata*	Adult medusae	Gastric cavity	Alcudia Bay, Balearic Sea	Culturing, NGS—454 pyrosequencing	[42]
					Adult medusae	Gastric cavity	Alcudia Bay, Balearic Sea	NGS—Illumina MiSeq platform, 2 × 250 bp, paired end	[43]
	Cubozoa	Carybdeida	Tripedaliidae	*Tripedalia cf. cystophora*	Adult medusae	Whole body	Indonesian marine lakes	NGS—454 technology V3–V4 16S rRNA region	[41]
	Hydrozoa	Anthoathecata	Bougainvilliidae	*Nemopsis bachei*	Adult medusae	Whole body	North Atlantic coastal waters	16S rRNA gene clone libraries	[36]
			Tubulariidae	*Tubularia indivisa*	Adult medusae	Tentacles	Scottish waters (Orkney)	Culturing DGGE, sequencing bands	[38]
		Leptothecata	Phialellidae	*Phialella quadrata*	Adult medusae	Whole body	Shetland Isles	Nested PCR with specific bacterial primers	[32]
					Adult medusae	Whole body	Ireland	RT PCR with specific bacterial primers	[34]
		Siphonophorae	Diphyidae	*Muggiaea atlantica*	Adult medusae	Whole body	Ireland	RT PCR with specific bacterial primers	[34]

Table 1. *Cont.*

Jellyfish Taxonomy					Study Design				
Phylum/ Subphylum	Class	Order	Family	Species	Life Stage	Body Part (Adult Medusae)	Sampling Location	Methodology to Study Associated Microbiota	Publication
Ctenophora	Tentaculata	Lobata	Bolinopsidae	*Mnemiopsis leidyi*	Adult specimen	Whole body	Tampa Bay, Florida, USA	16S rRNA gene clone libraries, T-RFLP	[44]
					Adult specimen	Whole body guts	Gullmar fjord, west coast of Sweden	NGS—454 pyrosequencing	[23]
					Adult specimen	Whole body	Helgoland roads, German Bight	ARISA of ITS region	[45]
				Bolinopsis infundibulum	Adult specimen	Whole body	Helgoland roads, German Bight	ARISA of ITS region	[45]
		Cydippida	Pleurobrachiidae	*Pleurobrachia pileus*	Adult specimen	Whole body	Helgoland roads, German Bight	ARISA of ITS region	[45]
	Nuda	Beroida	Beroidae	*Beroe ovata*	Adult specimen	Whole body	Helgoland roads, German Bight	ARISA of ITS region	[45]
					Adult specimen	Whole body	Tampa Bay, Florida, USA	16S rRNA gene clone libraries, T-RFLP	[44]

2.1. Microbiome Associated with Specific Jellyfish Taxa

Different taxonomic groups of jellyfish were studied for their associated microbiome, but maybe even more important, the vast diversity of jellyfish as hosts remain to be explored.

2.1.1. Medusozoa

Within Medusozoa, most studies were performed on Scyphozoa, some on Hydroza and, to our knowledge, only one study on Cubozoa was published so far (Table 1).

2.1.2. Scyphozoa

Inside the Scyphozoa class, also known as the "true jellyfish", only members from Rhizostomeae and Semaeostomeae orders have been investigated for their associated microbes so far.

2.1.3. Semaeostomeae

Within Semaeostomeae members of Ulmaridae (*Aurelia aurita*, or moon jellyfish, [35–37]), Cyaneidae (*Cyanea capillata*, known as lion's mane jellyfish, and *Cyanea lamarckii*, or blue fire jellyfish [38,39]) and Pelagiidae (*Pelagia noctiluca*, known as mauve stinger [33], *Chrysaora plocamia*, a South American sea nettle [40], and *Chrysaora hysoscella*, compass jellyfish [39]) were studied for their associated microbial communities.

2.1.4. Ulmaridae

One of the most studied jellyfish is *Aurelia aurita*, also known as moon jellyfish. To our knowledge, it is the only jellyfish species from this family that has been investigated for its associated microbiome, and at the same time the most comprehensively investigated of all jellyfish species. Nevertheless, we have to point out the unclear taxonomy of genus *Aurelia*. Mayer [46] and Kramp [47] described 12 and 6 *Aurelia* species, respectively, based on morphological characteristics of the medusa; later only *A. aurita* and *A. labiata* were recognized as distinct species [48]. Despite more than 100 years of *Aurelia* research, the taxonomy of this genus is still unclear [49], and recent molecular analysis indicated that *A. aurita* is represented by several cryptic species [50,51]. The World Register of Marine Species (WoRMS) currently (as of 17 December 2018) recognizes nine *Aurelia* species. It is therefore not always clear which of these species are investigated for their associated microbiome, but for this review we will retain the species name given by the authors of the respective articles.

The first and most detailed study of the microbiome of *A. aurita* was performed by Weiland-Bräuer et al. [35]. These authors investigated the microbiome associated with different life stages of *A. aurita* (polyp, strobila, ephyrae, juvenile, and adult medusae), investigated different compartments of the adult medusae (mucus versus gastric cavity), and compared the microbiome of the polyp stage of specimens from different geographic locations. In this study, fluorescence in situ hybridization (FISH) was used to determine the distribution of specific groups of bacteria on the polyps. The taxonomic composition of the microbiome of *A. aurita* was assessed using the next generation sequencing technique 454 pyrosequencing of the V1–V2 hypervariable region of the 16S rRNA gene.

While several interesting findings were reported in the study of Weiland-Brauer et al. [35], here we highlight only those most important for our review. Different stages of the strobilation event, strobila and ephyra, and juvenile medusa harbored a similar microbiome, which significantly differed from the microbiome of the non-strobilating perennial polyp. At the same time, the richness of the bacterial community was approximately the same among all examined life stages. The microbiome analysis revealed that *Gammaproteobacteria*, *Alphaproteobacteria*, *Bacteroidetes*, and *Actinobacteria* dominate all life stages, with different relative contributions of each bacterial group depending on the stage. Based on 16S rRNA gene amplicon sequencing and FISH coupled with confocal laser scanning microscopy, the authors reported that the entire epithelial surface of the polyp stage of *A. aurita* was covered by bacteria. They proposed that colonization of the polyp occurs on the mucus coating its epithelial surface, mainly by *Gammaproteobacteria* (mainly *Crenothrix*), while bacteria detected inside and between cells of the polyp tissue were most likely a novel *Mycoplasma* strain (class *Molicutes*), suggested to be potential endosymbionts of *A. aurita* polyps. However, sequencing approaches failed to detect this strain in polyps. Other dominant colonizers of polyps were *Bacteroidetes* (in particular *Lacinutrix*, a member of the *Flavobacteriaceae* family) and *Alphaproteobacteria* (dominated by *Phaeobacter*, a member of the *Rhodobacteriaceae* family). The next developmental phase of *A. aurita*, the strobilating polyp, was associated with *Gammaproteobacteria* (*Crenotrichaceae* and *Vibrionaceae*) and *Actinobacteria* (*Nocardiaceae*). Newly released ephyrae were dominated by *Gammaproteobacteria* (*Crenotrichaceae* and *Pseudoalteromonadaceae*), *Alphaproteobacteria* (*Rhodobacteraceae*) and *Actinobacteria* (*Microbacteriaceae*). Juvenile medusae were dominated by *Alphaproteobacteria* (*Rhodobacteraceae*), *Flavobacteriaceae*, and *Gammaproteobacteria* (*Vibrionaceae*). The differences in bacterial community composition between gastric cavity and umbrella mucus were also determined, revealing that both compartments were dominated by unclassified *Mycoplasma*. A small number of bacterial sequences from the gastric cavity and mucus were affiliated with *Rhodobacteraceae* (*Alphaproteobacteria*). Bacterial diversity differed between the gastric cavity and mucus, with mucus exhibiting greater variability in richness. The drawback of this study is the relatively low number of samples analyzed. One should be careful in extrapolating these results to the entire *A. aurita* taxonomic group (or even beyond), in particularly, since this study also shows the effect of the host's natural environment on the composition of jellyfish-associated microbiome.

The other two available studies of the microbiome of *A. aurita*, from the North Atlantic Ocean [36] and the Northern Adriatic [37], investigated only the adult medusa stage. Both Daley et al. [36] and Kos Kramar et al. [37] constructed and sequenced bacterial 16S rRNA gene clone libraries, which means that compared to the next-generation sequencing approach used by Weiland-Bräuer et al. [35], the number of sequences obtained and analyzed was lower, hence the entire diversity of bacteria was most likely not captured, particularly that of the rare community members. At the same time, the bacterial 16S rRNA gene clone libraries resulted in long and good-quality sequences, allowing the classification to lower taxonomic levels. However, even these two studies using clone libraries cannot be compared entirely, as they used different bacterial primers to amplify different regions of 16S rRNA. Besides, Kos Kramar et al. [37] also applied a culturing approach to investigate the cultivatable part of the jellyfish-associated microbial community.

Daley et al. [36] analyzed seven adult specimens, all collected during two sampling days in spring of the same year. In contrast, Kos Kramar et al. [37] collected 20 adult individuals during two sampling events in the same year, one in spring and one in early summer, investigating changes of the *A. aurita*-associated microbiome in relation to the age of the jellyfish population. Each jellyfish sample pool was then split in half for a culture-based and culture-independent approach to analyze the composition of the *A. aurita*-associated microbial community [37]. For both studies, it is difficult to tell whether the analyzed sample pool is truly representative of the studied jellyfish population, but both sampling designs did not account for interannual variability or spatial patchiness.

The conclusions drawn from these two studies differ. While Daley et al. [36] showed that *Aurelia* is associated with a consortium of bacteria composed of *Mycoplasmatales* (*Tenericutes*, *Mollicutes*) and many unclassified bacteria, several of them distantly related to *Mycoplasma*, in line with the results of Weiland-Brauer et al. [35], Kos Kramar et al. [37] did not detected any bacteria affiliated with this taxonomic group. Also, while in the North Atlantic *A. aurita* harbored only few *Gammaproteobacteria* (including *Psychrobacter* spp.), *A. aurita* collected in the Northern Adriatic were dominated by the gammaproteobacterial families *Vibrionaceae*, *Pseudoalteromonadaceae*, *Xanthomonadaceae*, and *Pseudomonadaceae*. Within the *Alphaproteobacteria*, different families were associated with Aurelia from the two studied systems: in the North Atlantic *Rickettsiales*, and in the Northern Adriatic *Rhodobacteraceae* (mostly *Phaeobacter* and *Ruegeria*). In the Northern Adriatic, *Betaproteobacteria* and *Actinobacteria* were found associated with *A. aurita*, but were not detected in *A.*

aurita from the North Atlantic while *Cyanobacteria* were associated with *A. aurita* in both the Northern Adriatic and the North Atlantic.

The body part-specificity of the *A. aurita* microbiome was also studied in the Northern Adriatic [37]. *Betaproteobacteria* dominated in the gastral cavity, while *Alphaproteobacteria* and *Gammaproteobacteria* dominated in the "outer" body parts, mostly ex-umbrella mucus, in accordance with Weiland-Brauer et al. [35]. Bacterial strains associated with polycyclic aromatic hydrocarbons (PAHs) and plastic degradation, such as *Stenotrophomonas, Pseudomonas, Burkholderia, Achromobacter*, and *Cupriavidus* [52–56], were present in *A. aurita*'s gastric cavity, indicating an adaptation to anthropogenic pollution. Furthermore, the bacterial community associated with *A. aurita* changed at the transition from the peak to the senescent phase of the jellyfish bloom, and was characterized by an increase of *Gammaproteobacteria*, especially *Vibrionaceae* and *Alteromonadaceae* [37]. Those authors speculated on the different roles that *Vibrionaceae* might play during different stages of jellyfish, from possible commensal opportunistic visitors at the peak of jellyfish bloom to consumers of moribund jellyfish biomass at the end of the jellyfish lifespan. This is in accordance with studies of the degradation of jellyfish biomass by ambient microbial communities, where *Vibrionaceae* were found to dominate the jellyfish biomass-degrading assemblages [25–29]. Furthermore, the authors suggested that *Vibrionaceae* could be exploiting the nutrient-rich niche provided by *A. aurita*. Under proper conditions, such as a disturbed defense mechanism of the jellyfish and elevated water temperatures at the end of jellyfish bloom, upregulating the determinants of *Vibrio*'s virulence, such as motility, resistance to antimicrobial compounds, hemolysis, and cytotoxicity [57], they could outcompete other bacteria and become highly dominant in *A. aurita*.

2.1.5. Cyaneidae

The first insights into the microbiome of Scyphozoa were published by Schuett and Doepke [38] focusing on the pathogenic potential of endobiotic bacteria associated with the tentacles of *Cyanea capillata* and *Cyanea lamarckii*. At the same time, they examined one Hydrozoa (discussed below in the Hydrozoa section) and one sea anemone (not discussed in this review) using culturing and culture-independent approaches.

The data from culturing and culture-independent approaches showed that most of the bacteria associated with both Cyaneidae species were *Gammaproteobacteria*, exhibiting >97% similarity to the *Vibrio* group and closely affiliated with *V. splendidus/V. tasmaniensis/V. lentus/V. kanaloa*. Bacteria affiliated with *Pseudoalteromonas tetradonis/P. elyacovii/P. haloplanctis, Photobacterium profundum, Shewanella violacea, Shewanella sairae/S. marinintestina, Sulfitobacter pontiacus*, and *Arcobacter butzleri* were found to be exclusively associated with *C. capillata* tentacles. In contrast, *Moritella viscosa* and *Mesorhizobium tianshanense* were only found associated with *C. lamarckii*. The authors unfortunately did not provide information on the relative abundance of these specific bacterial groups. As no remarkable similarities between communities associated with specific jellyfish were detected, the authors suggested that the associated microbiome could be host-specific to some extent. However, as there was no attempt to analyze the bacterial community of the ambient seawater, we believe that the information available is not sufficient to support this conclusion.

The culture-independent approach should have described the diversity of the bacterial community associated with *C. lamarckii* and *C. capillata* to a great extent. However, the authors reported that the number of denaturing gradient gel electrophoresis (DGGE) bands, representing the richness of the bacterial community, was lower than the number of bands obtained using a preculturing step (see Reference [38] for details on study design). One possible explanation is that the abundance of bacteria on jellyfish tentacles is low and the preculturing step increased their biomass and consequently the DNA extraction yield. The authors reported many high-quality sequences exhibiting low similarity to any of the sequences available in the National Center for Biotechnology Information (NCBI) database at that time (<97% similarity). Therefore, as they were not able to show an affiliation of the phylogeny of these bacteria using the tools available at the time, they suggested that these bacteria should be considered as potentially new species. Their preculturing step, however, was probably selective toward a cultivatable jellyfish-associated microbiota, which might represent only a minor fraction of the total microbial community. In addition, as information on the number of jellyfish individuals, the amount of analyzed tentacles, and the number of replicates is not provided, it is difficult to evaluate how representative these results are. From the description of sampling, the analyzed individuals were all collected at one location, during one season, and over three consecutive years. It is not clear, however, which jellyfish were collected when and whether specific species were collected over different years. Thus, the interannual variability could not be assessed.

The second study on *C. lamarckii* was published by Hao et al. [39]. This study provides valuable insights into the microbial community associated with different *Cyanea* body parts, revealing significant differences between bacterial communities associated with the umbrella and other parts, i.e., gonads, mouth, arm, and tentacles. The bacterial diversity was higher on the umbrella than on the mouth, arms, tentacles, and gonads. Altogether these results suggest a certain degree of body part specificity in the *Cyanea*-associated microbiome. The life stage specificity of the associated bacterial community was also investigated. Significant differences between bacterial communities associated with different jellyfish life stages were apparent, with the community's richness increasing from larval stage to adult medusa. Furthermore, it was found that the type of food fed to the polyps affected the structure of the associated bacterial community.

The number of specimens analyzed in this study was high (n = 44) and samples were collected frequently within a short time interval (twice per week), accounting for the variability within the population, but only during one season. At the same time, spatial, seasonal, or interannual variability was not assessed. This is important, especially in light of reports on interannual fluctuations and large spatial variability for Scyphomedusae in the study region [58]. Furthermore, jellyfish analyzed in this study were collected using 500 μm mesh trawls, introducing potential cross-contamination with microbes associated with other biological material accumulated in the cod end of the net. The data on microbial community associated with *C. lamarckii* tentacles from the two available studies cannot be compared, unfortunately, since the taxonomic composition of the microbial community was not analyzed by Hao et al. [39] and the data obtained by DGGE [38] and automated ribosomal intergenic spacer analysis (ARISA) [39] are not comparable due to the different resolutions of these two fingerprinting techniques.

2.1.6. Pelagiidae

Within the Pelagiidae family, jellyfish from genus *Chrysaora* have been most extensively studied, while only one study on the microbiota of *Pelagia noctulica* was published, focusing on a single specific bacterium, the fish pathogen *Tenacibaculum maritimum* (see Section 3.6 and Reference [33]). We did not find any studies on jellyfish from the genera *Mawia* and *Sanderia*.

The microbiome of *Chrysaora* was analyzed by Lee et al. [40], focusing on the benthic life stages of *Chrysaora plocamia*. In particular, forms involved in propagation through cyst formation, i.e., the mother polyp, its dormant cysts (podocysts), and polyps recently excysted from podocysts, were studied. The microbial community was analyzed using Illumina MiSeq sequencing of the V1–V2 region of the 16S rRNA gene. *Chrysaora plocamia* was collected along the Chilean–Peruvian Humboldt Current System (along the coast of Northern Chile) at different sampling times, accounting for temporal/seasonal variability. Polyps were grown in the lab from planulae subtracted from oral arms of collected jellyfish and produced podocysts, often followed by strobilation. No comparison with the bacterial community of the ambient seawater community was made, however, which would have allowed us to understand the degree of specialization of jellyfish-associated microbiome.

At a broader taxonomic level, the microbiomes of all life stages of *C. plocamia* were dominated by *Gammaproteobacteria* (mainly *Alteromonadales*, *Legionellales*, and *Methylococcales*), followed by *Alphaproteobacteria* (mostly *Rhizobiales*), *Bacteroidetes* (*Flavobacteriaceae*), and *Planctomycetes* [40]. Only *Betaproteobacteria* were significantly more abundant in the polyps of *Chrysaora* as compared to isolated podocyst. The microbiome of *Chrysaora* clustered according to the life stage rather than by sampling location and time, with polyp bacterial communities being significantly different from their podocyst and excyst. Thus, Lee et al. [40] speculated that the polyp-specific microbiome might be essential for their sessile lifestyle and important for the initiation of later developmental stages. Furthermore, the microbiome of the polyps was more similar to the next developmental stage, the excyst, than to the podocyst, which, as Lee et al. [40] argued, demonstrates an early stage in the successive restructuring of microbial communities from podocysts to polyps. Bacteria that were more abundant in the cyst stages than in polyps included chemolithoautotrophs of the genera *Nitrospira, Nitrospina, Thiogranum,* and *Desulfovermiculus* and nitrogen fixers of the order *Rhizobiales*. It has been suggested that the podocyst capsule harbors a specific beneficial assemblage of microbes to sustain the viability of podocysts [40]. This specific assemblage consists of bacteria affiliated with *Bacteroidetes, Planctomycetes, Alpha-, Beta-,* and *Gammaproteobacteria* and *Chloroflexi*.

One of the most interesting results of Lee et al. [40] was that within the jellyfish-associated microbiome, half of the detected bacteria were closely related to microbial communities found associated with seaweed, sponges, and sea squirts, all of which interestingly have a benthic lifestyle. Bacteria also had closest relatives known to be drivers of major elemental cycles, including methanotrophs. Interestingly, a number of these bacteria were also present in the microbiome of *A. aurita* as determined by Weiland-Brauer et al. [35]. However, it is important to note that for the comparison, bacteria associated with benthic and not medusa stages from the *Aurelia* study should be considered. Whether or not this was taken into account is not clear. Furthermore, Lee et al. [40] speculated that these shared bacteria might play an essential role in the scyphozoan life cycle, especially since the two jellyfish species originate from different marine regions. However, the microbiome of *Chrysaora* exhibited higher diversity than *A. aurita*, potentially explained by the fact that some of the studied life stages of *Aurelia* originated from polyps grown in the laboratory, most likely in an environment with lower microbial diversity than under *in situ* conditions. However, we would like to stress that Lee et al. [40] used 1 μm prefiltered seawater as the medium for their laboratory experiments, which could have introduced ambient seawater bacteria and altered the diversity of the jellyfish-associated microbial community. Also, the studies used different sequencing platforms, sequencing depths, and primers to amplify bacterial 16S rRNA genes, hence comparisons between these two jellyfish species should be done with caution.

Another representative of the *Chrysaora* genus, *Chrysaora hysoscella*, collected in the German Bight was studied in Reference [39]. The pluses and minuses of this study are discussed above (see the paragraph on *Cyanea lamarckii*). For this jellyfish species no significant differences between the microbiomes associated with different body parts were found. The microbiome of the umbrella exhibited the highest variability and richness, while the gonads revealed the lowest richness of all body parts. The microbiomes associated with different life stages indicated a strong selective colonization process, with decreasing diversity from larvae to adult medusae. Unfortunately, the methodological approach used in this study did not provide any insight into the phylogenetic composition of the associated microbiome.

2.1.7. Rhizostomeae

Within Rhizostomeae, *Mastigias papua* (a member of the Mastigiidae family), called spotted or lagoon jellyfish, was studied [41] and a member of the Cepheidae family, *Cotylorhiza tuberculata*, known as fried egg jellyfish, was investigated for its microbiome [42,43].

The first study of the microbiome of Rhizostomeae was on the *Cotylorhiza tuberculata* gastric cavity, adopting a culture-independent molecular approach in combination with classical culturing of aerobic heterotrophs [42]. However, no body compartment or life stage specificity was tested or comparison with the diversity of ambient seawater communities made. Both the culturing and culture-independent approaches revealed similar results: reduced diversity of the gastric cavity-associated microbial community, with four major groups of microorganisms detected. The dominant bacteria were

Spiroplasma, with the closest relative *S. poulsonii* (*Spiroplasmataceae, Mollicutes*), followed by *Thalassospira* (*Rhodospirillaceae, Alphaproteobacteria*), affiliated with the fish pathogen of the genus *Tenacibaculum* (with the closest relative *T. soleae*), and *Synechococcus*. Cortes-Lara et al. [42] suggested that bacteria affiliated with *Spiroplasma, Thalassospira*, and *Tenacibaculum* might be part of the jellyfish's digestive system, while *Synechococcus* could have been ingested by *C. tuberculata*. Within the culturable fraction, almost 80% of the bacteria were *Vibrionaceae* (*V. xuii* and *V. harveyi*). However, even though Cortes-Lara et al. [42] reported high similarity of the microbiomes among the studied specimens, one should note that this study was based on only four individuals sampled at just one location in one season. Thus, these conclusions might not be representative of the entire population or the jellyfish species.

A subsequent study [43], also focusing on the gastric microbiome of *C. tuberculata*, this time using a shotgun metagenomic approach, revealed the entire genomic repertoire of the dominant members of the bacterial community. Applying fluorescence *in situ* hybridization using fluorescently labeled oligonucleotide probes to target specific bacterial groups and epifluorescence microscopy (CARD-FISH) allowed for localization and identification of specific bacteria within adult medusae. Viver et al. [43] proposed a simple model of microbial-animal digestive associations and hypothesized on the role of the microbiome of *C. tuberculata* in jellyfish ecology. *C. tuberculata* individuals were collected during the same sampling campaign and location as in the study by Cortes-Lara et al. [42], but over an additional two consecutive years to account for interannual community variability.

One of the most important findings of this study was that the microbiome of the gastric cavity of *C. tuberculata* was dominated by eukaryotic cells (*Onychodromopsis*-like and *Symbiodinium* cells) and their infecting bacteria. In line with the previous study [42], low diversity of the gastric cavity-associated microbial community was found, dominated by a few bacterial species, such as a *Simkania*-like lineage of the phylum *Chlamydia, Tenacibaculum*-like, *Spiroplasma*-like, and *Mycoplasma*-like bacteria. The association between the jellyfish and its low-diversity microbiome was temporarily stable and possibly related to food ingestion and protection from pathogens. Microscopic analysis contradicted to some extent the results of the metagenome analysis, emphasizing the importance of coupling microscopy-based methods with omic approaches to study the jellyfish-associated microbiome. Based on their findings, Viver et al. [43] proposed three candidate taxa: *Simkania*-like lineage, *Candidatus* Syngnamydia medusae sp.nov. (affiliated with the candidate genus "*Candidatus* Syngnamydia" [59], in particular *Candidatus* Syngnamydia salmonis [60]), *Spiroplasma*-like bacteria, *Candidatus* Medusoplasma gen nov, *Candidatus* Medusoplasma mediterranei sp.nov., and *Tenacibaculum*-like bacteria, *Candidatus* Tenacibaculum medusae.

Within Rhizostomeae, the microbiome of *Mastigias cf. papua* was also studied [41]. This study employed 16S rRNA gene 454 pyrosequencing to examine the bacterial community associated with adult medusae collected during three sampling campaigns at three sampling locations during one season. Major groups of the microbiome of *Mastigias cf. papua* were *Gammaproteobacteria* (with *Endozoicimonaceae* being most abundant), followed by *Mollicutes, Spirochaetes*, and *Alphaproteobacteria* (with the orders *Kiloniellales* and *Rhodobacterales*).

2.1.8. Cubozoa

The only study that ever made an attempt to investigate the microbiome of Cubozoa was published by Clearly et al. [41] (for study details, see previous paragraph on Mastigiidae family), in particular with *Tripedalia cf. cystophora*. *Endozoicimonaceae* (*Gammaproteobacteria*) were the most abundant microbes associated with the adult medusae, followed by *Spirochaetes, Kiloniellales* (*Alphaproteobacteria*), and the low-abundance *Mollicutes* and *Firmicutes*.

2.1.9. Hydrozoa

Among cnidarians, *Hydrozoa* is the only taxon with freshwater species and formation of colonies combining both polyp-zoids and medusa-zoids; however, one stage or the other, more frequently medusa, is reduced or absent. They are found in nearly every marine habitat type and their diversity is significantly higher than that of Scyphomedusae, the former comprising more than 3800 species. Chronologically, to our knowledge, the first report of bacteria associated with Cnidaria was published in 1978 by Margulis et al. [31], who reported a large number of rod-shaped Gram-negative bacteria with a single polar flagellum, found in gastrodermal cells of healthy freshwater green hydras *Hydra viridis*. Afterward, the microbiota associated with the freshwater *Hydra* was extensively studied (reviewed in Reference [19]). This review focuses on the marine hydromedusan species. Marine Hydrozoa *Phialella quadrata* (*Leptothecata*) and *Muggiaea atlantica* (*Siphonophorae*) were investigated by Ferguson et al. [32] and Fringuelli et al. [34] and recognized as a vector and potential natural reservoir of *Tenacibaculum maritimum*, a bacterial pathogen frequently found in farmed fish. *Tubularia indivisia* (*Anthothecata*) was investigated by Schuett and Doepke ([38]; see first paragraph on Cyaneidae for details of this study), showing that its tentacles are associated with potential pathogenic endobiotic bacteria, such as *Cobetia marina*, *Colwellia aestuarii*, *Endozoiciminas elysicola*, *Vibrio aestuarianus*, *Bacillus subtilis*, and *Ilyobacter psychrophilus*. Daley et al. [36] investigated the microbiome of *Nemopsis bachei* (Anthothecata) parallel to the *A. aurita* analysis (see paragraph on Ulmaridae for details of this study). Bacteria found to be associated with *N. bachei* were affiliated with *Gammaproteobacteria* (*Vibrionales*, *Oceanospirillales*, *Enterobacteriales*, and *Alteromonadales*), with *Alphaproteobacteria* (*Rickettsiales* and *Rhizobiales*), with *Flavobacteria* (e.g., *Tenacibaculum maritimum*), with *Cyanobacteria* (e.g., *Synechoccous*), and with *Firmicutes*.

2.1.10. Ctenophora

The WoRMS lists over 200 Ctenophora species, but only a few have been examined for their associated microbial communities: among the *Tentaculata* class, *Mnemiopsis leidyi* [26,44,45,61], *Bolinopsis infundibulum* [45], and *Pleurobrachia pileus* (*Cydippida* order) [45], and only one species of the *Nuda* class, *Beroe ovata* [44,45].

Although Ctenophora are known to be parasitized by a variety of eukaryotes, including amoebae, dinoflagellates, and sea anemones [62–65], probably the first report on bacteria associated with Ctenophora was published by Moss et al. [61]. They observed rod-shaped bacteria in the ciliary structure (g-cilium) of the food grove of the lobate ctenophore *Mnemiopsis mccradyi* (now accepted as *Mnemiopsis leidyi* morphotype) that they described for the first time. Later, bacterial communities associated with *Mneniopsis leidyi* and its natural predator *Beroe ovata* were studied using terminal restriction fragment length polymorphism (T-RFLP), cloning, and sequencing of 16S rRNA genes by Daniels and Breitbart [44]. These authors found that the composition of bacterial communities associated with Ctenophora varied over time (sampling over three consecutive years and different seasons at the same location). *M. leidyi*-associated bacterial communities exhibited some degree of seasonal specificity, with families of *Proteobacteria* and *Bacteroides* being associated with specific months, while no such pattern was observed for *Beroe ovate*-associated bacterial communities. In general, the diversity of the Ctenophora-associated bacterial community was lower than the diversity in ambient seawater assemblages and included some known pathogens of sea anemones. *M. leidyi* was dominated by *Marinomonas* sp. (*Oceanospirillales*, *Gammaproteobacteria*), previously described in corals and sponges, which were not detected in the water column or in the microbiome of *Beroe ovata*, while the microbiome of *B. ovata* was dominated by *Rhodospirillaceae* (*Alphaproteobacteria*), followed by *Tenacibaculum aiptasiae* a5, originally described as a pathogen of sea anemones, which was not detected in *M. leidyi*. Interestingly, both ctenophores harbored cultured representatives of *Alphaproteobacteria* capable of hydrocarbon degradation, *Thalassospira*, and *Alcanivorax*. *M. leidyi* also contained *Nisaea*, a bacterial genus known to mediate denitrification and nitrite reduction. In both ctenophores, *Mycoplasma* and *Spiroplasma* sp., representatives of the phylum *Tenericutes* were present in low abundance, but were not detected in the surrounding water. Generally, many sequences associated with *M. leidyi* were closely related to bacteria previously described in various marine invertebrates, including corals, sponges, sea anemones, and bivalves, but also some *Betaproteobacteria*, previously reported from freshwater.

Dinasquet et al. [26] investigated the response of bacterioplankton to the presence of *Mnemiopsis leidyi* and analyzed the microbiota of its mesoglea and gastral space. As this analysis supplemented some core experiments and hence was not the main aim of the paper, only one sample of each compartment was analyzed. The resolution of the microbial community composition was high; however, as sequencing of 16S rRNA bacterial genes was performed with next generation sequencing technology 454 pyrosequencing. The gut community was different from the mesoglea community and dominated by *Bacteroidetes* (*Flavobacteriaceae*, with the most dominant bacteria affiliated with *Tenacibaculum*) and *Alphaproteobacteria* (*Rhodobacteraceae*), but also contained *Cyanobacteria* and *Actinobacteria*. Since the examined individuals were starved prior to the analysis, these bacteria could be part of the gut microbiome and were most likely not ingested with prey.

Hao et al. [45] investigated the microbiomes of several species of *Ctenophora*: *M. leidyi*, *Beroe* sp., *B. infundibulum*, and *Pleurobrachia pileus* collected from Helgoland Roads in the German Bight. Ctenophores were collected three times per week for two consecutive years, 10 individuals of each species each time (in total 496 specimens were analyzed), accounting for seasonal and interannual variability [45]. Total prokaryotic genomic DNA was extracted and analyzed using ARISA and 16S rRNA amplicon pyrosequencing. The four ctenophore species harbored species-specific bacterial communities. The largest difference in microbiome composition was found between *M. leidyi*, a highly invasive species in the region, and the other three species. The seasonal variability in microbiome composition was only analyzed for *M. leidyi*, as it was the only species present throughout all sampling seasons. The *M. leidyi* microbiomes collected in the summer clustered together, clearly separated from the microbiome composition associated with *M. leidyi* collected in autumn and winter. Altogether four major bacterial phyla were detected, *Proteobacteria*, *Tenericutes*, *Actinobacteria*, and *Firmicutes*, with *Proteobacteria* being the dominant phylum in all four ctenophore species. While *M. leidyi* and *P. pileus* were dominated by *Gammaproteobacteria*, *Beroe* sp. was dominated by *Alphaproteobacteria*, and *B. infundibulum* harbored roughly equal numbers of *Alpha-* and *Gammaproteobacteria*. The major groups of *Alphaproteobacteria* in *Beroe* sp. were affiliated with *Rhodospirillaceae* (*Thalassospira*). The microbiome of *B. infundibulum* was dominated by *Oceanospirillaceae* (*Marinomonas*), while in *P. pileus*, *Pseudoalteromonadaceae* and *Moraxellaceae* were most abundant. *Beroe* and *M. leidyi* were also associated with a small percentage of *Vibrionaceae*.

2.2. Critical Overview of Methodological Approaches Used to Study Jellyfish Microbiome

To depict the characteristics of the microbiome systems of jellyfish, the available datasets and the conclusions of different studies need to be compared. The methodologies used were different, making it occasionally difficult to directly compare the outcomes of these studies. Of particular relevance is that the taxonomic composition of the microbial communities was determined by applying different methods, but also different sampling approaches were used that might affect the relevance of the resulting data. The purpose here is not to criticize individual methodological approaches applied by different studies, since many of them used state-of-the-art techniques at the time they were performed. Rather, we want to point out that direct comparisons frequently cannot be made and general conclusions cannot be drawn when only limited comparable datasets are available.

Considering microscopy-based methods of enumerating microbes, the methodological spectrum spans from light and electron microscopy to epifluorescence and confocal laser scanning microscopy. Early studies applied light and electron microscopy to detect jellyfish-associated microbes and describe their morphology and location within the jellyfish host [31,33,38]. Later, epifluorescence and confocal laser scanning microscopy combined with specific oligonucleotide probes labeled with specific fluorochromes were applied, allowing localization of specific microbial taxa at varying phylogenetic resolutions ranging from kingdom- to strain-specific within specific jellyfish life stages and body compartments [35,43].

The range of techniques available to determine the phylogenetic composition of jellyfish microbiomes spans from culturing to culture-independent molecular biology-based techniques. Classical culturing, mostly applied in early studies, provides insight into only the cultivatable part of the microbial community, which presumably currently represents only 1–5% of the total community [66]. The first culture-independent techniques applied were DNA fingerprinting techniques, such as ARISA, DGGE, and T-RFLP (Table 1). These techniques allow recording of differences in the microbial community structure with varying resolution at low cost, but cover only those populations with a DNA content of 0.1–1% of the total community DNA. These fingerprinting techniques offer only limited possibility to identify the phylogeny of specific members of the microbial community. For example, the DGGE method has high resolution, in terms of detecting 1 bp differences among individual sequences and allows many samples to be analyzed and compared concurrently. However, to obtain information on specific phylogenetic groups of the bacterial community, one has to excise individual DGGE bands—each representing, in theory, different bacterial species/groups and clone and sequence them. In this process, especially the band-excision step is critical, as many bands are not separated well enough, which means that it is not possible to excise some and cross-contamination can occur during the procedure. Clone libraries using the 16S rRNA gene were also widely used as a culture-independent technique to study the phylogenetic composition of individual microbial communities. This approach allowed for in-depth phylogenetic analysis, as the obtained sequences were usually of good quality and long, allowing for classification even to the genus level. This approach, however, does not sufficiently describe the diversity of a given population, as it fails to detect the rare community members. Despite all these limitations and shortcomings, we believe that the early sequencing efforts provided valuable insights into dominant groups of bacteria associated with jellyfish, contributing important information to our understanding of jellyfish microbiomes. Due to these shortcomings and the labor-intensive nature of these methods, next-generation sequencing (NGS) techniques have been applied over the last decade to study the composition of jellyfish microbiomes with high resolution and efficiency. Next generation sequencing approaches have been rapidly developing since they were first introduced to marine microbial ecology by Sogin et al. [67], describing the rare biosphere using 454 pyrosequencing. Currently, the most frequently used NGS platform in microbial ecology is the Illumina MiSeq or HiSeq. Although the resulting data in different studies can be compared most easily, almost all publications on jellyfish microbiome (Table 1) have used different prokaryotic DNA extraction protocols, and different kits to construct DNA libraries, and frequently, different regions of the 16S rRNA gene were amplified with different sets of primers. All of these limits the comparability of the available sequencing datasets.

Nearly all of the studies conducted so far performed only bacterial 16S rRNA gene amplicon sequencing to obtain information on the phylogenetic diversity of the jellyfish microbiome. Therefore, hypotheses on the function and role of the microbiome associated with jellyfish and the relationships and interactions within host–microbiome associations are merely speculations based on our knowledge of specific microorganisms. The one exception to this is the study on the gastric cavity of *Cotylorhiza tuberculata*, in which a metagenome approach was used providing insight into the genetic potential of the microbiome [43]. To our knowledge no study on the metatranscriptome, metaproteome, or metabolome level of the jellyfish microbiome has been performed so far.

Studies have analyzed the jellyfish microbiome at different levels, investigating the degree of microbiome specialization, species specificity of the microbiome, jellyfish life stage and body compartment specificity, and jellyfish population specificity (Table 1). However, to our knowledge, a holistic approach covering all possible aspects in a single study has never been used. For example, the degree of specialization was addressed only occasionally by comparing the degree of similarity between the jellyfish microbiome and ambient water microbial communities [26,35,36,41,44]. Concurrent analyses of the jellyfish microbiome and the host's natural background, i.e., the pool of ambient water microbial community, would also provide insights into possible transmission mechanisms of the jellyfish microbiome (i.e., horizontal vs. vertical transmission).

Another important point to consider is the fact that most of the studies examined different jellyfish species collected from diverse marine systems, where jellyfish are continuously present or occur seasonally. A comparison of microbiomes associated with the same jellyfish species collected from different marine systems would enable us to understand the population specificity of the associated microbiome. However, this comparison is currently only possible for a few jellyfish families, i.e., Cyaneidae, Pelagiidae, Phialellidae, and Bolinopsidae, and in one single case at the genus level, i.e., for *Aurelia aurita* (Table 1). In addition, there are several studies on different jellyfish species from the same system, which allow us to speculate on the effect of the microbial community of the ambient water on the jellyfish microbiome [34,36,38,39,41,44,45]. The spatial patchiness and variability within specific systems have rarely been addressed [40,43], as mostly jellyfish collected from a single sampling location were analyzed. Studies have rarely accounted for the variability within jellyfish populations, as most are based on only a small number of specimens. Furthermore, only few studies have accounted for the interannual [40,43] and seasonal [37,44,45] variability of jellyfish–microbe associations. With a single exception [42,43], no studies of specific jellyfish species from the same marine system are available where different groups of researchers have studied them. Consequently, no critical assessment of the reproducibility of the results can be made. Most of the studies were performed on jellyfish collected from their natural habitat, usually during bloom conditions, but some studies were conducted using jellyfish grown in the laboratory [35,40].

Jellyfish were collected and analyzed at different stages of their life cycle, but this comprehensive approach was applied only rarely [35,39,40]. To our knowledge, there is just a single study that investigated the microbial community associated with each life stage of a specific jellyfish (*Aurelia aurita*, [35]). However, even in this case, not all body compartments were analyzed, and in the future, the methodological approach should be improved to avoid possible cross-contamination.

3. Characteristics of the Jellyfish-Associated Microbiome

The screening and critical overview of all published studies to date on the jellyfish-associated microbiome revealed some potential characteristics and patterns of jellyfish-associated microbial communities.

3.1. What is the Degree of Specialization of the Jellyfish-Associated Microbiome?

Only a few attempts have been made to investigate whether there is a jellyfish-specific microbiome. Are these microorganism generalists and thus also present in the water column or on organic detrital particles? Are they generalist symbionts (i.e., can be found in association with other similar organisms) or specialists (i.e., are found only in association with jellyfish)?

The approach applied by these studies was to compare the diversity of the jellyfish-associated microbiome with the bacterial community in the ambient water. The conclusion from these studies is that the jellyfish-associated bacterial community is significantly different in terms of composition and lower in diversity than the bacterial community in the ambient water. This was reported in all the studies on *A. aurita*'s microbiome from the North Atlantic coastal waters [36], Kiel Bight, Baltic Sea, English Channel, North Sea [35], and Northern Adriatic Sea [37]. In addition, this was also reported for the ctenophore *M. leidyi*, although only in one study so far [44].

The jellyfish-associated microbiome is frequently also found associated with other host organisms, such as corals, sponges, sea squirts, and sea horses, which sometimes also share similar lifestyle, morphological, and/or biochemical characteristics [37,40,41]. This indicates a certain degree of specialization of the jellyfish-associated microbiome, suggesting that the jellyfish microbiota could be symbiotic generalists. Some studies concluded that at least certain members of the jellyfish-associated microbiome are specific for certain jellyfish taxa. For example, a comparison of the microbiomes of *Chrysaora* and *Aurelia* revealed a certain number of shared bacterial species, suggesting that they are specific for scyphozoans in general [40]. However, one should be careful when drawing general conclusions from a limited number of investigations performed with different methods. There is evidence, however, that some members of jellyfish-associated microbiomes are rarely present in ambient water microbial assemblages and are more frequently found in association with similar organisms, probably due to their ability to attach and thrive under the specific environmental conditions (e.g., high viscosity, high nutrient, and low oxygen concentrations, presence of toxins and other antimicrobial compounds) that these organisms have to offer.

The observed low-diversity microbiome of jellyfish is in contrast to the trend observed for corals and sponges, where the diversity of associated microbiomes is usually higher than that in the ambient water [11,68–70]. One possible explanation is the production of antimicrobial compounds by jellyfish or their associated microbiomes. As discussed by Cleary et al. [41], *A. aurita* produces an antimicrobial peptide, named aurelin, that is active against Gram-positive and -negative bacteria [71]. Likewise, extracts of *Cassiopeia* spp. showed strong antimicrobial activity against bacteria [72]. Antibacterial polyketides were also isolated from a fungal symbiont (*Paecilomyces variotii*) of the jellyfish *Nemopilema nomurai* [73]. Thus, survival in this specific environment might be possible for only some bacterial taxa, leading to reduced diversity within the jellyfish-associated microbiome. In conclusion, the degree of specialization of the jellyfish-associated microbiome needs to be further studied, expanding to different taxonomic groups of jellyfish, marine systems, and biogeographic provinces.

3.2. Is the Jellyfish-Associated Microbiome Jellyfish Population-Specific?

The fact that studied jellyfish were collected in different marine systems ranging from the North Atlantic Ocean to marine lakes in Indonesia (Table 1) might help to determine whether the microbiome is specific to certain jellyfish populations.

Jellyfish are known to be able to inhabit very different marine systems around the world, from shallow coastal seas to deep-sea environments, and, even more, seem to be able to quickly acclimatize to changing physical conditions (e.g., temperature, salinity [74,75]) and adapt to emerging features of marine habitats such as lack of predators/competitors caused by overfishing [76] or use of increasing marine sprawl, such as platforms as a surface for their polyp phase [77]. This could also be reflected in their associated microbiome. From the research conducted so far, the question of whether there is population specificity of the microbiome can only be addressed for *A. aurita*, as it was analyzed from different marine systems (North Atlantic [36], Kiel Bight, Baltic Sea, Southern English Channel, North Sea [35], and Northern Adriatic Sea [37]), and for *M. leidyi* from its native environment, Tampa Bay (Florida, USA) [44], and to areas where it was recently introduced, such as Gullmar fjord (Sweden) [26] and Helgoland Roads (German Bight) [45]. Interestingly, the results for the two jellyfish phyla contrast. Whereas in scyphozoan *A. aurita* the microbiome seems to be population-specific and hence varies with sample location, the ctenophore *M. leidyi* harbored a similar microbiota regardless of the sample location (e.g., *Marinomonas* was detected in ctenophore from all studied systems).

If we assume that the jellyfish-associated microbiome is determined to some extent by the environment, many more jellyfish ecosystems should be explored to better understand the selective pressure. In particular, one would expect differences in the microbiome of open-ocean (or even deep-sea) vs. coastal jellyfish, which are subjected to different degrees of anthropogenic impact. In particular, coastal environments are common entry points of pathogenic bacteria into marine ecosystems via coastal runoff, wastewater treatment plant discharge, or other kinds of human activities. Furthermore, as pathogenic bacteria are known to prefer organic-rich environments and are capable of surface attachment, they could easily hitchhike with bypassing jellyfish or, even more likely, attach to and thrive on polyps commonly found on biofouling at pillars in ports, on platforms, or on other structures. As known filter feeders, some of the organisms that dominate in biofouling communities, such as mussels and oysters, can accumulate pathogens and microbes that produce toxic compounds, which could be transmitted to polyps and from there to adult medusae, which can drift large distances. In this way, jellyfish can be seen as vectors for pathogens and other allochthonous bacteria from coastal to open water. This could be important in light of the recently modelled dynamics of jellyfish populations showing that a population of jellyfish can drift far away from its source polyp population area [77]. Another important unexplored aspect of the transmission route of the jellyfish microbiome is ballast water, potentially introducing invasive/nonnative species of both jellyfish and microbiota into new marine environments.

3.3. Is there a Jellyfish Taxa-Specific Microbiome?

Several studies have compared the composition of microbial communities associated with different jellyfish taxa collected from the same environment (Table 1). In a study on an Indonesian marine lake, representatives from different medusozoan classes were collected, i.e., a jellyfish from the Mastigiidae family, representative of the Scyphozoa class, and a jellyfish from the Tripedaliidae family, representative of the Cubozoa class [41]. This study indicated that the microbiome is jellyfish taxa-specific [41]. Another study performed on scyphozoan jellyfish of the order Semaeostomeae, *Cyanea lamarckii*, a representative of Cyaneidae family, and *Chrysaora hysoscella*, a representative of Pelagiidae family, both collected in the German Bight, showed that there was a clear difference between the microbiome associated with the life stages of the two species, indicating specificity of the microbiome for certain jellyfish taxa [39]. In a study on two species of Cyaneidae, *Cyanea capillata* and *Cyanea lamarckii*, collected in Scottish coastal waters, their microbiomes were found to be species-specific [38].

3.4. Is the Jellyfish-Associated Microbiome Specific to Different Life Stages?

Selecting jellyfish as a host seems to be an advantage or survival strategy for at least some bacteria. However, due to dramatic changes of lifestyle in the life cycle of jellyfish (benthic/attached vs. pelagic/swimming), morphology (i.e., surface architecture), and (bio)chemical characteristics (i.e., different expression patterns of antimicrobial compounds during different life stages), its associated microbiome needs both resiliency and plasticity. Each life stage might represent a unique niche allowing for specific bacteria to grow, and consequently, life stage specificity of the microbiome might be possible. From a jellyfish perspective, different life stages are associated with different necessities and requirements, which could drive corresponding shifts in the structure of associated microbiomes, as hypothesized by Lee et al. [40].

Microbial communities associated with different life stages of specific jellyfish species have been only rarely investigated [35,39,40]. Comparing the results of these studies, it seems that the composition of the jellyfish-associated microbiome changes with the life stage, particularly, during the transition from the benthic to the pelagic stages. The observed shifts in the community composition during the different developmental stages raise questions on the functional role of the microbiome in general and the development of jellyfish, as discussed by Weiland-Brauer et al. [35]. Certain members of the microbiome might have specific metabolic functions, which might play a role during specific development stages of the jellyfish [35]. However, available data on the diversity of the microbiome associated with different jellyfish life stages provide information only on the genetic potential or potential metabolism of specific bacterial groups. Hence, one can only speculate on the functions of specific bacteria in different life stages of jellyfish.

Our current knowledge of the bacterial colonization of scyphozoans or other jellyfish larvae, the establishment of the microbiome, and the compositional stability at different life stages, especially during strobilation, is limited. In jellyfish, however, bacteria have been shown to be important for larval settlement: it was shown that the settlement of pedal stolons of scyphopolyps of *A. aurita* was induced by *Micrococcaceae*, presumably via its effective substances acylgalactosidyldiglyceride and monogalactosidyldiglyceride [78,79], while in *Cassiopea andromeda*, swimming buds and planulae were induced to settle and metamorphose by a compound released from a *Vibrio* species during growth [80–83]. Regulation of the jellyfish life cycle, in particular the transition from one life stage to another and the induction of metamorphosis, is still only poorly understood. Recently, three major studies shed light on the regulatory mechanisms of the life cycle of jellyfish. One study focused on transcriptome profiling of the life stages of *A. aurita* [84], another study focused specifically on molecules critical in controlling the polyp-to-jellyfish transition [85], and the third study determined the genome of *A. aurita* [86], representing the first fully sequenced genome of the medusa stage of a cnidarian and providing important insight into the evolution of animal complexity. These analyses of life stage transcriptomes showed that different transcript expression profiles are related to each specific life stage and that not only shifts in ambient temperature, but also other signaling factors could initiate and regulate the strobilation and metamorphosis processes. It was shown that the nuclear hormone receptors, including the retinoic acid signaling cascade, are core elements of the regulation machinery of the life cycle of *A. aurita*, identifying strobilation inducer and precursor, novel CL390 protein, and its minimal pharmacophore, 5-methoxy-2-methylindole [85]. We argue that additional strobilation regulators might be present, and it is tempting to speculate on the role that ambient or associated microbes play in this process, especially as this was not tested in any of the studies conducted so far and the exclusive production of these compounds by the host has not been shown. One might speculate that similar or antagonistic molecules are produced by the jellyfish-associated microbiome in particular, since it was shown that bacteria by themselves or via their extracellular vesicles induce metamorphosis in marine invertebrates ([87] and references therein). The identification of strobilation and proliferation agents also has direct biotechnological application, as it might represent an avenue to control jellyfish blooms [85].

In addition, it was speculated that for the benthic life stage, the associated microbiome is particularly important for producing vitamins, amino acids, and secondary metabolites and possibly for inducing subsequent developmental stages [35]. Among bacteria associated with *A. aurita* polyps, *Phaeobacter* is known as a producer of compounds that inhibit fouling of surfaces of the host and substrates in general by interfering with cues for the settlement of invertebrate larvae and spores of algae [88–90]. Also, the polyp-associated bacteria of the genus *Rhodococcus*, the only exclusively marine genus within *Actinobacteria*, produce extracellular enzymes and are an exceptionally rich source of secondary metabolites, particularly compounds with the potential to serve as novel antibiotics and anticancer drugs (reviewed in Reference [91]). Polyp-associated bacteria of the genus *Vibrio* are known to produce quorum-sensing signals, and their antagonistic behavior was proposed to stimulate the settlement of spores of other organisms [88,92]. In a study of *Chrysaora plocamia*, polyps were proposed to form the base for a plethora of asexual reproduction strategies, and it is likely that they represent a reservoir of microbial members essential for the initiation, development, and survival of the subsequent life forms [40], while the authors hypothesized that the microbiota associated with *C. plocamia* podocyst are conserved within the podocyst capsule, where they help to maintain the viability of podocysts and potentially facilitate the activation of excystment [40].

Taking all this information together, it seems that a major restructuring of the polyp-associated microbiome takes place at the transition from the benthic to the pelagic stage, also in terms of reduced diversity. Different pelagic stages seem to be more similar to each other than to the benthic stage. The observed changes could be a response to different lifestyles, metabolisms, morphologies and biochemical characteristics of jellyfish, but could also be due to different environmental conditions that the microbiome experiences in the benthic vs. pelagic stage of the host.

3.5. Is the Jellyfish-Associated Microbiome Body Part-Specific?

In jellyfish, different types of cells are present in different body compartments [93]. Body compartments are characterized by different morphological and biochemical features, as well as by their degree of contact with the ambient water. Thus, different body compartments might be preferentially colonized by specific microbes.

Few studies have investigated the compartment specificity of the jellyfish microbiome. In *A. aurita*, the differences between the microbiomes of the mucus and the gastric cavity were investigated [35]; in *Cyanea lamarckii* and *Chrysaora hysoscella*, the composition of the microbiomes of the tentacles, umbrella, mouth arm, and gonads was studied [39]; and in ctenophore *M. leidyi*, differences in the microbiome composition between the gut and tissue were investigated [26]. In other studies, jellyfish were analyzed as whole individuals, even when they were dissected prior to analysis (Table 1).

3.5.1. Microbiome Associated with Outer Body Parts and Its Potential Role

For all investigated species, compartment specificity of the jellyfish-associated microbiome has been reported, with outer body parts, e.g., umbrella and mucus, usually harboring more diverse, rich, and variable microbial communities than the inner body compartments, e.g., gonads and gastric cavity [26,35,37,39]. The degree of similarity between the microbiomes associated with outer body compartments and ambient water bacterial assemblages was higher than that between the microbiomes of inner body compartments. The jellyfish's outer body compartments are in direct contact with microbial communities of the ambient water, thus the microbiome of the outer compartments is probably directly recruited from those communities, indicating the possibility of horizontal transfer of the jellyfish's ecto-microbiome.

The epidermis and gastrodermis of the jellyfish contain numerous types of unicellular mucus-producing gland cells, leading to the formation of a thin, constantly renewed mucus layer covering the entire external surface of the medusa [94,95]. Under conditions such as stress and moribundity, but also during reproduction and digestion, mucus release rates are higher than under nonstress conditions [95]. Mucus contains toxins and nematocysts, thus serves as an important chemical defense mechanism of jellyfish and plays a major role in surface cleansing [95,96]. Jellyfish also produce toxins and antimicrobial compounds, such as the peptide aurelin in the mesoglea of *A. aurita* [71]. The mesoglea is an extracellular matrix situated between the epidermal and gastrodermal layers [97] containing collagen and collagen-like proteins associated with mucopolysaccharides [98,99]. Secreted mucus and the mesoglea are mainly composed of proteins, lipids, and, to a lesser extent, carbohydrates in different ratios [100,101], representing an attractive niche for bacteria, especially those with a competitive advantage and specialized for settling from the ambient water. The specific physiochemical characteristics of mucus and the physiology of the host represent selective pressure and determine the abundance and diversity of metabolically active bacteria [102]. Hence, one can hypothesize that jellyfish can actively or passively select their bacterial associates. Whether bacteria directly adhere to external cell layers of jellyfish or are only associated with the thin mucus layer remains to be resolved.

Based on the two studies that actually focused on the composition of the microbiome associated with jellyfish mucus, *Gammaproteobacteria*, particularly *Pseudoalteromonas* and *Vibrio*, are abundant, but to some extent also *Alphaproteobacteria* (*Phaeobacter*, *Rugeria*, and *Roseovarius*) [35,37]. These bacteria were previously recognized as important players in the host defense against pathogens and fouling organisms from the surrounding seawater [88,92,103,104] because of their ability to produce antimicrobial compounds when attached to live or inert surfaces [90,104–107]. Considering that the host would recruit microbes that are beneficial for its development or contribute to its well-being [92], one might speculate that the mucus-associated microbiome serves as a defense mechanism to protect the jellyfish from hostile microbes and other organisms in the ambient water. It was even proposed that the antimicrobial compounds and even toxins found in jellyfish mucus could originate from the associated microbiome (by Schuett and Doepke [38], who found tetrodotoxin-producing bacteria on jellyfish tentacles). At the same time, the mucus-inhabiting microbiome could benefit from constant nutrient input and/or other compounds from the host.

3.5.2. Microbiome Associated with Inner Body Compartments and Its Potential Role

The inner body parts of jellyfish, such as the gonads and gastric cavity, are isolated from the surrounding environment and have different morphological and biochemical characteristics, allowing for development of very specific bacterial groups that are otherwise rarely found in the ambient water. This is in line with reports on lower diversity, richness, and variability of the microbiome of the inner body compartments of jellyfish [26,35,37,39]. Reduced microbial diversity has been found in the gastric cavity of *Cotylorhiza tuberculata* [42,43]. Since a considerable fraction of the bacterial community present in the inner compartments of jellyfish is rarely or never found in the ambient water, these bacteria may have lost the ability to live independently and may be acquired by the host via vertical transmission as parental heritage through reproductive cells and larvae [43]. Based on metagenomic analysis, the role of these bacteria is possibly related to food digestion and protection from pathogens [43].

3.6. The Composition, Potential Role, and Biotechnological Potential of the Jellyfish-Associated Microbiome

Overall, it is not surprising that the jellyfish-associated microbiome is a consortium of bacteria (see below) that prefer a particle-attached lifestyle, are capable of degrading complex organic compounds, and are known to be found in association with other marine organisms. These bacteria are known for commensalism, symbiosis, and parasitism, or are even pathogens of marine organisms. Frequently, they are capable of producing quorum-sensing signal molecules, antagonistic compounds, and/or factors that interfere with quorum sensing of other microbes (which involves sensing the abundance of other bacteria, expressing of virulence factors, and interfering with the chemical communication of other bacteria, exhibiting antagonistic behavior). All of these features have either a direct or indirect application in blue biotechnology (Table 2). However, so far there has been no comprehensive study investigating the biotechnological potential of bacteria associated with jellyfish. Therefore, the listed features (Table 2) are based on the literature on the biotechnological potential of these bacterial strains or their closest relatives isolated from other marine organisms or substrates. Bacteria associated with jellyfish are also known for their diverse metabolisms and involvement in the cycling of carbon, nitrogen, sulfur, and phosphorus, which prompted several authors to speculate on their role in supplying specific compounds to jellyfish, implying that there is a symbiotic relationship between jellyfish and the associated microbiome (discussed in detail below). For example, it was proposed that nitrifying bacteria could play an important role in the life cycle of jellyfish [40], as they are known to harbor key enzymes involved in the conversion of ammonia to hydroxylamine and further to nitric oxide, the latter known as an important messenger molecule to regulate metamorphosis in marine invertebrates [108], regulate swimming of the jellyfish *Aglantha digitale* [109], and facilitate the discharge of nematocytes in the sea anemone *Aiptasia diaphana* [110]. All of these features have the potential to be exploited by blue biotechnology (Table 2). Several jellyfish-associated bacteria were previously also associated with the processing of more peculiar substances, such as PAHs, plastics, and xenobiotics found in the ocean, with possible benefits for the host [37,40]. For example, the presence of PAH-degrading bacteria agrees with the high tolerance of *A. aurita* to crude oil exposure and its ability to accumulate PAHs [111], suggesting that the PAH-degrading microbial community associated with *Aurelia* facilitates the survival of jellyfish in polluted coastal systems. This also agrees with the findings of Kos Kramar et al. [37], who detected PAH and plastic-degrading bacteria within the gastric cavity of *A. aurita* collected in the Northern Adriatic. The potential of these microbes and/or the compounds they produce to be used for biotechnological applications is obvious (Table 2). Marine archaea are known for their biotechnological potential (reviewed in References [112,113]); however, only one study so far made an attempt to investigate the archaeal community potentially associated with jellyfish [42]. However, amplification of archaeal 16S rRNA failed [42].

The most frequently detected members of the jellyfish microbiome are affiliated with the following representatives: *Alpha-* and *Gammaproteobacteria*, *Bacteroidetes*, *Tenericutes*, and *Cyanobacteria*. Among the less frequently, but still repeatedly reported were bacteria affiliated with *Betaproteobacteria*, *Spirochaetes*, *Actinobacteria*, *Firmicutes*, *Chlamyidiae*, *Chloroflexi*, *Planctomycetes*, *Nitrospirae*, and *Nitrospinae*. As we have shown that the associated microbiome is probably not exclusively associated with jellyfish and is to some extent affected by the physiochemical boundary conditions, the term "transient microorganisms" would probably best describe the jellyfish-associated microbiome [114].

Table 2. Overview of dominant bacteria found to be associated with jellyfish and their attributed features with biotechnological potential.

Bacteria Associated with Jellyfish		Features with Biotechnological Potential	Jellyfish
Class	Representative Families		
Gammaproteobacteria	*Vibrionaceae* *Pseudoalteromonadaceae* *Alteromonadaceae* *Oceanospirillaceae* *Shewanellaceae* *Crenotrichaceae* *Methylococcalaceae* *Endozoicimonadaceae* *Moraxellaceae* *Legionelaceae*	- readily culturable - extremophiles - quorum-sensing factors/signals - degradation of high-molecular-weight compounds - antagonistic behavior - symbionts and/or pathogens - possibly stimulate the settlement of spores of other organisms (*Vibrio*) - producers of *ToxR* gene that mediate virulence expression (*Photobacterium profundum*) - producers of highly bioactive compounds (extracellular enzymes, exopolysaccharides, compounds involved in antimicrobial antifouling, with algicidal activity and various pharmaceutically relevant activities, e.g., *Pseudoa termonas*) - multifunctional polyphenol oxidases, involved in secondary metabolism, biodegradation processes, breakdown of dimethylsulfoniopropionate (e.g., *Marinomonas*) - producers of antibacterial compounds - tetrodotoxin producers (*Pseudoalteromonas tetraodonis*) - septicemic, necrotic hemolytic, and cytolytic activity - produce polyunsaturated fatty acids (*Shewanella*)	- widespread among all studied jellyfish taxa; outer jellyfish body parts
Alphaproteobacteria	*Rhodospirillaceae* *Rhodobacteriaceae* *Kiloniellaceae*	- degradation of aromatic hydrocarbons; production of biosurfactants; utilization of hydrocarbons, carbohydrates, organic acids, or amino acids; and degradation of polycyclic aromatic hydrocarbons (e.g., *Thallasospira*) - readily culturable - found in association with living organisms and/or organic matter particles - found in sediment and microbial mats - inhibitory compound producers inhibit fouling of surfaces of host/substrate by interfering with cues for settlement of invertebrate larvae or spores of algae (e.g., *Phaeobacter*) - active antibiotic producers	- widespread among all studied jellyfish taxa
Flavobacteria	*Flavobacteriaceae*	- readily culturable - production/expression of various extracellular hydrolytic enzymes - degradation of complex organic materials - some pathogen representatives - some psychrophilic representatives - found in sewage-polluted waters, human guts (e.g., indicators of water quality) - carbohydrate and protein catalysis (e.g., *T. maritimum*)	- *A. aurita*, *P. noctiluca* and *C. plocamia*, *C. tuberculata gastric cavity* - *M. Leidyi* and *B. ovata* - Hydrozoa
Mollicutes	*Spiroplasmataceae* *Mycoplasmataceae*	- parasitic and commensal life-style - pathogens of humans, mammals, reptiles, fish, plants, and other arthropods and have been reported in different invertebrates, also cnidarians, such as corals - anaerobic fermenter metabolism	- *A. aurita* - *C.tuberculata gastric cavity* - Mastiigiidae - Cubozoa - ctenophores

Table 2. *Cont.*

Bacteria Associated with Jellyfish		Features with Biotechnological Potential	Jellyfish
Class	Representative Families		
Spirochaetes		- gut microbiota of marine animals - pathogens	- Mastigiidae - Cubozoa
Actinobacteria		- repertoire of enzymes to break down polysaccharides, proteins, and fats - rich source of secondary metabolites, antimicrobial and anticancer drugs	- A. aurita - M. leidyi
Firmicutes		- pathogens - bind and oxidize metals; bioremediation - powerful chitinolytic activity - source of toxins	- Cubozoa - ctenophores
Chlamydiae	*Simkania*-like bacteria	- pathogenic capability	- C. tuberculata's gastric cavity
Nitrospirae and *Nitrospinae*		- potentially involved in regulating metamorphosis in marine invertebrates - potentially involved in regulating swimming of jellyfish - potential to facilitate discharge of nematocytes in *Aiptasia diaphana* - formation of bullet-shaped magnetite magnetosomes	- C. plocamica
Betaproteobacteria	*Burkholderia* *Achromobacter* *Cupriavidus*	- associated with polycyclic aromatic hydrocarbon degradation - associated with plastic degradation - associated with xenobiotic degradation - bioremediation and biopesticidal properties, - ability to synthesize wide range of antimicrobial compounds	- A. aurita - C. plocamica - M. leiydi
Cyanobacteria		- can survive extreme temperatures and salinities - production of secondary metabolites including exopolysaccharides, vitamins, toxins, enzymes, and pharmaceuticals - used in aquaculture, wastewater treatment, food, fertilizers - source of biologically active compounds with antiviral, antibacterial, antifungal, and anticancer activity - potential biofuel producers - removal of heavy metals from water, degrading oil components	- C. tuberculata's gastric cavity - A. aurita - M.leidyi

3.6.1. Gamma- and Alphaproteobacteria

Within the *Proteobacteria*, *Gammaproteobacteria* are reported to be associated with every comprehensively studied jellyfish taxon so far, where they either dominated or at least represented a very substantial part of the jellyfish-associated microbiome. The exception to this seems to be the jellyfish gut microbiome, where either only a small fraction of *Gammaproteobacteria* was detected [42,43] or none at all [26]. Within *Gammaproteobacteria*, different families were regularly reported as a part of jellyfish-associated microbiome: *Vibrionaceae*, *Pseudoalteromonadaceae*, *Alteromonadaceae*, *Oceanospirillaceae*, *Shewanellaceae*, *Crenotrichaceae*, *Methylococcalaceae*, *Endozoicimonadaceae*, *Moraxellaceae*, *Xanthomonadaceae*, and *Legionelaceae*. In general, *Gammaproteobacteria* in marine environments are associated with the ability to attach to surfaces and to degrade high-molecular-weight compounds. Within this class, there are many readily culturable bacteria, such as *Vibrio*, *Alteromonas*, *Pseudoalteromonas*, *Marinomonas*, *Shewanella*, and *Oceanospirillum* [115]. This suggests that most of the jellyfish-associated *Gammaproteobacteria* could be readily cultured, facilitating their possible exploitation for biotechnological applications (Table 2). Furthermore, these bacteria usually exhibit rapid growth and a feast-or-famine lifestyle with quorum-sensing playing an important role. In addition, many of these bacteria are found in biofilms and/or in association with other marine organisms (as symbionts or pathogens) and occupy micro-niches with specific environmental conditions (i.e., temperature, oxygen, nutrient availability). All of the listed features have direct and/or indirect potential for application in biotechnology (Table 2). For example, *Marinomonas* was detected as an abundant member of the microbiome of the ctenophore *M. leidyi* in several marine systems. In terms of their biotechnological potential, these bacteria contain multifunctional polyphenol oxidases that are able to oxidize a wide range of substrates, are producers of antibacterial compounds, and are involved in biodegradation processes [116,117].

Marinomonas also contain genes for the breakdown of dimethylsulfoniopropionate, indicative of their role in the cycling of sulfur [44]. *Pseudoalteromonas* is known to produce a variety of highly bioactive compounds, including extracellular enzymes, exopolysaccharides, and compounds involved in antimicrobial antifouling, with algicidal activity and various pharmaceutically relevant activities [118]. *Vibrio* is known as a producer of quorum-sensing signals and for its antagonistic behavior, but also for its proposed role in stimulating the settlement of spores of other organisms [119].

Within the jellyfish-associated *Alphaproteobacteria*, representatives of the *Rhodospirillaceae*, *Rhodobacteriaceae*, and *Kiloniellaceae* families and the order *Rhizobiales* are documented (Table 2). Among them, members of *Rhodobacteriaceae* can easily be cultivated and are commonly found in association with living organisms and detrital particles, in sediment and microbial mats, playing an important role in carbon and sulfur cycles. The *Phaeobacter* genus of the *Rhodobacteriaceae* family is known as a producer of inhibitory compounds that prevent or inhibit fouling of surfaces by interfering with cues for the settlement of invertebrate larvae or spores of algae [88–90], a feature with potential for biotechnological application. One of the most reported *Rhodospirillaceae* associated with jellyfish are bacteria affiliated with *Thalassospira* [42,44,45]. Bacteria of the *Thalassospira* genus might be involved in carbon cycling by providing an additional source of fixed carbon for jellyfish and exhibit chemotaxis to phosphate (as suggested in Reference [44]). Bacteria from this genus were found in microbial consortia degrading aromatic hydrocarbons [120,121] as part of the microbiome of sabellids (Polychaeta, Annelida) in crude oil enrichments with potential production of biosurfactants [122], features with biotechnological potential. Bacteria of the *Kiloniellaceae* family are also producers of antibiotic compounds [123], with direct application in biotechnology.

3.6.2. Bacteroidetes, Flavobacteria, Flavobacteriaceae

Jellyfish-associated bacteria of the *Bacteroidetes* phylum are affiliated with the *Flavobacteriaceae* family, and were found in association with Medusozoa, within which with Semaeostomeae, in particular with Ulmaridae (*A. aurita*, [35,36]), with Pelagiidae (*P. noctiluca* and *C. plocamia*, [33,40]), with Rhizostomeae (*C. tuberculata*, [42,43]), and with Ctenophora, where they were not always detected in the same species, suggesting that their presence within the ctenophore microbiome might depend on food and/or the environment [26,44,45]. *Flavobacteriaceae* were also detected in all studied Hydrozoa from different systems. Therefore, *Flavobacteria* represent an important part of the jellyfish-associated microbiome, as they are widespread within jellyfish. However, the association seems to be dependent on the host's natural environment to some degree. This bacterial group is known to be easily cultivated and its most distinctive properties are gliding motility and the expression of various extracellular hydrolytic enzymes to degrade complex organic materials, with potential application in biotechnology (Table 2). Some members are pathogenic and some psychrophilic, both characteristics with biotechnological potential. They can be found in the human gut and, in sewage-polluted waters, but also in seawater, where they persist for a long time, thus were proposed as indicators of water quality [124], also with potential biotechnological application. Among the most frequently reported members of the *Flavobacteria* found in association with different jellyfish species are bacteria affiliated with the genus *Tenacibaculum. Tenacibaculum maritimum* is a known fish pathogen that causes tenacibaculosis, a disease considered to be an important threat to aquaculture worldwide [125]. *Tenacibaculum maritimum* was the first specific bacterium associated with jellyfish to be extensively studied, after the first report that it might infect fish gills damaged by jellyfish venom [126]. Consequently, 20 years ago, jellyfish, in particular *Phialella quadrata, Cyanea capillata*, and *Pelagia noctiluca*, were recognized as a possible cause of mass mortality of fish farmed in sea cages [30,127,128]. Transmission of this bacterium via the ambient water and direct transmission from host to host have been proposed as possible routes of infection, in addition to ingestion through food [129]. However, as the survival of *T. maritimum* in seawater is rather limited [130], its natural reservoir remains unclear. Ferguson et al. [32] provided the first evidence of *Phialella quadrata* carrying filamentous bacteria affiliated with *T. maritimum* and proposed that jellyfish is a vector and a carrier of this fish pathogen. These authors suggested that the presence of this proteolytic enzyme-producing bacterium in the jellyfish mouth could support the pre-digestion of the prey of jellyfish and that it could be specific to jellyfish, playing an important role in both immune defense and their nutrition [32]. In a subsequent study, the presence of *T. maritimum* in the mouth of *Pelagia noctiluca* was shown [33]. Supported by the results of both studies, Delannoy et al. [33] proposed that some cnidarians might represent a natural host for *T. maritimum*. However, the environmental reservoir of *T. maritimum* has not been determined yet. Also, its survival in seawater and the suitable niche for its growth in ambient water remain unclear. Quantitative real-time PCR was applied to detect *T. maritimum* in both of the tested species, *Phialella quadrata* and *Muggiaea atlantica* [34]. Bacteria affiliated with the genus *Tenacibaculum* were also found in association with *Cotylorhiza tuberculata* [42,43], with hydromedusa (*N. bachei* [36]), and with the ctenophores *M. leidyi* [26] and *B. ovata* [44]. Based on the large number of genes indicative of carbohydrate and protein metabolism, it has been suggested that *Tenacibaculum*-like bacteria are polymer degraders in jellyfish and their high abundance in the mesogleal axis of the gastric filaments indicate their role in the digestion of ingested food items such as copepods [43]. The possibility for the biotechnological application of this bacterial species as a diagnostic tool and beyond is obvious (Table 2).

3.6.3. Tenericutes

Within the *Tenericutes* phylum, the class *Mollicutes* has frequently been associated with jellyfish. In particularly, in *A. aurita* it has been suggested that *Mollicutes* is a potential endosymbiont [35,36]. In addition, *Mollicutes* was detected in both studied representatives of Rhizostomeae, in the gastric cavity of *Cotylorhiza tuberculata* [34,43] and associated with the Mastiigiidae family. *Mollicutes* was also reported as part of the microbiome of Cubozoa [41] and frequently associated with ctenophores [44,45]. Most of the detected bacteria within the *Mollicutes* class were affiliated with either the *Spiroplasmataceae* or *Mycoplasmataceae* family. The unique characteristics of *Mollicutes* are the lack of a cell wall, small size and simple cell structure, reduced genome, and simplified metabolic pathways, all indicative of a parasitic lifestyle corresponding to their preferred habitat in jellyfish, the gastric cavity, and their suggested endosymbiotic relationship with jellyfish. These bacteria are widespread commensals or pathogens of humans, mammals, reptiles, fish, plants, and arthropods and have been reported also in cnidarians, such as corals [131,132], and are known for their antimicrobial resistance, all characteristic with the potential to be exploited for biotechnology. Their frequent detection as part of the jellyfish microbiome indicates that they are important members of the jellyfish-associated microbial consortium. However, it is important to note that these microorganisms are also one of the most common sources of cross-contamination in academic and biopharmaceutical production laboratories. For the *Spiroplasma*-like bacteria identified as the dominant microbes in the gastric cavity of *C. tuberculata*, the estimated genome was smaller than any other currently known genome of *Spiroplasmas*, which might be indicative of their intracellular lifestyle as predicted anaerobic fermenters. The apparently fit status of the analyzed jellyfish suggests that the intracellular *Spiroplasma*-like bacteria are commensals of *C. tuberculata* rather than pathogens [43]. Features of these bacteria with biotechnological potential are summarized in Table 2; however, the full biotechnological potential of *Mollicutes* associated with jellyfish remains to be explored as a major problem in research with these bacteria is the difficulty of cultivating them in vitro.

3.6.4. Minor Members of the Jellyfish Microbiome

Betaproteobacteria that were detected as a part of jellyfish microbiome affiliated with *Burkholderia*, *Achromobacter*, and *Cupriavidus* [37,40,44] were previously associated with degradation of PAHs, plastics, and xenobiotics in the marine environment [88–90]. Direct application of these features in biotechnology is apparent (Table 2). *Spirochaetes* found in association with Rhizostomeae, in particular with the Mastigiidae family, and with Cubozoa are known for their unique morphology, with tightly coiled spirals, and their motility. They are very widespread in marine habitats, but not much is known about their ecological role or their biotechnological potential. However, they might play an important role in the gut microbiota of marine animals. *Spirochaetes* were reported in two jellyfish species in one stud, in which both species were collected at the same sampling location, an Indonesian marine lake [41]. This raises questions about a location-specific pattern and the methodological approach used by this particular study [41]. In any case, these bacteria include aerobic and anaerobic species, well-known pathogens (syphilis and Lyme disease), and mutualists, for example, inhabiting the guts of cows and termites ([41] and references therein), all features with biotechnological potential. Also, *Actinobacteria* were not common in the jellyfish microbiome; however, they were found in association with *A. aurita* [35,37] and with *M. leidyi* [26,45]. *Actinobacteria* is a bacterial taxon with the greatest biotechnological potential known [133]. *Firmicutes* probably represents a minor part of the jellyfish microbiome as well. However, since they are otherwise rarely recorded as part of natural marine microbial assemblages, their presence within the jellyfish microbiome might indicate that they play an important role in this consortium. They were recorded in association with Cubozoa [41] and with different ctenophores [45]. The biotechnological applicability of these bacteria (e.g., in remediation of polluted marine sediments, as probiotics in aquaculture) and their features (e.g., pathogenicity, resistance to high temperatures, irradiation, desiccation, wide range of fermentation pathways, source of toxins) are known (reviewed in Reference [6]). Within *Chlamydiae*, *Simkania*-like bacteria were found to be the most dominant microorganism in the gastric cavity of *C. tuberculata*. Based on their metagenome, Viver et al. [43] speculated that they are putative endosymbionts of a ciliate that was probably symbiotic with the jellyfish. These authors speculated that the ciliate might be involved in controlling the free-living microbial population within the gastric cavity through grazing, therefore reducing competition between the host and the specific bacterial population for food (i.e., copepods). Furthermore, the authors speculated that the role of the *Simkania*-like endosymbiont of a ciliate, for which metabolic modelling predicts an aerobic heterotrophic lifestyle, could be understood as a nested symbiosis (supported by size and genetic repertoire exhibiting a versatile lifestyle) or even potential pathogenic capability (supported by the repertoire of genes indicative of virulence factors, and their relationship could be mutually beneficial for the host), features with biotechnological potential.

4. Conclusions and Future Research Directions

Different taxonomic groups of jellyfish were studied for their associated microbiomes; but this review also reveals that the vast diversity of jellyfish as hosts remains to be explored (Figure 1). For instance, to our knowledge, not a single member of the Coronatae order was examined for its microbiome, while within Rhizostomeae, an entire suborder of Daktyliophorae remains unexplored and within Semaeostomeae, some members of the Drymonematidae family were never investigated for their microbiome. The WoRMS recognizes 69 genera of Scyphozoa comprising 191 species that populate different habitats, from tropical to polar regions, thus potentially harboring different microbes with unexplored biotechnological potential (Figure 1). Also, the Hydrozoa orders Actinulidae, Limnomedusae, Narcomedusae, and Trachymeduase remain to be investigated, and within Ctenophora several orders of the Tentaculata class remain to be explored.

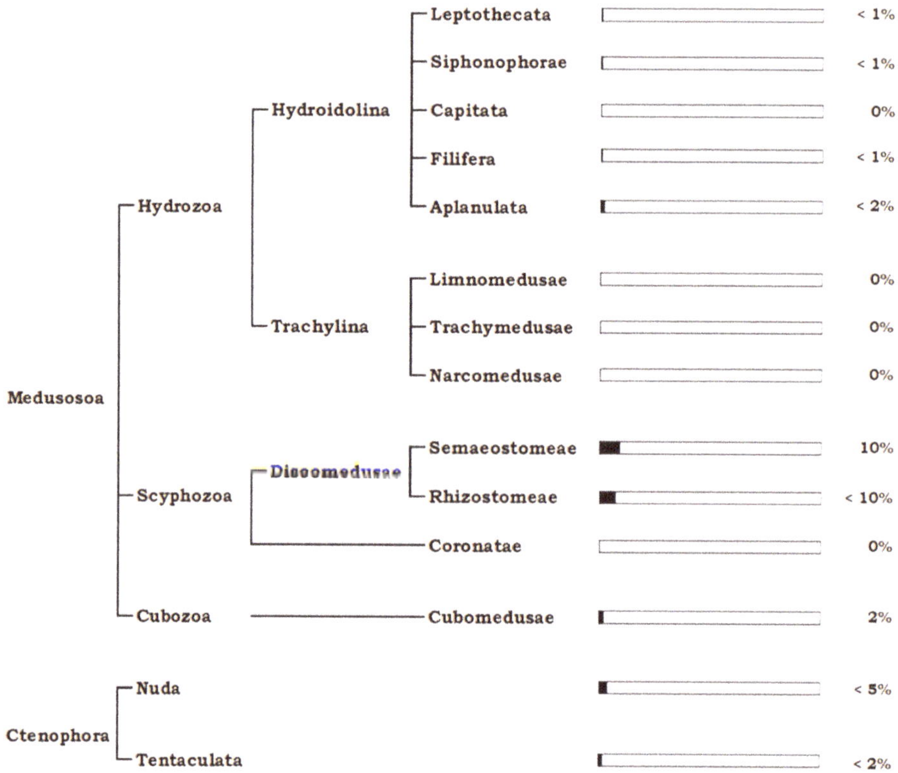

Figure 1. Relative amounts of investigated species in gelatinous taxa. Number of species per taxon was assembled from World Register of Marine Species (WoRMS) database (accessed December 2018). In the Hydrozoa class, only the species with a pelagic stage in their lifecycle were considered, following the species list of Reference [134].

Based on existing data, we tried to depict some basic characteristics of the jellyfish microbiome. It seems that the jellyfish microbiome is distinct from the bacterial community of the ambient water, comprising bacteria known for their preference for a surface-attached lifestyle and in association with marine organisms. This implies a certain degree of specialization of the microbiome of jellyfish, which are potentially generalists or possibly generalist symbionts. To some extent, it appears that the microbiome is jellyfish species-specific. However, in some instances, there is evidence that the jellyfish microbiome also depends on the background microbial community in the ambient water, possibly for recruiting members. This would also suggest a horizontal transmission of the microbiome to the outer body parts of the jellyfish. Accordingly, the microbiome associated with the outer body parts seems to exhibit a higher degree of similarity to the bacterial community in the ambient water and is also more diverse and variable than the microbiome associated with the inner compartments. It seems that the relationship between the microbiome of the inner body compartments and the jellyfish could be symbiotic, and that in this case, the mechanism of transmission could be vertical. The microbiomes of different life stages of jellyfish seem to vary, indicating a significant restructuring of the microbiome from the benthic to the pelagic stage of the jellyfish life cycle.

Our review of different methodological approaches used to study the jellyfish microbiome and the difficulties in comparing the available datasets calls for the establishment of more standardized and holistic sampling and analytic approaches. Also, as currently all hypotheses on the role and function of the jellyfish microbiome are rather speculative, we believe that metatranscriptomic, metaproteomics, and metabolomics should be applied, coupled with state-of-the-art microscopy techniques, to study the microbiome and the mechanisms underlying the associations to provide insight into the potential roles the microbiome might play in the ecology of jellyfish. Furthermore, we propose that investigations should be scaled down to the molecular level, i.e., the level at which microbial-mediated processes take place. Finally, regarding the biotechnological potential of the jellyfish-associated microbiome, the features of specific bacteria found to be associated with jellyfish that have potential application in blue biotechnology are, in fact, known characteristics of these bacterial strains or their closest relatives isolated from other marine organisms or substrates. To the best of our knowledge, no comprehensive study of biotechnological potential of the jellyfish-associated bacteria has been conducted to date. With this review, we provide insight into the jellyfish-associated microbiome and highlight its biotechnological potential, hoping to draw the attention of the blue biotechnology sector to explore jellyfish as a potentially great source of biotechnologically interesting microbes and the compounds they produce. Altogether, this will allow to fully exploiting the biotechnological potential locked in the jellyfish microbiome association.

Author Contributions: Conceptualization, T.T., A.M., V.T.; Investigation, T.T. and V.T.; Writing—original draft preparation, T.T.; Writing—review and editing, T.T., V.T., A.M., K.K., T.K., G.J.H.; Visualization, T.T., T.K., and K.K.; Supervision, V.T. and G.J.H.; Project administration, T.T., G.J.H., V.T.; Funding acquisition, T.T., G.J.H., V.T.

Funding: T.T. received funding from the European Union's Horizon 2020 Research and Innovation Programme under the Marie Sklodowska-Curie grant agreement No 793778. V.T. and K.K. were funded by the Ministry of Higher Education, Science and Technology of the R Slovenia (ARRS P1-0237). T.K. was funded by project "Raziskovalci-2.0-NIB-529024" (contract number C3330-17-529024) by the Ministry of Higher Education, Science and Technology of the R Slovenia. G.J.H. was supported by the Austrian Science Fund (FWF) project ARTEMIS (project number: P 28781-B21).

Acknowledgments: T.T. would like to acknowledge Monika Bright for valuable discussions regarding some basic concepts discussed in the review.

Conflicts of Interest: The authors declare no conflicts of interest. The funders had no role in the design of the study; in the collection, analyses, or interpretation of data; in the writing of the manuscript, or in the decision to publish the results.

References

1.   Bull, A.T. *Microbial Diversity and Bioprospecting*; American Society for Microbiology (ASM) Press: Washington, DC, USA, 2004.

2.   Azam, F.; Malfatti, F. Microbial structuring of marine ecosystems. *Nat. Rev. Microbiol.* **2007**, *5*, 782–791. [CrossRef] [PubMed]

3.   Gulder, T.A.M.; Moore, B.S. Chasing the treasures of the sea—Bacterial marine natural products. *Curr. Opin. Microbiol.* **2009**, *12*, 252–260. [CrossRef] [PubMed]

4. Antranikian, G. Industrial relevance of thermophiles and their enzymes. In *Thermophiles: Biology and Technology at High Temperatures*; Robb, F., Grogan, D., Eds.; GRC Press: Boca Raton, FL, USA, 2007; pp. 113–189.

5. Kennedy, J.; Marchesi, J.R.; Dobson, A.D.W. Marine metagenomics: Strategies for discovery of novel enzymes with biotechnological applications from marine environments. *Microb. Cell Factories* **2008**, *7*, 27. [CrossRef] [PubMed]

6. Munn, C.B. *Marine Microbiology: Ecology and Applications*; Garland Science, CRC Press: Abingdon, UK, 2011.

7. Fenical, W. Marine Pharmaceuticals: Past, Present and Future. *Oceanography* **2006**, *19*, 112–119. [CrossRef]

8. Newman, D.J.; Hill, R.T. New drugs from marine microbes: The tide is running. *J. Ind. Microbiol. Biotechnol.* **2006**, *33*, 539–544. [CrossRef] [PubMed]

9. Williams, P.G. Panning for chemical gold: Marine bacteria as a source of new therapeutics. *Trends Biotechnol.* **2009**, *27*, 45–52. [CrossRef] [PubMed]

10. Choudhary, A.; Naughton, L.M.; Montánchez, I.; Dobson, A.D.W.; Rai, D.K. Current Status and Future Prospects of Marine Natural Products (MNPs) as Antimicrobials. *Mar. Drugs* **2017**, *15*, 272. [CrossRef]

11. Taylor, M.W.; Radax, R.; Steger, D.; Wagner, M. Sponge-associated microorganisms: Evolution, ecology, and biotechnological potential. *Microbiol. Mol. Mol. Rev.* **2007**, *71*, 295–347. [CrossRef]

12. Thomas, T.R.A.; Kavlekar, D.P.; LokaBharathi, P.A. Marine drugs from sponge-microbe association—A review. *Mar. Drugs* **2010**, *8*, 1417–1468. [CrossRef]

13. Penesyan, A.; Kjelleberg, S.; Egan, S. Development of novel drugs from marine surface associated microorganisms. *Mar. Drugs* **2010**, *8*, 438–459. [CrossRef]

14. Blockley, A.; Elliott, D.R.; Roberts, A.P.; Sweet, M. Symbiotic Microbes from Marine Invertebrates: Driving a New Era of Natural Product Drug Discovery. *Diversity* **2017**, *9*, 49. [CrossRef]

15. Haygood, M.G.; Schmidt, G.; Davidson, E.W.; Faulkner, D.J. Microbial symbionts of marine invertebrates: Opportunities for microbial biotechnology. *J. Mol. Microbiol. Biotechnol.* **1999**, *1*, 33–43. [PubMed]

16. Brinkmann, C.M.; Marker, A.; Kurtböke, I. An Overview on Marine Sponge-Symbiotic Bacteria as Unexhausted Sources for Natural Product Discovery. *Diversity* **2017**, *9*, 40. [CrossRef]

17. Rizzo, C.; Lo Giudice, A. Marine Invertebrates: Underexplored Sources of Bacteria Producing Biologically Active Molecules. *Diversity* **2018**, *10*, 52. [CrossRef]

18. Stabili, L.; Parisi, M.G.; Parrinello, D.; Cammarata, M. Cnidarian Interaction with Microbial Communities: From Aid to Animal's Health to Rejection Responses. *Mar Drugs* **2018**, *16*, 296. [CrossRef] [PubMed]

19. Bosch, T.C.G. Cnidarian-Microbe Interactions and the Origin of Innate Immunity in Metazoans. *Ann. Rev. Microbiol.* **2013**, *67*, 499–518. [CrossRef] [PubMed]

20. Condon, R.H.; Duarte, C.M.; Pitt, K.A.; Robinson, K.L.; Lucas, C.H.; Sutherland, K.R.; Mianzan, H.W.; Bogeberg, M.; Purcell, J.E.; Decker, M.B.; et al. Recurrent jellyfish blooms are a consequence of global oscillations. *Proc. Natl. Acad. Sci. USA* **2013**, *110*, 1000–1005. [CrossRef]

21. Condon, R.H.; Steinberg, D.K.; del Giorgio, P.A.; Bouvier, T.C.; Bronk, D.A.; Graham, W.M.; Hugh, W.; Ducklow, H.W. Jellyfish blooms result in a major microbial respiratory sink of carbon in marine systems. *Proc. Natl. Acad. Sci. USA* **2011**, *108*, 10225–10230. [CrossRef]

22. Riemann, L.; Titelman, J.; Båmstedt, U. Links between jellyfish and microbes in a jellyfish dominated fjord. *Mar. Ecol. Prog. Ser.* **2006**, *325*, 29–42. [CrossRef]

23. Dinasquet, J.; Granhag, L.; Riemann, L. Stimulated bacterioplankton growth and selection for certain bacterial taxa in the vicinity of the ctenophore *Mnemiopsis leidyi*. *Front. Microbiol.* **2012**, *3*, 302. [CrossRef]

24. Manzari, C.; Fosso, B.; Marzano, M.; Annese, A.; Caprioli, R.; D'Erchia, A.M.; Gissi, C.; Intranuovo, M.; Picardi, E.; Santamaria, M.; et al. The influence of invasive jellyfish blooms on the aquatic microbiome in a coastal lagoon (Varano, SE Italy) detected by an Illumina-based deep sequencing strategy. *Biol. Invas.* **2015**, *17*, 923–940. [CrossRef]

25. Blanchet, M.; Pringault, O.; Bouvy, M.; Catala, P.; Oriol, L.; Caparros, J.; Ortega-Retuerta, E.; Intertaglia, L.; West, N.; Agis, M.; et al. Changes in bacterial community metabolism and composition during the degradation of dissolved organic matter from the jellyfish *Aurelia aurita* in a Medeterranean coastal lagoon. *Environ. Sci. Pollut. Res. Int.* **2014**, *22*, 13638–13653. [CrossRef] [PubMed]

26. Dinasquet, J.; Kragh, T.; Schroter, M.L.; Sondergard, M.; Riemann, L. Functional and compositional succession of bacterioplankton in response to a gradient in bioavailable dissolved organic carbon. *Environ. Microbiol.* **2013**, *15*, 2616–2628. [CrossRef] [PubMed]

27. Tinta, T.; Kogovšek, T.; Malej, A.; Turk, V. Jellyfish Modulate Bacterial Dynamic and Community Structure. *PLoS ONE* **2012**, *7*, e39274. [CrossRef] [PubMed]

28. Tinta, T.; Kogovšek, T.; Turk, V.; Shiganova, T.A.; Mikaelyan, A.S.; Malej, A. Microbial transformation of jellyfish organic matter affects the nitrogen cycle in the marine water column—A Black Sea case study. *J. Exp. Mar. Biol. Ecol.* **2016**, *475*, 19–30. [CrossRef]

29. Tinta, T.; Malej, A.; Kos, M.; Turk, V. Degradation of the Adriatic medusa *Aurelia* sp. by ambient bacteria. *Hydrobiologia* **2010**, *645*, 179–191. [CrossRef]

30. Bruno, D.W.; Ellis, A.E. Mortalities in farmed Atlantic salmon associated with the jellyfish *Phialella quadrata*. *Bull. Eur. Ass. Fish Pathol.* **1985**, *5*, 64.

31. Margulis, L.; Thorington, G.; Berger, B.; Stolz, J. Endosymbiotic bacteria associated with the intracellular green algae of *Hydria Viridis*. *Curr. Microbiol.* **1987**, *1*, 227–232. [CrossRef]

32. Ferguson, H.W.; Delannoy, C.M.J.; Hay, S.; Nicolson, J.; Sutherland, D.; Crumlish, M. Jellyfish as vectors of bacterial disease for farmed salmon (*Salmo salar*). *J. Vet. Diagn. Investig.* **2010**, *22*, 376–382. [CrossRef]

33. Delannoy, C.M.J.; Houghton, J.D.R.; Fleming, N.E.C.; Ferguson, H.W. Mauve Stingers (*Pelagia noctiluca*) as carriers of the bacterial fish pathogen *Tenacibaculum maritimum*. *Aquaculture* **2011**, *311*, 255–257. [CrossRef]

34. Fringuelli, E.; Savage, P.D.; Gordon, A.; Baxter, E.J.; Rodger, H.D.; Graham, D.A. Development of a quantitative real-time PCR for the detection of *Tenacibaculum maritimum* and its application to field samples. *J. Fish Dis.* **2012**, *35*, 579–590. [CrossRef]

35. Weiland-Bräuer, N.; Neulinger, S.C.; Pinnow, N.; Künzel, S.; Baines, J.F.; Schmitz, R.A. Composition of bacterial communities associated with *Aurelia aurita* changes with compartment, life stage, and population. *Appl. Environ. Microbiol.* **2015**, *81*, 6038–6052. [CrossRef] [PubMed]

36. Daley, M.C.; Urban-Rich, J.; Moisander, P.H. Bacterial associations with the hydromedusa *Nemopsis bachei* and scyphomedusa *Aurelia aurita* from the North Atlantic Ocean. *Mar. Biol. Res.* **2016**, *12*, 1088–1100. [CrossRef]

37. Kos Kramar, M.; Tinta, T.; Lučić, D.; Malej, A.; Turk, V. Bacteria Associated with Moon Jellyfish during Bloom and Post-bloom Periods in the Gulf of Trieste (northern Adriatic). *PLoS ONE* **2019**, *14*, e0198056. [CrossRef] [PubMed]

38. Schuett, C.; Doepke, H. Endobiotic bacteria and their pathogenic potential in cnidarian tentacles. *Helgol. Mar. Res.* **2010**, *64*, 205–212. [CrossRef]

39. Hao, W.; Gerdts, G.; Holst, S.; Wichels, A. Bacterial communities associated with scyphomedusae at Helgoland Roads. *Mar. Biodiv.* **2018**. [CrossRef]

40. Lee, D.L.; Kling, J.D.; Araya, R.; Ceh, J. Jellyfish life stages shape associated microbial communities, while a core microbiome is maintained across all. *Front. Microbiol.* **2018**, *9*, 1534. [CrossRef]

41. Cleary, D.F.R.; Becking, L.E.; Polónia, A.R.M.; Freitas, R.M.; Gomes, N.C.M. Jellyfish-associated bacterial communities and bacterioplankton in Indonesian Marine lakes. *FEMS Microbiol. Ecol.* **2016**, *92*, 1–14. [CrossRef]

42. Cortés-Lara, S.; Urdiain, M.; Mora-Ruiz, M.; Prieto, L.; Rosselló-Móra, R. Prokaryotic microbiota in the digestive cavity of the jellyfish *Cotylorhiza tuberculata*. *Syst. Appl. Microbiol.* **2015**, *38*, 494–500. [CrossRef]

43. Viver, T.; Orellana, L.H.; Hatt, J.K.; Urdiain, M.; Diaz, S.; Richter, M.; Anton, J.; Avian, M.; Amann, R.; Konstantinidis, K.T.; et al. The low diverse gastric microbiome of the jellyfish *Cotylorhiza tuberculata* is dominated by four novel taxa. *Environ. Microbiol.* **2017**, *19*, 3039–3058. [CrossRef]

44. Daniels, C.; Breitbart, M. Bacterial communities associated with the ctenophores *Mnemiopsis leidyi* and *Beroe ovata*. *FEMS Microbiol. Ecol.* **2012**, *82*, 90–101. [CrossRef] [PubMed]

45. Hao, W.; Gerdts, G.; Peplies, J.; Wichels, A. Bacterial communities associated with four ctenophore genera from the German Bight (North Sea). *FEMS Microbiol. Ecol.* **2015**, *91*, 1–11. [CrossRef]

46. Mayer, A.G. *Medusae of the World. Volume III. The Scyphomedusae*; Carnegie Institution of Washington: Washington, DC, USA, 1910; 735p.

47. Kramp, P.L. Synopsis of the *Medusae* of the World. *J. Mar. Biol. Assoc. UK* **1961**, *40*, 7–382. [CrossRef]

48. Russell, F.S. *Medusae of the British Isles. Vol II. Pelagic Scyphozoa, with a Supplement to Vol. 1*; Cambridge University Press: Cambridge, UK, 1970.

49. Gambill, M.; Jarms, G. Can *Aurelia* (*Cnidaria, Scyphozoa*) species be differentiated by comparing their scyphistomae and ephyrae. *Eur. J. Taxon.* **2014**, *107*, 1–23.

50. Dawson, M.N.; Jacobs, D.K. Molecular Evidence for Cryptic Species of *Aurelia aurita* (*Cnidaria, Scyphozoa*). *Biol. Bull.* **2001**, *200*, 92–96. [CrossRef] [PubMed]

51. Dawson, M.N. Macro-morphological variation among cryptic species of the moon jellyfish, *Aurelia* (*Cnidaria: Scyphozoa*). *Mar. Biol.* **2003**, *143*, 369–379. [CrossRef]

52. Ryan, R.P.; Monchy, S.; Cardinale, M.; Taghavi, S.; Crossman, L.; Avison, M.B.; Berg, G.; Van der Lelie, D.; Dow, J.M. The versatility and adaptation of bacteria from the genus *Stenotrophomonas*. *Nat. Rev. Microbiol.* **2009**, *7*, 514–525. [CrossRef]

53. Hong, Y.H.; Ye, C.C.; Zhou, Q.Z.; Wu, X.Y.; Yuan, J.P.; Peng, J.; Deng, H.; Wang, J.H. Genome sequencing reveals the potential of *Achromobacter* sp. HZ01 for bioremediation. *Front. Microbiol.* **2017**, *8*, 1507. [CrossRef]

54. Maravić, A.; Skočibušić, M.; Šprung, M.; Šamanić, I.; Puizina, J.; Pavela-Vrančić, M. Occurrence and antibiotic susceptibility profiles of *Burkholderia cepacia* complex in coastal marine environment. *Int. J. Environ. Health Res.* **2012**, *22*, 531–542. [CrossRef]

55. Deng, M.C.; Li, J.; Liang, F.R.; Yi, M.; Xu, X.M.; Yuan, J.P.; Peng, J.; Wu, C.F.; Wang, J.H. Isolation and characterization of a novel hydrocarbon-degrading bacterium *Achromobacter* sp. HZ01 from the crude oil-contaminated seawater at the Daya Bay, southern China. *Mar. Pollut. Bull.* **2014**, *83*, 79–86. [CrossRef]

56. Harayama, S.; Kishira, H.; Kasai, Y.; Shutsubo, K. Petroleum biodegradation in marine environments. *J. Mol. Microbol. Biotechnol.* **1999**, *1*, 63–70.

57. Dang, H.; Lovell, C.R. Microbial Surface Colonization and Biofilm Development in Marine Environments. *Microbiol. Mol. Biol. Rev.* **2016**, *80*, 91–138. [CrossRef] [PubMed]

58. Hay, S.J.; Hislop, J.R.G.; Shanks, A.M. North Sea *Scyphomedusae*; summer distribution, estimated biomass and significance particularly for 0-group gadoid fish. *Neth. J. Sea Res.* **1990**, *25*, 113–130. [CrossRef]

59. Fehr, A.; Walther, E.; Schmidt-Posthaus, H.; Nufer, L.; Wilson, A.; Svercel, M.; Richter, D.; Segner, H.; Pospischil, A.; Vaughan, L. *Candidatus Syngnamydia venezia*, a novel member of the phylum *Chlamydiae* from the broad nosed pipefish *Syngnathus typhle*. *PLoS ONE* **2013**, *8*, e70853. [CrossRef] [PubMed]

60. Nylund, S.; Steigen, A.; Karlsbakk, E.; Plarre, H.; Andersen, L.; Karlsen, M.; Watanabe, K.; Nylund, A. Characterization of "*Candidatus Syngnamydia salmonis*" (*Chlamydiales*, *Simkaniaceae*), a bacterium associated with epitheoliocystis in Atlantic salmon (*Salmo salar*). *Arch. Microbiol.* **2015**, *197*, 17–25. [CrossRef] [PubMed]

61. Moss, A.G.; Estes, A.M.; Muellner, L.A.; Morgan, D.D. Protistan epibionts of the ctenophore *Mnemiopsis mccradyi* Mayer. *Hydrobiologia* **2001**, *451*, 295–304. [CrossRef]

62. Hay, S. Marine Ecology: Gelatinous Bells May Ring Change in Marine Ecosystems. *Curr. Biol.* **2006**, *16*, 679–682. [CrossRef]

63. Reitzel, A.M.; Sullivan, J.C.; Brown, B.K.; Chin, D.W.; Cira, E.K.; Edquist, S.K.; Genco, B.M.; Joseph, O.C.; Kaufman, C.A.; Kovitvongsa, K.; et al. Ecological and developmental dynamics of a host-parasite system involving a sea anemone and two ctenophores. *J. Parasitol.* **2007**, *93*, 1392–1402. [CrossRef]

64. Smith, K.; Dodson, M.; Santos, S.; Gast, R.; Rogerson, A.; Sullivan, B.; Moss, A.G. *Pentapharsodinium tyrrhenicum* is a parasitic dinoflagellate of the ctenophore *Mnemiopsis leidyi*. *J. Phycol.* **2007**, *43*, 119.

65. Kogovšek, T.; Vodopivec, M.; Raicich, F.; Uye, S.; Malej, A. Comparative analysis of the ecosystems in the northern Adriatic Sea and the Inland Sea of Japan: Can anthropogenic pressures disclose jellyfish outbreaks? *Sci. Total Environ.* **2018**, *626*, 982–994. [CrossRef]

66. Stewart, E.J. Growing Unculturable Bacteria. *J. Bacteriol.* **2012**, *194*, 4151–4160. [CrossRef] [PubMed]

67. Sogin, M.L.; Morrison, H.G.; Huber, J.A.; Welch, D.M.; Huse, S.M.; Neal, P.R.; Arrieta, J.M.; Herndl, G.J. Microbial diversity in the deep sea and the underexplored "rare biosphere". *Proc. Natl. Acad. Sci. USA* **2006**, *103*, 12115–12120. [CrossRef] [PubMed]

68. Rohwer, F.; Seguritan, V.; Azam, F.; Knowlton, N. Diversity and distribution of coral-associated bacteria. *Mar. Ecol. Prog. Ser.* **2002**, *243*, 1–10. [CrossRef]

69. Webster, N.S.; Taylor, M.W.; Behnam, F.; Lücker, S.; Rattei, T.; Whalan, S.; Horn, M.; Wagner, M. Deep sequencing reveals exceptional diversity and modes of transmission for bacterial sponge symbionts. *Environ. Microbiol.* **2010**, *12*, 2070–2082. [CrossRef]

70. Lee, O.O.; Wang, Y.; Yang, J.; Lafi, F.F.; Al-Suwailem, A.; Qian, P.Y. Pyrosequencing reveals highly diverse and species-specific microbial communities in sponges from the Red Sea. *ISME J.* **2011**, *5*, 650–664. [CrossRef] [PubMed]

71. Ovchinnikova, T.V.; Balandin, S.V.; Aleshina, G.M.; Tagaev, A.A.; Leonova, Y.F.; Krasnodembsky, E.D.; Men'shenin, A.V.; Kokryakov, V.N. Aurelin, a novel antimicrobial peptide from jellyfish *Aurelia aurita* with structural features of defensins and channel-blocking toxins. *Biochem. Biophys. Res. Commun.* **2006**, *348*, 514–523. [CrossRef] [PubMed]

72. Bhosale, S.H.; Nagle, V.L.; Jagtap, T.G. Antifouling potential of some marine organisms from India against species of *Bacillus* and *Pseudomonas*. *Mar. Biotechnol.* **2002**, *4*, 111–118. [CrossRef] [PubMed]

73. Liu, J.; Li, F.; Kim, E.L.; Li, J.L.; Hong, J.; Bae, K.S.; Chung, H.Y.; Kim, H.S.; Jung, J.H. Antibacterial polyketides from the jellyfish-derived fungus *Paecilomyces variotii*. *J. Nat. Prod.* **2011**, *74*, 1826–1829. [CrossRef]

74. Mills, C.E. Jellyfish blooms: Are populations increasing globally in response to changing ocean conditions? *Hydrobiologia* **2001**, *451*, 55–68. [CrossRef]

75. Lucas, C.H. Reproduction and life history strategies of the common jellyfish, *Aurelia aurita*, in relation to its ambient environment. *Hydrobiologia* **2001**, *451*, 229–246. [CrossRef]

76. Purcell, J.E.; Arai, M.N. Interactions of pelagic cnidarians and ctenophores with fish: A review. *Hydrobiologia* **2001**, *451*, 27–44. [CrossRef]

77. Vodopivec, M.; Peliz, A.J.; Malej, A. Offshore marine constructions as propagators of moon jellyfish dispersal. *Environ. Res. Lett.* **2017**, *12*, 084003. [CrossRef]

78. Schmahl, G. Induction of stolon settlement in the scyphopolyps of *Aurelia aurita* (*Cnidaria, Scyphozoa, Semaeostomeae*) by glycolipids of marine bacteria. *Helgol. Meeresunters.* **1985**, *39*, 117–127. [CrossRef]

79. Schmahl, G. Bacterially induced stolon settlement in the scyphopolyp of *Aurelia aurita* (*Cnidaria, Scyphozoa*). *Helgol. Meeresunters.* **1985**, *39*, 33–42. [CrossRef]

80. Hofmann, D.K.; Neumann, R.; Henne, K. Strobilation, budding and initiation of scyphistoma morphogenesis in the rhizostome *Cassiopea andromeda* (Cnidaria, Scyphozoa). *Mar. Biol.* **1978**, *47*, 161–176. [CrossRef]

81. Hofmann, D.K.; William, K.F.; Fleck, J. Checkpoints in the life-cycle of *Cassiopea* spp.: Control of metagenesis and metamorphosis in a tropical jellyfish. *Int. J. Dev. Biol.* **1996**, *40*, 331–338. [PubMed]

82. Neumann, A. Bacterial induction of settlement and metamorphosis in the planula larvae of *Cassiopea andromeda* (*Cnidaria: Scyphozoa, Rhizostomeae*). *Mar. Ecol. Prog. Ser.* **1979**, *1*, 21–28. [CrossRef]

83. Neumann, R.; Schmahl, G.; Hofmann, D.K. Bud formation and control of polyp morphogenesis in *Cassiopea andromeda* (*Scyphozoa*). In *Developmental and Cellular Biology of Coelenterates*; Tardent, P., Tardent, R., Eds.; Elsevier, North-Holland Biomedical Press: Amsterdam, The Netherlands, 1980; pp. 217–223.

84. Brekhman, V.; Malik, A.; Haas, B.; Sher, N.; Lotan, T. Transcriptome profiling of the dynamic life cycle of the scypohozoan jellyfish *Aurelia aurita*. *BMC Genom.* **2015**, *16*, 74. [CrossRef]

85. Fuchs, B.; Wang, W.; Graspeuntner, S.; Li, Y.; Insua, S.; Herbst, E.M.; Dirksen, P.; Marei Bohm, A.M.; Hemmrich, G.; Sommer, F.; et al. Regulation of Polyp-to-Jellyfish Transition in *Aurelia aurita*. *Curr. Biol.* **2014**, *24*, 263–273. [CrossRef]

86. Gold, D.A.; Katsuki, T.; Li, Y.; Yan, X.; Regulski, M.; Ibberson, D.; Holstein, T.; Steele, R.E.; Jacobs, D.K.; Greenspan, R.J. The genome of the jellyfish *Aurelia* and the evolution of animal complexity. *Nat. Ecol. Evol.* **2018**, *3*, 96–104. [CrossRef]

87. Freckelton, M.L.; Nedved, B.T.; Hadfield, M.G. Induction of Invertebrate Larval Settlement; Different Bacteria, Different Mechanisms? *Sci. Rep.* **2017**, *7*, 42557. [CrossRef] [PubMed]

88. Dobretsov, S.; Qian, P.Y. The role of epibotic bacteria from the surface of the soft coral *Dendronephthya* sp. in the inhibition of larval settlement. *J. Exp. Mar. Biol. Ecol.* **2004**, *299*, 35–50. [CrossRef]

89. Dang, H.; Li, T.; Chen, M.; Huang, G. Cross-ocean distribution of *Rhodobacterales* bacteria as primary surface colonizers in temperate coastal marine waters. *Appl. Environ. Microbiol.* **2008**, *74*, 52–60. [CrossRef] [PubMed]

90. Porsby, C.H.; Nielsen, K.F.; Gram, L. *Phaeobacter* and *Ruegeria* Species of the *Roseobacter* Clade Colonize Separate Niches in a Danish Turbot (*Scophthalmus maximus*)-rearing farm and antagonize *Vibrio anguillarum* under different growth conditions. *Appl. Environ. Microbiol.* **2008**, *74*, 7356–7364. [CrossRef] [PubMed]

91. Elsayed, Y.; Refaat, J.; Abdelmohsen, U.R.; Fouad, M.A. The Genus *Rhodococcus* as a source of novel bioactive substances: A review. *J. Pharma. Phytochem.* **2017**, *6*, 83–92.

92. Shnit-Orland, M.; Kushmaro, A. Coral mucus-associated bacteria: A possible first line of defense. *FEMS Microbiol. Ecol.* **2009**, *67*, 371–380. [CrossRef] [PubMed]

93. Lesh-Laurie, G.E.; Suchy, P.E. *Scyphozoa* and *Cubozoa*. In *Placozoa, Porifera, Cnidaria, and Ctenophora, Volume II*; Harrison, F.W., Westfall, J.A., Eds.; Wiley-Liss: New York, NY, USA, 1991; pp. 185–266.

94. Heeger, T.; Möller, H. Ultrastructural observations on prey capture and digestion in the scyphomedusa *Aurelia aurita*. *Mar. Biol.* **1987**, *96*, 391–400. [CrossRef]

95. Patwa, A.; Thiéry, A.; Lombard, F.; Lilley, M.K.S.; Boisset, C.; Bramard, J.; Bottero, J.Y.; Philippe Barthélémy, P. Accumulation of nanoparticles in "jellyfish" mucus: A bio-inspired route to decontamination of nano-waste. *Sci. Rep.* **2015**, *5*, 11387. [CrossRef]

96. Shanks, A.; Graham, W. Chemical defense in a scyphomedusa. *Mar. Ecol. Prog. Ser.* **1988**, *45*, 81–86. [CrossRef]

97. Shaposhnikovaa, T.; Matveevb, I.; Naparac, T.; Podgornayab, O. Mesogleal cells of the jellyfish *Aurelia aurita* are involved in the formation of mesogleal fibres. *Cell. Biol. Int.* **2005**, *29*, 952–958. [CrossRef]

98. Hoeger, U. Biochemical composition of ctenophores. *J. Exp. Mar. Biol. Ecol.* **1983**, *72*, 251–261. [CrossRef]

99. Hsieh, Y.H.P.; Rudloe, J. Potential of utilizing jellyfish as food in Western countries. *Trends Food Sci. Technol.* **1994**, *5*, 225–229. [CrossRef]

100. Ducklow, H.W.; Mitchell, R. Composition of mucus released by coral reef coelenterates. *Limnol. Oceanogr.* **1979**, *24*, 706–714. [CrossRef]

101. Stabili, L.; Schirosi, R.; Parisi, M.G.; Piraino, S.; Cammarata, M. The Mucus of *Actinia equina* (*Anthozoa, Cnidaria*): An Unexplored Resource for Potential Applicative Purposes. *Mar. Drugs* **2015**, *13*, 5276–5296. [CrossRef] [PubMed]

102. Wahl, M.; Goecke, F.; Labes, A.; Dobretsov, S.; Weinberger, F. The second skin: Ecological role of epibiotic biofilms on marine organisms. *Front. Microbiol.* **2012**, *3*, 292. [CrossRef] [PubMed]

103. Harder, T.; Lau, S.C.K.; Dobretsov, S.; Fang, T.K.; Qian, P.Y. A distinctive epibiotic bacterial community on the soft coral *Dendronephthya* sp. and antibacterial activity of coral tissue extracts suggest a chemical mechanism against bacterial epibiosis. *FEMS Microbiol. Ecol.* **2003**, *43*, 337–347. [CrossRef] [PubMed]

104. Holmström, C.; Egan, S.; Franks, A.; McCloy, S.; Kjelleberg, S. Antifouling activities expressed by marine surface associated *Pseudoalteromonas* species. *FEMS Microbiol. Ecol.* **2002**, *41*, 47–58. [CrossRef]

105. Long, R.A.; Azam, F. Antagonistic Interactions among Marine Pelagic Bacteria. *Appl. Environ. Microbiol.* **2001**, *67*, 4975–4983. [CrossRef]

106. Bruhn, J.B.; Nielsen, K.F.; Hjelm, M.; Hansen, M.; Bresciani, J.; Schulz, S.; Gram, L. Ecology, Inhibitory Activity, and Morphogenesis of a Marine Antagonistic Bacterium Belonging to the *Roseobacter* Clade. *Appl. Environ. Microbiol.* **2005**, *71*, 7263–7270. [CrossRef]

107. Gram, L.; Melchiorsen, J.; Bruhn, J.B. Antibacterial Activity of Marine Culturable Bacteria Collected from a Global Sampling of Ocean Surface Waters and Surface Swabs of Marine Organisms. *Mar. Biotechnol.* **2010**, *12*, 439–451. [CrossRef]

108. Bishop, C.D.; Brandhorst, B.P. On nitric oxide signalling, metamorphosis, and the evolution of biphasic life cycles. *Evol. Dev.* **2003**, *5*, 542–550. [CrossRef] [PubMed]

109. Moroz, L.L.; Meech, R.W.; Sweedler, J.V.; Mackie, G.O. Nitric oxide regulates swimming in the jellyfish *Aglantha digitale*. *J. Comp. Neurol.* **2004**, *471*, 26–36. [CrossRef] [PubMed]

110. Salleo, A.; Musci, G.; Barra, P.; Calabrese, L. The discharge mechanism of acontial nematocytes involves the release of nitric oxide. *J. Exp. Biol.* **1996**, *199 Pt 6*, 1261–1267.

111. Almeda, R.; Wambaugh, Z.; Chai, C.; Wang, Z.; Liu, Z.; Buskey, E.J. Effects of crude oil exposure on bioaccumulation of polycyclic aromatic hydrocarbons and survival of adult and larval stages of gelatinous zooplankton. *PLoS ONE* **2013**, *8*, e74476. [CrossRef] [PubMed]

112. Schiraldi, C.; Giuliano, M.; De Rosa, M. Perspectives on biotechnological applications of archaea. *Archaea* **2002**, *1*, 75–86. [CrossRef] [PubMed]

113. Egorova, K.; Antranikian, G. Industrial relevance of thermophilic *Archaea*. *Curr. Opin. Microbiol.* **2005**, *8*, 649–655. [CrossRef]

114. Ainsworth, T.D.; Krause, L.; Bridge, T.; Torda, G.; Raina, J.B.; Zakrzewski, M.; Gates, R.D.; Padilla-Gamiño, J.L.; Spalding, H.L.; Smith, C.; et al. The coral core microbiome identifies rare bacterial taxa as ubiquitous endosymbionts. *ISME J.* **2015**, *9*, 2261–2274. [CrossRef]

115. Giovannoni, S.J.; Rappe, M. Evolution, diveristy and molecular ecology of marine prokaryotes. In *Microbial Ecology of the Ocean*; Kirchman, D.L., Ed.; Wiley-Liss: New York, NY, USA, 2000; pp. 47–84.

116. Solano, F.; Sanchez-Amat, A. Studies on the phylogenetic relationships of melanogenic marine bacteria: Proposal of *Marinomonas mediterranea* sp. nov. *Int. J. Syst. Bacteriol.* **1999**, *49*, 1241–1246. [CrossRef]

117. Sanchez-Amat, A.; Lucas-Elo, P.; Fernández, E.; García-Borrón, J.C.; Solano, F. Molecular cloning and functional characterization of a unique multipotent polyphenol oxidase from *Marinomonas mediterranea*. *Biochim. Biophys. Acta* **2001**, *1547*, 104–116. [CrossRef]

118. Bowman, J.P. Bioactive compound synthetic capacity and ecological significance of marine bacterial genus pseudoalteromonas. *Mar. Drugs* **2007**, *5*, 220–241. [CrossRef]

119. Thompson, F.L.; Iida, T.; Swings, J. Biodiversity of *Vibrios*. *Microbiol. Mol. Biol. Rev.* **2004**, *68*, 403–431. [CrossRef] [PubMed]

120. Kodama, Y.; Stiknowati, L.I.; Ueki, A.; Ueki, K.; Watanabe, K. *Thalassospira tepidiphila* sp. nov., a polycyclic aromatic hydrocarbon-degrading bacterium isolated from seawater. *Int. J. Syst. Evol. Microbiol.* **2008**, *58*, 711–715. [CrossRef] [PubMed]

121. Gallego, S.; Vila, J.; Tauler, M.; Nieto, J.M.; Breugelmans, P.; Springael, D.; Gri-foll, M. Community structure and PAH ring-hydroxylating dioxygenasegenes of a marine pyrene-degrading microbial consortium. *Biodegradation* **2014**, *25*, 543–556. [CrossRef] [PubMed]

122. Rizzo, C.; Michaud, L.; Hörmann, B.; Gerçe, B.; Syldatk, C.; Hausmann, R.; DeDomenico, E.; Lo Giudice, A. Bacteria associated with sabellids (*Polychaeta*: *Annelida*; as a novel source of surface active compounds. *Mar. Pollut. Bull.* **2013**, *70*, 125–133. [CrossRef] [PubMed]

123. Wiese, J.; Thiel, V.; Gartner, A.; Schmaljohann, R.; Imhoff, J.F. *Kiloniella laminariae* gen. nov., sp. nov., an *Alphaproteobacterium* from the marine macroalga *Laminaria saccharina*. *Int. J. Syst. Evol. Micrrobiol.* **2009**, *59*, 350–356. [CrossRef] [PubMed]

124. Converse, R.R.; Blackwood, A.D.; Kirs, M.; Griffith, J.F.; Noble, R.T. Rapid Q-PCR-based assay for fecal *Bacteroidetes* spp. as tool for assessing fecal contamination in recreational water. *Water Res.* **2009**, *43*, 4828–4837. [CrossRef] [PubMed]

125. Toranzo, A.E.; Magarinos, B.; Romalde, J.L. A review of the main bacterial fish diseases in mariculture systems. *Aquaculture* **2005**, *246*, 37–61. [CrossRef]

126. Handlinger, J.; Soltani, M.; Percival, S. The pathology of *Flexibacter maritimus* in aquaculture species in Tasmania, Australia. *J. Fish. Dis.* **1997**, *20*, 159–168. [CrossRef]

127. Roberts, R.J. Miscellaneous non-infectious diseases. In *Fish Pathology*; Roberts, R.J., Ed.; WB Saunders: London, UK, 2001; pp. 367–379.

128. Queruel, P. Envenimations par la méduse *Pelagia noctiluca* sur nos côtes méditerranéennes. *Presse Méd.* **2000**, *29*, 188.

129. Mitchell, S.O.; Rodger, H.D. A review of infectious gill disease in marine salmonid fish. *J. Fish. Dis.* **2011**, *34*, 411–432. [CrossRef]

130. Avendanõ-Herrera, R.; Toranzo, A.E.; Magarinõs, B. *Tenacibaculosis* infection in marine fish caused by *Tenacibaculum maritimum*: A review. *Dis. Aquat. Organ.* **2006**, *71*, 255–266. [CrossRef] [PubMed]

131. Kellogg, C.A.; Lisle, J.T.; Galkiewicz, J.P. Culture-independent characterization of bacterial communities associated with the cold-water coral *Lophelia pertusa* in the northeastern Gulf of Mexico. *Appl. Environ. Microbiol.* **2009**, *75*, 2294–2303. [CrossRef] [PubMed]

132. Neulinger, S.C.; Gartner, A.; Jarnegren, J.; Ludvigsen, M.; Lochte, K.; Dullo, W.C. Tissue-associated "Candidatus Mycoplasma corallicola" and filamentous bacteria on the cold-water coral *Lophelia pertusa* (*Scleractinia*). *Appl. Environ. Microbiol.* **2009**, *75*, 1437–1444. [CrossRef] [PubMed]

133. Bull, A.T.; Stach, J.E. Marine actinobacteria: New opportunities for natural product search and discovery. *Trends Microbiol.* **2007**, *11*, 491–499. [CrossRef] [PubMed]

134. Bouillon, J.; Boero, F. Synopsis of the families and genera of the *Hydromedusae* of the world, with a list of the worldwide species. *Thalassia Salentina* **2000**, *24*, 47–296.

Article

Antitumor Anthraquinones from an Easter Island Sea Anemone: Animal or Bacterial Origin?

Ignacio Sottorff [1,2], Sven Künzel [3], Jutta Wiese [1], Matthias Lipfert [4], Nils Preußke [4], Frank D. Sönnichsen [4] and Johannes F. Imhoff [1,*]

[1] GEOMAR Helmholtz Centre for Ocean Research Kiel, Marine Microbiology, 24105 Kiel, Germany; isottorff@geomar.de (I.S.); jwiese@geomar.de (J.W.)

[2] Facultad de Ciencias Naturales y Oceanográficas, Universidad de Concepción, 4070386 Concepción, Chile

[3] Max Planck Institute for Evolutionary Biology, 24306 Plön, Germany; kuenzel@evolbio.mpg.de

[4] Otto Diels Institute for Organic Chemistry, University of Kiel, 24118 Kiel, Germany; mlipfert@oc.uni-kiel.de (M.L.); npreusske@oc.uni-kiel.de (N.P.); fsoennichsen@oc.uni-kiel.de (F.D.S.)

* Correspondence: jimhoff@geomar.de; Tel.: +49-431-600-4450

Received: 31 January 2019; Accepted: 26 February 2019; Published: 5 March 2019

Abstract: The presence of two known anthraquinones, Lupinacidin A and Galvaquinone B, which have antitumor activity, has been identified in the sea anemone (*Gyractis sesere*) from Easter Island. So far, these anthraquinones have been characterized from terrestrial and marine Actinobacteria only. In order to identify the anthraquinones producer, we isolated Actinobacteria associated with the sea anemone and obtained representatives of seven actinobacterial genera. Studies of cultures of these bacteria by HPLC, NMR, and HRLCMS analyses showed that the producer of Lupinacidin A and Galvaquinone B indeed was one of the isolated Actinobacteria. The producer strain, SN26_14.1, was identified as a representative of the genus *Verrucosispora*. Genome analysis supported the biosynthetic potential to the production of these compounds by this strain. This study adds *Verrucosispora* as a new genus to the anthraquinone producers, in addition to well-known species of *Streptomyces* and *Micromonospora*. By a cultivation-based approach, the responsibility of symbionts of a marine invertebrate for the production of complex natural products found within the animal's extracts could be demonstrated. This finding re-opens the debate about the producers of secondary metabolites in sea animals. Finally, it provides valuable information about the chemistry of bacteria harbored in the geographically-isolated and almost unstudied, Easter Island.

Keywords: Easter Island; Actinobacteria; anthraquinones; symbionts; sea anemone; marine invertebrates; spectroscopy; chromatography

---

1. Introduction

Since the discovery of Easter Island, compelling explorations have characterized the flora [1] and fauna [2–4] of this geographically isolated location. However, little has been done to understand the chemistry harbored in this territory. The best-known finding is the discovery of rapamycin in a soil dwelling actinobacterial representative, which serves as an immunosuppressive drug [5,6]. Beyond that, little progress has been made in exploiting the chemical diversity harbored by marine invertebrates and microorganisms dwelling in this territory, despite the high degree of endemism found [7].

Marine invertebrates are immensely diverse, well distributed in the world oceans [8], and widely known to contain medicinally relevant molecules [9–11]. While these metabolites play different roles in nature, e.g., they act as chemical defense, chemical communication or reproductive signaling molecules, they also find application as human medicines [12]. During the last decades, much effort has been made to identify and characterize the chemicals contained in marine invertebrates. This has resulted in the discovery of astonishing chemicals with novel biological activities and chemical scaffolds [12–14].

Further, the immense progress in DNA sequencing technologies, the development of bioinformatics, and an improvement in analytical techniques enables the identification of the source of the chemicals. Thus, several molecules contained in marine invertebrates have now been shown to actually be of microbial origin [15,16]. It is expected that with the increasing availability of metagenomic information more identifications of the real producers of these metabolites will be made and will establish the metabolite relevance for the interaction of host, symbiont, and the environment.

Marine sponges represent a classic example of marine invertebrates that harbor microbes producing secondary metabolites. They have been studied in detail to determine the origin of the metabolites [17,18]. Another example is the producer of the approved anticancer drug Yondelis (Ecteinascidin-743). This compound was first assigned to the tunicate *Ecteinascidia turbinata*, but later identified as the product of a microbe, Candidatus *Endoecteinascidia frumentensis* [16].

Anthraquinones have been characterized in different marine invertebrates, for example crinoids [19] and sponges [20]. Anthraquinones have broad biological activity and are substances of pharmaceutical relevance for revealing antitumor [21], antibacterial [22], antifungal [23] and epigenetic modulator activities [24]. Two recently isolated anthraquinones, Lupinacidin A (**1**) and Galvaquinone B (**2**), have been reported to be produced by Actinobacteria belonging to the genera *Streptomyces* and *Micromonospora*. Lupinacidin A (**1**) was firstly reported as a specific inhibitor on murine colon 26-L5 carcinoma cells [25], and Galvaquinone B (**2**) showed moderate cytotoxicity against non-small-cell lung cancer cells Calu-3 and H2887, in addition of epigenetic modulatory activity [24].

Herein, we report the identification of two known anthraquinone molecules, Lupinacidin A (**1**) and Galvaquinone B (**2**), contained in an Easter Island sea anemone, *Gyractis sesere*. Interestingly, so far these molecules have been only characterized from microbial origin. Therefore, we undertook the isolation and culture of the Actinobacteria associated with this marine invertebrate. The culture, chemical and genomic evaluation of these bacteria showed that the real producer of the metabolites was an Actinobacterium belonging to the genus *Verrucosispora* and not the sea anemone.

2. Results and Discussion

2.1. Sea Anemone Dereplication

Samples of the sea anemone *Gyractis sesere* [26], also known as *Actiniogeton rapanuiensis* [27], were collected in the intertidal zone of Easter Island, South Pacific Ocean, and extracted with chloroform to give 8 mg of a yellowish residue. The HRLCMS analysis of this crude extract showed the presence of two previously identified antitumor anthraquinones (Figure 1) [24,25], Lupinacidin A (**1**) ([M + H]⁺ m/z 341.1378) and Galvaquinone B (**2**) ([M + H]⁺ m/z 369.3510), which were confirmed by complete NMR-spectroscopic characterization.

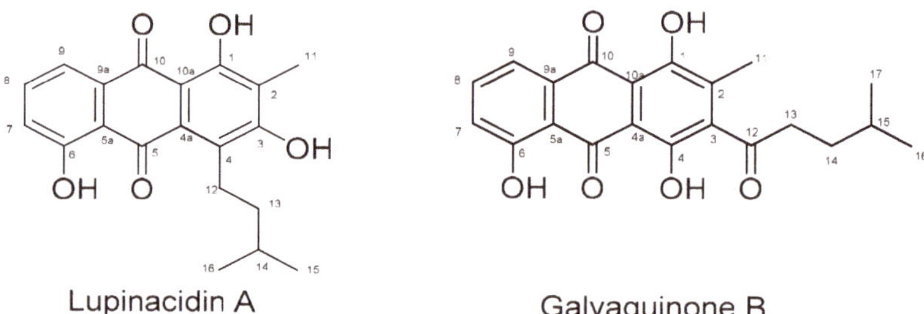

Figure 1. Identified molecules from the Easter Island sea anemone *Gyractis sesere*.

Interestingly, the crude extract of the sea anemone (Figure 2) shows five resonances above 10 ppm; a region which is characteristic for hydroxyl protons. The resonances at δ 14.18 and 12.96 ppm are assigned to the groups at C-1 and C-6 of Lupinacidin A (**1**), as their vicinity and consequential hydrogen bonding to the ketogroups slows down their exchange. Similarly, the resonances at δ 13.49, 12.50, and 12.14 ppm originate from the three hydroxylprotons in Galvaquinone B (**2**).

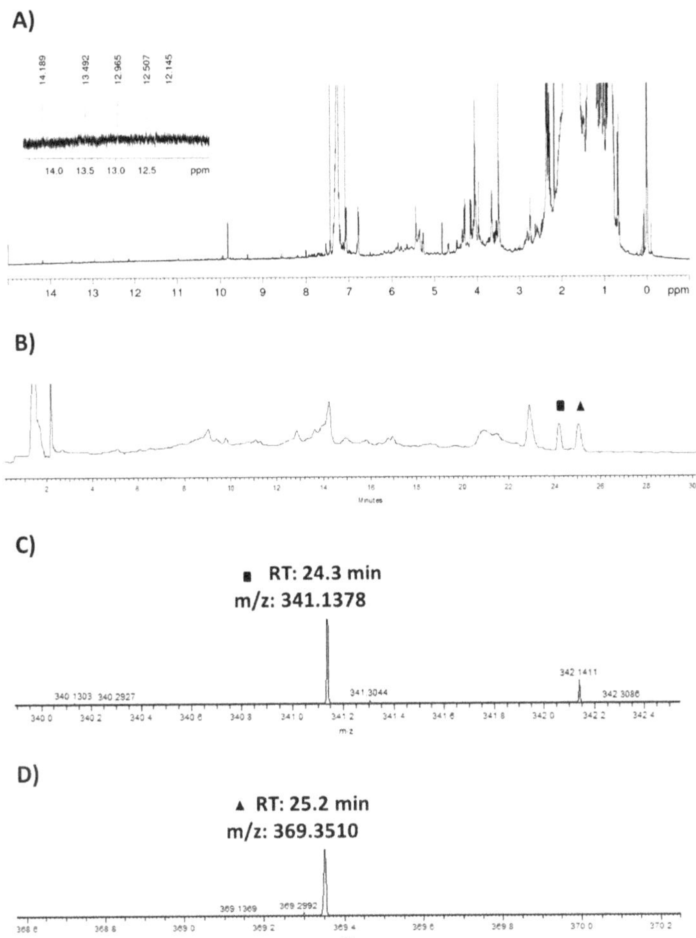

Figure 2. Chemical analysis of the crude extract of the sea anemone *Gyractis sesere*. (**A**) ^{1}H NMR spectra of the crude extract of marine anemone *Gyractis sesere* acquired in CDCl$_3$, 600 MHz. Highlighted in the zoomed area are the frequencies of characteristic resonances originating from hydroxyl exchangeable protons in vicinity to ketogroups. (**B**) UV chromatogram (254 nm) of the crude extract of the sea anemone *Gyractis sesere* highlighting the specific peaks for ■RT: 24.3 min, and ▲: RT: 25.2 min. (**C**) High resolution mass for ■ (m/z [M + H]$^+$ 341.1378) and (**D**) high resolution mass for ▲ (**2**) (m/z [M + H]$^+$ 369.3510). *RT: Retention Time.

Other main peaks found in the sea anemone crude extract were peaks at RT 14.2 min with a HRMS [M + H]$^+$ m/z 295.19009 and at RT 23 min with a HRMS [M + H]$^+$ m/z 256.26312. Both exact masses were evaluated using the MarinLit database, however their HRMS did not match any known compound to-date.

2.2. Bacterial Metabolites and Harbored Bacteria

Lupinacidin A (**1**) and Galvaquinone B (**2**) have so far only been characterized in actinobacterial representatives, specifically from the genera *Streptomyces* [24,28] and *Micromonospora* [25], raising the question of the origin of these compounds in the sea anemone extract. Thus, we cultivated the Actinobacteria harbored by this sea anemone to determine if the anthraquinone producer was a bacterium or the sea anemone. Isolation media and the respective obtained strains are specified in Supplementary Table S1. Ten strains were identified through analysis of the 16S rRNA gene sequences as members of the genera *Micromonospora, Streptomyces, Verrucosispora, Dietzia, Arthrobacter, Rhodococcus,* and *Cellulosimicrobium* (Figure 3). Remarkably, *Gyractis sesere* harbors a high number of Actinobacteria genera, in total seven; the most abundant genus being *Micromonospora* with three different species, followed by *Arthrobacter* with two different species. Other actinobacterial genera were present with only one species each. Outstandingly, the only isolate belonging to the *Verrucosispora* genus was the most abundant single Actinobacterium in the sea anemone.

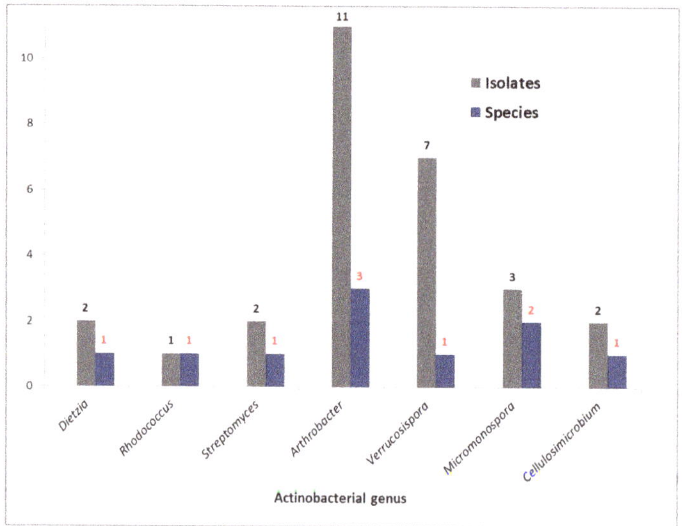

Figure 3. Genera and number of Actinobacteria species strains isolated from the sea anemone *Gyractis sesere.*

2.3. Bacterial Growth

To evaluate the production of the anthraquinones by the isolated bacteria, we selected the actinobacterial representatives for growth experiments that were most closely related to known producers of the genera *Streptomyces, Micromonospora* and *Verrucosispora*. In addition, we also grew the *Dietzia* representative due to the lack of comprehensive information about its secondary metabolite production. On the other hand, *Rhodococcus, Arthrobacter,* and *Cellulosimicrobium* were omitted here because of their known poor production of secondary metabolites.

The growth yield of the selected Actinobacteria was in the range of 20 to 100 mg crude extract. The chromatograms of HPLC analyses of the crude extracts were compared in order to facilitate the metabolic comparison between grown bacteria, the sea anemone and the pure substances (Figure 4).

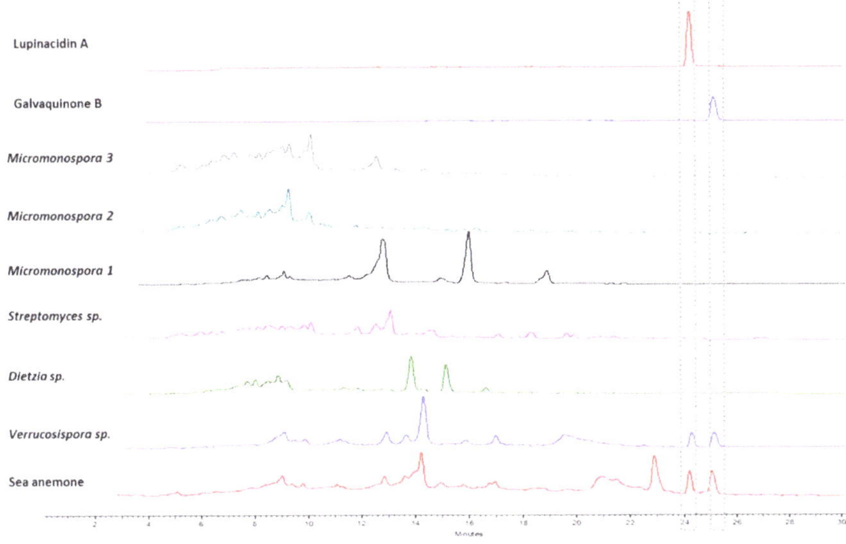

Figure 4. HPLC chromatograms of the crude extracts of the sea anemone *Gyractis sesere*, its respective actinobacterial isolates, and the purified anthraquinones, Lupinacidin A (**1**) and Galvaquinone B (**2**). Approximate retention times of Lupinacidin A (**1**) and Galvaquinone B (**2**) are highlighted by boxes.

By comparison of chromatograms and retention times, we observed that Lupinacidin A (**1**) and Galvaquinone B (**2**) were only present in the crude extract of the sea anemone *Gyractis sesere* and *Verrucosispora sp.* SN26_14.1, and not in the other actinobacterial representatives. Clearly, *Verrucosispora* must be the producer of the anthraquinones. Further, it is obvious that the metabolites of *Verrucosispora sp.* strain SN26_14.1 are dominant in the marine invertebrate. The chromatograms of *Verrucosispora* and sea anemone extracts are nearly identical and differ only slightly in the region of retention time 20–23 min. Notably the chromatogram of the sea anemone extract also does not show any peaks that suggest the presence of metabolites of any other of the cultivated bacteria. Together, this strongly suggests *Verrucosispora* appeared to be the most abundant microbe in the sea anemone biomass during the collection.

The subtle difference in the metabolite profiles between *Verrucosispora* and sea anemone extract in the retention time region 20–23 min appears to be to metabolites produced by the sea anemone itself. Overall, the amount appears to be surprisingly small. This may however be caused by the isolation methodology (chloroform extraction), that prioritizes lipophilic substances and selects against the isolation of polar compounds such as peptides.

2.4. Actinobacterial Producer

To confirm and replicate the production of these metabolites, we undertook a scale up culture of *Verrucosispora* sp. SN26_14.1. Thus, 10 L of the Actinobacterium culture were grown, and extracted through the use of amberlite XAD-16 resin, yielding 1 g of crude extract with a brownish coloration. This extract was subjected to stepwise flash chromatography using iso-octane and ethyl acetate gradients, which produced a total of ten fractions. The fractions were evaluated through HPLC to find the fractions containing Lupinacidin A (**1**) and Galvaquinone B (**2**). The chromatogram evaluation showed that only the orange colored fraction two, which was eluted with 90% iso-octane and 10% ethyl acetate, contained 78 mg of metabolites enriched with Lupinacidin A (**1**) and Galvaquinone B (**2**). The purification of Lupinacidin A (**1**) and Galvaquinone B (**2**) was achieved through HPLC using normal and reverse phase chromatography.

Lupinacidin A (**1**) was isolated as a yellow powder, with a yield of 11 mg from a 10 L culture, which suggested to be an intermediate yield compared with *Streptomyces* and *Micromonospora* producers [24,25,28]. High resolution APCI-MS gave an [M + H]$^+$ adduct of m/z 341.1378, which results in a molecular formula of $C_{20}H_{20}O_5$. The calculation of the degree of unsaturation indicated 11 degrees. ^{1}H NMR showed the characteristic exchangeable protons of (**1**) at δ at 14.18 and 12.96 ppm, in addition to the three neighboring aromatic proton signals δ 7.26, 7.62, and 7.79 ppm that showed the expected coupling pattern for three neighboring aromatic protons in a para-ortho, ortho-meta relationship (two duplets, and one duplet of duplets). The ^{13}C NMR experiment showed 20 carbons of which two represented ketone signals (δ 190.2 and 186.9 ppm), 12 aromatic carbons, and six aliphatic carbons (see Table 1). Two-dimensional NMR experiments, Homonuclear COrrelated SpectroscopY (COSY), Heteronuclear Single Quantum Correlation (HSQC), and Heteronuclear Multiple Bond Correlation (HMBC), helped to confirm the identity of the molecules. These data were in agreement with the published information [24,25,28].

Table 1. Spectroscopic NMR data of Lupinacidin A (**1**) and Galvaquinone B (**2**).

Position	Lupinacidin A				Galvaquinone B			
	$δ_C$	$δ_H$ Mult (J in Hz)	HMBC	COSY	$δ_C$	$δ_H$ Mult (J in Hz)	HMBC	COSY
1	162.5				157.5			
2	117.5				137.1			
3	159.6				141.1			
4	130.4				153.8			
4a	127.9				116.2			
5	190.2				190.7			
5a	117.1				116.2			
6	162.6				162.7			
7	124.3	7.26, d (8.2)	5a, 6, 9	H-8	124.8	7.32, d (8.5, 1.3)	5a, 6, 9	H-8
8	136	7.62, dd (8.2, 7.5)	6, 9a	H-7, H-9	137.1	7.72, dd (8.5, 7.6)	6, 9, 9a	H-7, H-9
9	118.3	7.79, d (7.5)	5a, 7, 10	H-8	119.7	7.90, d (7.6, 1.3)	5a, 7, 10,	H-9
9a	133				133.4			
10	186.9				186.5			
10a	110.8				111.7			
11	8.4	2.27, s	1, 2, 3		13.2	2.25, s	1, 2, 3	
12	24.8	3.21, m (6.6)	2, 4a, 12, 14	H-13	204.9			
13	37.7	1.46, m (6.6)	4a, 12, 14, 15	H-12, H-14	42.4	2.85, m	12, 14, 15	H-14
14	28.4	1.80, m (6.6)	12, 13, 15, 16	H-13, H-15, H-16	31.9	1.63, m	14, 15, 16, 17	H-15
15	22.5	1.04, d (6.6)	13, 14, 16	H-14	27.6	1.63, m	14, 15, 16, 17	H-13, H-16, H-17
16	22.5	1.04, d (6.6)	13, 14, 15	H-14	22.4	0.93, d, (6.2)	14, 15, 17	H-15
17					22.4	0.93, d, (6.2)	14, 15, 16	H-15
1-OH		14.18	1, 2, 10a,			13.49, s	1, 2, 10a	
3-OH		5.62	2, 4a, 3					
4-OH						12.50, s	3, 4, 10a	
6-OH		12.96	5a, 6, 7			12.14, s	5a, 6, 8	

** ^{1}H NMR (600 MHz) Solvent: CDCl$_3$ (δ^{1}H, mult, J in Hz), *** ^{13}C NMR (125 MHz), Solvent: CDCl$_3$.

Galvaquinone B (**2**) was isolated as a red powder, with a yield of 7 mg from a 10 L culture, which is an intermediate yield compared with *Streptomyces* and *Micromonospora* producers [24,28]. High resolution APCI-MS gave an [M + H]$^+$ adduct of m/z 369.3510 and a molecular formula of $C_{21}H_{20}O_6$. The calculation of the degree of unsaturation indicated 12 degrees. Galvaquinone B (**2**) showed characteristic exchangeable proton signals at δ 13.49, 12.50, and 12.14 ppm, respectively. Aromatic signals were similar to those found in compound (**1**), showing three neighboring aromatic protons in a para-ortho, ortho-meta relationship (two duplets, and one duplet of duplets), but with a higher frequency (see Table 1). The ^{13}C NMR experiment showed 21 carbons of which three represented ketone signals (δ 205, 190.2 and 186.9 ppm), 12 aromatic carbons, and six aliphatic carbons (see Table 1). Two dimensional experiments (COSY, HSQC, HMBC) confirmed the identity of the molecules and were in agreement with the published data [24,28].

2.5. Phylogeny of the Producer

To determine whether the present isolate was a new species and to evaluate its evolutionary relationship, we performed a phylogenetic evaluation of the strain based of the 16S rRNA gene sequence (Figure 5). The evaluation of the 16S gene showed a high similarity (99%) to the next related type strain, *Verrucosispora maris* DSM 45365^T. However, analysis of the gyrase subunit B taxonomic marker (gyrB) showed 94.4% similarity to *Verrucosispora maris* DSM 45365^T. The construction of the phylogenetic tree with the closest relatives in terms of the 16S rRNA gene sequence as well as known producers of (**1**) and (**2**) confirmed that the producer strain SN26_14.1 belongs to the genus *Verrucosispora*, and that quite likely it represents a new species within the genus. This result represents the first report of anthraquinone production for the *Verrucosispora* genus. Interestingly, it appears that anthraquinones are more widespread metabolites in the *Actinobacteria* phylum, since compounds (**1**) and (**2**) have now been found in three actinobacterial genera, *Micromonospora*, *Streptomyces*, and *Verrucosispora* producers [24,25,28].

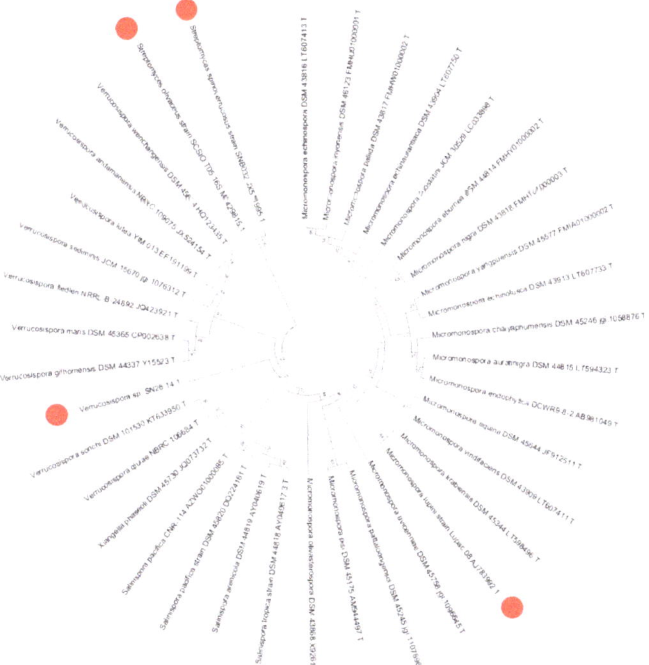

Figure 5. Phylogenetic tree based on 16S rRNA gene sequence of *Verrucosispora sp.* SN26_14.1. The tree was calculated using a neighbor-joining statistical method and Jukes–Cantor model. • Red dots highlight Lupinacidin A (**1**) and Galvaquinone B producers (**2**).

2.6. Biosynthesis

Recently, the biosynthetic machinery for the production of compound (**1**) and (**2**) was described as a type II polyketide synthase (PKS) that features a special Baeyer–Villiger type rearrangement, and was allocated to an Rsd gene cluster in *Streptomyces olivaceus* SCSIO T05 [28]. The Rsd biosynthetic gene cluster (BGC) showed great similarity to the Rsl BGC reported for the production of rishirilide A and B in *Streptomyces bottropensis* (also known as *Streptomyces. sp.* Gç C4/4) [29]. The BGC Rsd is responsible for the production of six molecules (rishirilide B, rishirilide C, Lupinacidin A (**1**), Lupinacidin D, Galvaquinone A and Galvaquinone B (**2**)), and among them compound (**1**) and (**2**) [28]. This raised the question of whether in *Verrucosispora,* compounds (**1**) and (**2**) follow the same biochemical assembly line as described for *Streptomyces*. Thus, the genome of *Verrucosispora sp.* SN26_14.1 was sequenced using Illumina MiSeq. Although the obtained short reads were not complemented with a long read sequencing technology as PacBio, we were still able to obtain a 6.9 Mb draft genome (NCBI Bioproject Access # PRJNA522941). This data was annotated with Prokka and analyzed with the Antismash online platform [30] to identify the secondary metabolite biosynthesis gene clusters. As shown in Figure 6, the draft genome of *Verrucosispora* sp. SN26_14.1 shared 60% of the genes of the Rsd gene cluster, as well as to an important percentage of the genes of the Rsl BGC. The similarity of the found genes ranged from 49% to 81%. The genetic architecture found in Vex BGC was quite similar to that of the Rsd and Rsl BGC. Remarkably, we could not detect any cyclase/aromatase and amidohydrolase sequences in our draft genome. Likely, this relates to the incompleteness of our sequence. Finally, it appears reasonable that *Verrucosispora* sp. SN26_14.1 follows the same biosynthetic machinery for the production of Lupinacidin A (**1**) and Galvaquinone B (**2**) as found for *Streptomyces* species (Figure 6).

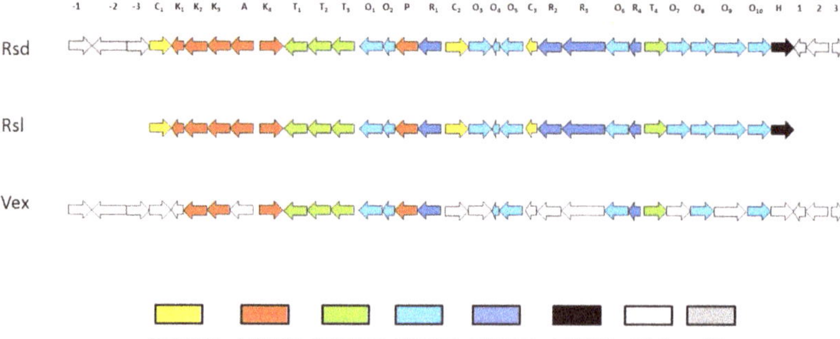

Figure 6. Biosynthetic gene cluster of anthraquinones producers. Rsd: *Streptomyces olivaceus* SCSIO T05 gene cluster [28], Rsl: *Streptomyces bottropensis* (*Streptomyces.* sp. Gc C4/4) gene cluster [29], Vex: *Verrucosispora* sp. SN26_14.1. **C1**: aromatase, **K1**: acyl carrier protein, **K2**: ketosynthase (beta), **K3**: ketosynthase (alpha), **A**: acyl transferase, **K4**: 3-oxoacyl-ACP synthase III, **T1**: ABC-transporter (substrate binding), **T2**: ABC-transporter (ATP binding), **T3**: ABC-transporter trans-membrane, **O1**: luciferase-like monooxygenase, **O2**: flavin reductase, **P**: phosphotransferase, **R1**: SARP family regulator, **C2**: second ring cyclase, **O3**: 3-oxoacyl-ACP reductase, **O4**: anthrone monooxygenase, **O5**: NADH: flavin oxidoreductase, **C3**: cyclase, **R2**: SARP regulatory protein, **R3**: LAL-family regulator, **O6**: luciferase-like monooxygenase, **R4**: MarR family transcriptional regulator, **T4**: drug resistance transporter, **O7**: putative NADPH quinone reductase, **O8**: putative NADPH: quinone oxidoreductase, **O9**: FAD-dependent oxidoreductase, **O10**: C9-keto reductase, **H**: amidohydrolase, **-3**: unknown function, **-2**: major facilitator superfamily protein, **-1**: Transcriptional regulatory protein, **1**: cupin, **2**: citrate/H$^+$ symporter, **3**: transcriptional regulator.

2.7. Antibiotic Activity Test

We performed a disc diffusion antibiotic test as a preliminary evaluation to determine if Lupinacidin A (**1**) and Galvaquinone B (**2**) have an inhibition effect on bacteria. As a positive control, we used streptomycin at a concentration of 25 µg/disc. The results showed that Lupinacidin A (**1**) and Galvaquinone B (**2**) did not produce any growth inhibition against the Gram-positive bacterium *Staphylococcus lentus* DSM 20352^T, and neither against the Gram-negative bacterium *Escherichia coli* DSM 498^T. In contrast, the positive control, streptomycin produced an inhibition halo of 22 mm for Gram-negative and 18 mm for Gram-positive bacteria.

3. Materials and Methods

3.1. Sample Collection

The sea anemone *Gyractis sesere* (also known as *Actiniogeton rapanuiensis*) was sampled from the coastal zone of Easter Island (27°08′45.1″S, 109°25′50.0″W) by the first author (Chilean citizen), in March 2016. The sampling site was outside the Isla de Pascua national park, and the sample was taken in agreement with regulations by the Chilean government. The sample was stored at 0 °C one hour after the sampling process.

3.2. Sea Anemone Dereplication

10 g of the sea anemone *Gyractis sesere* (wet weight) were thawed and homogenized with a mortar and pestle. When a creamy consistency was obtained, the tissue was transferred to a 250 mL beaker and 50 mL of chloroform was added. This extraction procedure was repeated three times. The obtained chloroform extract was concentrated until dryness under reduced pressure in a rotatory evaporator. The dried extract was resuspended in 1 L deionized water and transferred to a separation funnel, where it was partitioned with chloroform (3 × 300 mL). This process produced 8 mg of crude extract with a brownish coloration. Part of the crude extract (0.5 mg) was resuspended in methanol (HPLC grade) and injected in a HPLC (Merck Hitachi LaChrom Elite, Darmstadt, Germany) and in a HRLCMS Thermo Scientific™ Q Exactive™ Hybrid-Quadrupol-Orbitrap (Bremen, Germany), positive mode, and a 30 minute gradient of H_2O and acetonitrile supplemented with 0.1% of formic acid. The gradient developed as following: 0 min: 90% water, 10% acetonitrile, 25 min: 0% water, 100% acetonitrile, 28 min: 0% water, 100% acetonitrile, 30 min: 90% water, 10% acetonitrile. Mass spectroscopic data was evaluated with Xcalibur® (Thermo Fisher Scientific, San Jose, CA, USA), and compared with online databases (MarinLit, and Scifinder), and literature. The entire remaining sample was dissolved in deuterated chloroform (Eurisotop™, Saint-Aubin, France) and analyzed by ^{1}H NMR using a Bruker (Rheinstetten, Germany) Avance 600 MHz NMR spectrometer.

3.3. Bacterial Isolation

Approximately 1 cm^3 (2 g) of the sea anemone *Gyractis sesere* (wet weight) were thawed and homogenized with a sterile mortar and pestle. Subsequently, the homogenized tissue was mixed with 9 mL of Ringer's buffer $\frac{1}{4}$ strength [31] to produce a final solution of 1:10. This solution was incubated at 56 °C for 10 min with the aim of reducing the viability of non-actinobacterial microbes. After the incubation, 1 min of vortex was applied. The inoculation of the culture media was done by adding 50 µL of the dilution into 15 cm diameter Petri dishes containing the media. The inoculum was spread out on the plate with a triangular cell spreader made of glass. Finally, plates were incubated at 25 °C in darkness. Darkness was chosen as a filtering factor to eliminate potential microalgae contamination. Four different media were prepared for the isolation of Actinobacteria from the sea anemone *Gyractis sesere*. Medium SIMA1 (*Salinispora* isolation media A1) was selected from literature and slightly modified as follows: 2.5 g starch, 1 g yeast extract, 0.5 g peptone, 1 L deionized water, and 25 g Tropic Marin™ salt (Wartenberg, Germany), 15 g/L agar [32]. The other media (BCM, BTM, and BSEM) were generated for this study as follows: BCM, 3 g chitin, 0.5 g N-acetyl glucosamine, 0.2 g K_2HPO_4, 0.25 g KNO_3, 0.25 g casein, 5 mL of mineral solution, 4 mL vitamin solution, 1 L deionized water, 15 g/L Tropic Marin™ salt (Wartenberg, Germany), 12 g/L Gellan gum, pH = 7.35; BTM, 1 g trehalose, 0.25 g histidine, 0,25 g proline, 0.2 g $MgCl_2{\cdot}6H_2O$, 4 mL vitamin solution, 12 g/L Gellan gum, 1 L deionized water, 15 g Tropic Marin™ salt (Wartenberg, Germany), pH = 7.2; and BSEM, 0.1 g tyrosine, 0.1 g D-galactose, 4 mL vitamin solution, 5 mL mineral solution, 1 L Baltic Sea water, 16 g/L agar, pH = 7.4. Mineral salt solution contained 1 L distillated water, 50 mg $FeSO_4 \cdot 7H_2O$, 50 mg $ZnCl_2$, and 50 mg $CuSO_4$. Vitamin solution contained 1 L distillated water, 5 mg thiamine·HCl, 5 mg riboflavin, 5 mg niacin, 5 mg pyrodoxine HCl, 5 mg inositol, 5 mg Ca-pantothenate, 5 mg p-amino benzoic acid, and 2.5 mg biotin.

The media were autoclaved for 35 min at 121 °C. Subsequently, the culture media were supplemented with 50 mg/L of nalidixic acid (Sigma-Aldrich, St. Louis, MO, USA) and 100 mg/L of cycloheximide (Carl Roth GmbH, Karlsruhe, Germany) [33], and poured into petri dishes. Once the sample was inoculated onto the petri dish, they were incubated for six weeks. When bacterial colonies were visually evident, we proceeded with the purification of the bacteria until obtaining an axenic culture. The isolated bacteria were conserved using Cryobank™ (Mast Diagnostica GmbH, Reinfeld, Germany) bacterial storage system.

3.4. Molecular Characterization and Phylogenetic Analysis

DNA was extracted from bacterial cells by use of a DNA isolation kit, DNeasy™ (Qiagen, Hilden, Germany), following the manufacturer instructions. Subsequently, the 16S rRNA gene sequence was amplified with PCR and the use of general bacterial primers in a concentration of 10 pmol/µL, i.e., 27f and 1492r [34], 342f and 534r [35], 1387r [36] as well as 1525r [37]. PCR reagents were obtained from GE Healthcare Illustra™ PuReTaq Ready-To-Go™ PCR Beads (GE Healthcare, Glattbrugg, Switzerland) containing DNA polymerase, $MgCl_2$, and dNTPs. The PCR conditions were the same as reported by Staufenberger et al. [35]. Once the PCR amplification process was terminated, a quality check of the PCR products was performed by gel electrophoresis. The sequencing process was run at the Centre for Molecular Biology at Kiel University (IKMB). The 16S rRNA gene sequences were manually curated using Chromas pro software, version 1.7.6 (Technelysium Pty Ltd., Tewantin QLD, Australia), and saved in FASTA format. Sequences were aligned with nucleotide BLAST [38] and EZbiocloud [39]. Phylogenetic analysis involved the alignment of the sequences with related reference strains in the web platform SILVA-SINA [40]. MEGA was used to delete gap sites and to run bootstrapped phylogenetic trees using a neighbor-joining model [41].

3.5. Bacterial Growth for Secondary Metabolites Production

For the evaluation of the secondary metabolites production, we grew the Easter Island isolated strain *Verrucosispora* sp. SN26_14.1 in 10 × 2.5 L Thomson Ultra Yield® flasks (Thomson Instrument, Oceanside, CA, USA), which contained 1 L each of a modified starch-glucose-glycerol (SGG) liquid medium [31]. The composition of the production medium was: 5 g glucose, 5 g soluble starch, 5 g glycerol, 1.25 g cornsteep powder, 2.5 g peptone, 1 g yeast extract, 1.5 $CaCO_3$, and 1 L deionized water. The medium was also supplemented with 15 g/L Tropic Marin™ salt (Wartenberg, Germany). The pH was adjusted to 7.7 using 1 M HCl and NaOH. The culture was kept in orbital agitation at 240 RPM, 28 °C, for 14 days in darkness.

3.6. Chemical Extraction, Purification and Structure Elucidation

After the growth period, 20 g/L amberlite XAD-16 (Sigma-Aldrich, St. Louis, MO, USA) was added to each culture medium flask and mixed for one hour using orbital agitation with 120 rpm. Subsequently, the resin was separated through cheesecloth filtration [42], and the liquid was discarded. Afterwards, amberlite plus cheesecloth was mounted on a glass funnel, washed with 3 L of deionized water, and eluted with 1 L of acetone [42]. Acetone was then concentrated under reduced pressure until an aqueous residue was obtained. One liter of deionized water was added to the acetone residue, and it was brought to a separation funnel. The organic molecules were extracted using 3 × 1 L of ethyl acetate. The organic phase was concentrated under reduced pressure until dryness.

For the evaluation of the produced metabolites, we used HPLC-DAD (Merck Hitachi LaChrom Elite, Darmstadt, Germany) and a 30 min gradient of H_2O-acetonitrile supplemented with 0.1% of formic acid. The gradient was developed as following: 0 min: 90% water, 10% acetonitrile, 25 min: 0% water, 100% acetonitrile, 28 min: 0% water, 100% acetonitrile, 30 min: 90% water, 10% acetonitrile. The gravity SB™ C-18 column was obtained from Macherey-Nagel (Düren, Germany).

The purification of chemicals involved three different steps: 1) Flash chromatography using standard silica gel 60, pore size ~ 60 Å (Macherey-Nagel, Düren, Germany) as a stationary phase, mounted in a glass Buchner funnel (D = 70 mm, H = 180 mm). The mobile phase solvents were iso-octane and ethyl acetate. The chromatographic process was developed in a stepwise increase of polarity (10% each), starting with 100% iso-octane, and 0% of ethyl acetate, and ending in 0% iso-octane and 100% ethyl acetate, resulting in 10 different fractions. 2) The fraction that contained compound (**1**) and (**2**) was selected and worked in HPLC (Merck Hitachi LaChrom Elite, Darmstadt, Germany) using a normal phase NUCLEODUR® 100-5 column (4.6 × 250 mm) from Macherey-Nagel (Düren, Germany). The method used for the purification was a combination of isocratic and gradient solvent mix, with a flow rate of 1 mL/min, where A: iso-octane, B: ethyl acetate, and C: dichloromethane/methanol (50:50). The method was developed as following: 0 min: 100% A and 0% B, 3 min: 100% A and 0% B, 5 min: 95% A and 5% B, 9 min: 95% A and 5% B, 11 min: 0% A and 100% B, 13 min: 0% A and 100% B, 14 min: 10% A, 50% B, and 40% C, 16 min: 10% A, 50% B, and 40% C, 18 min: 50% A and 50% B, 19 min: 100% A and 0% B, 21 min: 100% A and 0% B. 3) The semi-purified compounds were purified through HPLC (Merck Hitachi LaChrom Elite, Darmstadt, Germany using a reverse phase C-18 column, 10 × 250 mm (YMC, Kyoto, Japan). The method used for the purification was a combination of isocratic and gradient solvent mix, with a flow rate of 2.5 mL/min. The method was developed as following: 0 min: 90% A and 10% B, 5 min: 20% A and 80% B, 9 min: 20% A and 80% B, 13 min: 0% A and 100% B, 19 min: 0% A and 100% B, 23 min: 90% A and 10% B, 25 min: 90% A, 10% B (A. water, B: acetonitrile).

After these purification steps, Lupinacidin A (**1**) and Galvaquinone B (**2**) were obtained with high purity to perform structural elucidation experiments. HRLCMS was performed with a Thermo Scientific™ Q Exactive™ Hybrid-Quadrupol-Orbitrap (Thermo Scientific, Bremen, Germany), positive mode, and a 30 min gradient of H_2O and acetonitrile supplemented with 0.1% of formic acid. The gradient was developed as follows: 0 min: 90% water, 10% acetonitrile, 25 min: 0% water, 100% acetonitrile, 28 min: 0% water, 100% acetonitrile, 30 min: 90% water, 10% acetonitrile. Mass spectroscopic data was evaluated with Xcalibur® (Thermo Fisher Scientific, San Jose, CA, USA), and the compared with online databases (MarinLit, and Scifinder), and literature.

Additionally, ^1H and ^{13}C NMR and two-dimensional NMR experiments (HMBC, HSQC, COSY) were acquired to characterize the main components of crude extract, and their chemical functionality. For this, compound (**1**) and (**2**) were redissolved in $CDCl_3$ (Eurisotop™, Saint-Aubin, France), and transferred to NMR tubes (178 × 5.0 mm). Experiments were acquired on a Bruker (Rheinstetten, Germany) Avance spectrometer operating at 600 MHz proton frequency equipped with a cryogenically cooled triple resonance z-gradient probe head using stand pulse sequences from the Bruker experiment library. Spectra were referenced against tetramethylsilane (Sigma-Aldrich, St. Louis, MO, USA) as internal standard.

3.7. Genome Sequencing

The samples were prepared with the Nextera® XT DNA sample preparation kit from Illumina (Illumina, San Diego, CA, USA) following the manufacturer's protocol. Afterwards the samples were pooled and sequenced on the Illumina MiSeq using the MiSeq® (Illumina, San Diego, CA, USA) Reagent Kit v3 600 cycles sequencing chemistry. The library was clustered to a density of approximately 1200 K/mm².

3.8. Genome Assembly

The quality control of reads was checked with FASTQC software [43] to evaluate the GC%, number of k-mers, sequence length, and total reads. Trimmomatic v0.36 [44] was used to filter low quality sequences and adapters. Filtered reads were assembled with SPAdes v3.11.0 [45] using default k-mer lengths. The obtained contigs were evaluated with QUAST tool [46] to select the best quality contig. Finally, Prokka [47] was used to annotate the draft genome.

3.9. Secondary Metabolites Gene Clusters Search

The online platform of Antismash [30] was used to detect the secondary metabolites gene clusters present in the draft genome.

3.10. Antibiotic Activity Test

To test the antibiotic activity, we used the disc diffusion method [48] as a primary indicator. Thus, compound (**1**) and (**2**) were tested to determine their activity on *Staphylococcus lentus* DSM 20352^T, and *Escherichia coli* DSM 498^T. These bacteria were cultured in GYM medium (4 g glucose, 4 g yeast extract, 10 g malt extract, 2 g $CaCO_3$, 1 L deionized water, pH = 7.2, and 12 g agar). Lupinacidin A (**1**), and Galvaquinone B (**2**) were transferred to a paper disc to reach a final concentration of 25 µg and 50 µg each in triplicate. Additionally, we used an antibiotic susceptibility disc of streptomycin (Oxoid$^®$, Columbia, MD, USA) as a positive indicator of antibiotic activity. The plates were inoculated with fresh culture of *Staphylococcus lentus* DSM 20352^T, and *Escherichia coli* DSM 498^T, and incubated at 37 °C for 24 h. After the incubation period, the inhibition zone was measured and registered.

4. Conclusions

We established that the Easter Island sea anemone *Gyractis sesere* contained two anthraquinones, Lupinacidin A (**1**) and Galvaquinone B (**2**), which were ultimately found to be produced by one of the Actinobacteria associated with this marine invertebrate, *Verrucosispora* sp. SN26_14.1. The production of the identified metabolites by the bacterial isolate apparently follows a recently characterized PKS type II pathway with a Baeyer−Villiger type rearrangement assembly line. Our finding adds a new actinobacterial genus to the producers of these anthraquinones, implying that these metabolites are not exclusive to the genera *Streptomyces* and *Micromonospora*. It was demonstrated, that culture-based approaches remain as effective tools for the isolation of polyketide producing Actinobacteria as sources for secondary metabolites of potential use in drug discovery. Our study confirms that cnidarians, and in specific sea anemones, can be a source of such pharmacologically relevant microorganisms. Finally, these findings re-open the debate about the real producers of secondary metabolites in sea animals and add another example of associated bacteria as producers of substances present in sea animals. In addition, the study provides information on the chemistry harbored in biota of the geographically isolated and almost unstudied, Easter Island.

Supplementary Materials: The following are available online at http://www.mdpi.com/1660-3397/17/3/154/s1. Information on NMR spectra, HRLCMS data, and secondary metabolite gene cluster.

Author Contributions: I.S., J.F.I. and J.W. planned the experiments, I.S. performed the experiments, analyzed and evaluated the data and wrote the first draft of the publication. J.F.I., F.D.S. and J.W. supervised the work and revised the manuscript. S.K. sequenced the genome and supplied the genome data. M.L. and N.P. acquired LCMS and NMR data. F.D.S. acquired and analyzed NMR data.

Funding: We thank the Deutscher Akademischer Austauschdienst (DAAD) for financial support under the stipend # PKZ91564794. We acknowledge financial support by the Land Schleswig-Holstein within the funding programme Open Access Publikationsfonds.

Acknowledgments: We thank Marion Höftmann and Gitta Kohlmeyer-Yilmaz for her valuable help with the NMR data acquisition. We thank Ute Hentschel Humeida (GEOMAR Helmholtz Centre for Ocean Research Kiel) for her support. We also thank Philip A. Thomas (ImagesByPT@PhilipT.com), for providing high quality photography of the sea anemone *Gyractis sesere*. I.S. thanks Millaray Sierra for her support during the research. N.P. thanks the Deutsche Bundesstiftung Umwelt (German Federal Environmental Foundation) for a predoctoral fellowship.

Conflicts of Interest: The authors declare no conflict of interest.

References

1. Dransfield, J.; Flenley, J.R.; King, S.M.; Harkness, D.D.; Rapu, S. A recently extinct palm from Easter Island. *Nature* **1984**, *312*, 750. [CrossRef]
2. Boyko, C.B. The endemic marine invertebrates of Easter Island: How many species and for how long? In *Easter Island: Scientific Exploration into the World's Environmental Problems in Microcosm*; Loret, J., Tanacredi, J.T., Eds.; Springer: Boston, MA, USA, 2003; pp. 155–175.
3. Skottsberg, C. *The Natural History of Juan Fernández and Easter Island*; Almqvist & Wiksells Boktryckeri: Uppsala, Sweden, 1920.
4. Kohn, A.J.; Lloyd, M.C. Marine polychaete annelids of Easter Island. *Int. Rev. Hydrobiol.* **1973**, *58*, 691–712. [CrossRef]
5. Sehgal, S.N. Rapamune® (RAPA, rapamycin, sirolimus): Mechanism of action immunosuppressive effect results from blockade of signal transduction and inhibition of cell cycle progression. *Clin. Biochem.* **1998**, *31*, 335–340. [CrossRef]
6. Vezina, C.; Kudelski, A.; Sehgal, S. Rapamycin (AY-22, 989), a new antifungal antibiotic. *J. Antibiot.* **1975**, *28*, 721–726. [CrossRef] [PubMed]
7. Allen, G.R. Conservation hotspots of biodiversity and endemism for Indo-Pacific coral reef fishes. *Aquat. Conserv.* **2008**, *18*, 541–556. [CrossRef]
8. Horton, T.; Kroh, A.; Ahyong, S.; Bailly, N.; Boyko, C.B.; Brandão, S.N.; Costello, M.J.; Gofas, S.; Hernandez, F.; Holovachov, O.; et al. *World Register of Marine Species (WoRMS)*; WoRMS Editorial Board: Ostend, Belgium, 2018.
9. Urda, C.; Fernández, R.; Pérez, M.; Rodríguez, J.; Jiménez, C.; Cuevas, C. Protoxenicins A and B, cytotoxic long-chain acylated xenicanes from the soft coral *Protodendron repens*. *J. Nat. Prod.* **2017**, *80*, 713–719. [CrossRef] [PubMed]
10. Nakamura, F.; Kudo, N.; Tomachi, Y.; Nakata, A.; Takemoto, M.; Ito, A.; Tabei, H.; Arai, D.; de Voogd, N.; Yoshida, M.; et al. Halistanol sulfates I and J, new SIRT1–3 inhibitory steroid sulfates from a marine sponge of the genus *Halichondria*. *J. Antibiot.* **2017**, *71*, 273. [CrossRef] [PubMed]
11. Aminin, D.; Menchinskaya, E.; Pisliagin, E.; Silchenko, A.; Avilov, S.; Kalinin, V. Anticancer activity of sea cucumber triterpene glycosides. *Mar. Drugs* **2015**, *13*, 1202–1223. [CrossRef] [PubMed]
12. Hay, M.E. Marine chemical ecology: What's known and what's next? *J. Exp. Mar. Biol. Ecol.* **1996**, *200*, 103–134. [CrossRef]
13. Mehbub, M.F.; Lei, J.; Franco, C.; Zhang, W. Marine sponge derived natural products between 2001 and 2010: Trends and opportunities for discovery of bioactives. *Mar. Drugs* **2014**, *12*, 4539–4577. [CrossRef] [PubMed]
14. Blunt, J.W.; Copp, B.R.; Keyzers, R.A.; Munro, M.H.G.; Prinsep, M.R. Marine natural products. *Nat. Prod. Rep.* **2014**, *31*, 160–258. [CrossRef] [PubMed]
15. Piel, J.; Hui, D.; Wen, G.; Butzke, D.; Platzer, M.; Fusetani, N.; Matsunaga, S. Antitumor polyketide biosynthesis by an uncultivated bacterial symbiont of the marine sponge *Theonella swinhoei*. *Proc. Natl. Acad. Sci. USA* **2004**, *101*, 16222–16227. [CrossRef] [PubMed]
16. Rath, C.M.; Janto, B.; Earl, J.; Ahmed, A.; Hu, F.Z.; Hiller, L.; Dahlgren, M.; Kreft, R.; Yu, F.; Wolff, J.J.; et al. Meta-omic characterization of the marine invertebrate microbial consortium that produces the chemotherapeutic natural product ET-743. *ACS Chem. Biol.* **2011**, *6*, 1244–1256. [CrossRef] [PubMed]
17. Unson, M.D.; Holland, N.D.; Faulkner, D.J. A brominated secondary metabolite synthesized by the cyanobacterial symbiont of a marine sponge and accumulation of the crystalline metabolite in the sponge tissue. *Mar. Biol.* **1994**, *119*, 1–11. [CrossRef]
18. Schmidt, E.W.; Obraztsova, A.Y.; Davidson, S.K.; Faulkner, D.J.; Haygood, M.G. Identification of the antifungal peptide-containing symbiont of the marine sponge *Theonella swinhoei* as a novel δ-Proteobacterium, "Candidatus *Entotheonella palauensis*". *Mar. Biol.* **2000**, *136*, 969–977. [CrossRef]
19. Feng, Y.; Khokhar, S.; Davis, R.A. Crinoids: Ancient organisms, modern chemistry. *Nat. Prod. Rep.* **2017**, *34*, 571–584. [CrossRef] [PubMed]
20. Khokhar, S.; Pierens, G.K.; Hooper, J.N.A.; Ekins, M.G.; Feng, Y.; Davis, R.A. Rhodocomatulin-type anthraquinones from the australian marine invertebrates *Clathria hirsuta* and *Comatula rotalaria*. *J. Nat. Prod.* **2016**, *79*, 946–953. [CrossRef] [PubMed]

21. Tietze, L.F.; Gericke, K.M.; Schuberth, I. Synthesis of highly functionalized anthraquinones and evaluation of their antitumor activity. *Eur. J. Org. Chem.* **2007**, *2007*, 4563–4577. [CrossRef]

22. Yang, K.-L.; Wei, M.-Y.; Shao, C.-L.; Fu, X.-M.; Guo, Z.-Y.; Xu, R.-F.; Zheng, C.-J.; She, Z.-G.; Lin, Y.-C.; Wang, C.-Y. Antibacterial anthraquinone derivatives from a sea anemone-derived fungus *Nigrospora* sp. *J. Nat. Prod.* **2012**, *75*, 935–941. [CrossRef] [PubMed]

23. Kim, Y.-M.; Lee, C.-H.; Kim, H.-G.; Lee, H.-S. Anthraquinones isolated from *Cassia tora* (Leguminosae) Seed Show an Antifungal Property against Phytopathogenic Fungi. *J. Agric. Food Chem.* **2004**, *52*, 6096–6100. [CrossRef] [PubMed]

24. Hu, Y.; Martinez, E.D.; MacMillan, J.B. Anthraquinones from a marine-derived *Streptomyces spinoverrucosus*. *J. Nat. Prod.* **2012**, *75*, 1759–1764. [CrossRef] [PubMed]

25. Igarashi, Y.; Trujillo, M.E.; Martínez-Molina, E.; Yanase, S.; Miyanaga, S.; Obata, T.; Sakurai, H.; Saiki, I.; Fujita, T.; Furumai, T. Antitumor anthraquinones from an endophytic actinomycete *Micromonospora lupini* sp. nov. *Bioorg. Med. Chem. Lett.* **2007**, *17*, 3702–3705. [CrossRef] [PubMed]

26. Haddon, A.C.; Shackleton, A.M. Description of some new species of Actiniaria from Torres Straits. *R. Dublin Soc.* **1893**, *8*.

27. Carlgren, O. Actiniaria und zoantharia von Juan Fernandez und der Osterinsel. In *The Natural History of Juan Fernandez and Easter Island*; Skottsberg, C., Ed.; Almquist & Wiksells Boktryckeri: Uppsala, Sweden, 1922; pp. 145–160.

28. Zhang, C.; Sun, C.; Huang, H.; Gui, C.; Wang, L.; Li, Q.; Ju, J. Biosynthetic Baeyer–Villiger chemistry enables access to two anthracene scaffolds from a single gene cluster in deep-sea-derived *Streptomyces olivaceus* SCSIO T05. *J. Nat. Prod.* **2018**, *81*, 1570–1577. [CrossRef] [PubMed]

29. Yan, X.; Probst, K.; Linnenbrink, A.; Arnold, M.; Paululat, T.; Zeeck, A.; Bechthold, A. Cloning and heterologous expression of three type II PKS gene clusters from *Streptomyces bottropensis*. *ChemBioChem* **2012**, *13*, 224–230. [CrossRef] [PubMed]

30. Weber, T.; Blin, K.; Duddela, S.; Krug, D.; Kim, H.U.; Bruccoleri, R.; Lee, S.Y.; Fischbach, M.A.; Müller, R.; Wohlleben, W.; et al. AntiSMASH 3.0—A comprehensive resource for the genome mining of biosynthetic gene clusters. *Nucleic Acids Res.* **2015**, *43*, W237–W243. [CrossRef] [PubMed]

31. Goodfellow, M.; Fiedler, H.-P. A guide to successful bioprospecting: Informed by actinobacterial systematics. *J. Microb.* **2010**, *98*, 119–142. [CrossRef] [PubMed]

32. Patin, N.V.; Duncan, K.R.; Dorrestein, P.C.; Jensen, P.R. Competitive strategies differentiate closely related species of marine Actinobacteria. *ISME J.* **2015**, *10*, 478. [CrossRef] [PubMed]

33. Thaker, M.N.; Waglechner, N.; Wright, G.D. Antibiotic resistance–mediated isolation of scaffold-specific natural product producers. *Nat. Protoc.* **2014**, *9*, 1469. [CrossRef] [PubMed]

34. Lane, D.J. 16S/23S rRNA sequencing. In *Nucleic Acid Techniques in Bacterial Systematics*; Stackebrandt, E., Goodfellow, M., Eds.; John Wiley and Sons: Chichester, UK, 1991; pp. 115–175.

35. Staufenberger, T.; Thiel, V.; Wiese, J.; Imhoff, J.F. Phylogenetic analysis of bacteria associated with *Laminaria saccharina*. *FEMS Microbiol. Ecol.* **2008**, *64*, 65–77. [CrossRef] [PubMed]

36. Ellis, R.J.; Morgan, P.; Weightman, A.J.; Fry, J.C. Cultivation-dependent and independent approaches for determining bacterial diversity in heavy-metal-contaminated soil. *Appl. Environ. Microb.* **2003**, *69*, 3223. [CrossRef]

37. Frank, J.A.; Reich, C.I.; Sharma, S.; Weisbaum, J.S.; Wilson, B.A.; Olsen, G.J. Critical evaluation of two primers commonly used for amplification of bacterial 16S rRNA genes. *Appl. Environ. Microb.* **2008**, *74*, 2461. [CrossRef] [PubMed]

38. Altschul, S.F.; Gish, W.; Miller, W.; Myers, E.W.; Lipman, D.J. Basic local alignment search tool. *J. Mol. Biol.* **1990**, *215*, 403–410. [CrossRef]

39. Yoon, S.-H.; Ha, S.-M.; Kwon, S.; Lim, J.; Kim, Y.; Seo, H.; Chun, J. Introducing EzBioCloud: A taxonomically united database of 16S rRNA gene sequences and whole-genome assemblies. *Int. J. Syst. Evol. Microbiol.* **2017**, *67*, 1613–1617. [CrossRef] [PubMed]

40. Quast, C.; Pruesse, E.; Yilmaz, P.; Gerken, J.; Schweer, T.; Yarza, P.; Peplies, J.; Glöckner, F.O. The SILVA ribosomal RNA gene database project: Improved data processing and web-based tools. *Nucleic Acids Res.* **2013**, *41*, D590–D596. [CrossRef] [PubMed]

41. Tamura, K.; Peterson, D.; Peterson, N.; Stecher, G.; Nei, M.; Kumar, S. MEGA5: Molecular evolutionary genetics analysis using maximum likelihood, evolutionary distance, and maximum parsimony methods. *Mol. Biol. Evol.* **2011**, *28*, 2731–2739. [CrossRef] [PubMed]

42. Cheng, Y.B.; Jensen, P.R.; Fenical, W. Cytotoxic and antimicrobial napyradiomycins from two marine-derived *Streptomyces* strains. *Eur. J. Org. Chem.* **2013**, 3751–3757. [CrossRef] [PubMed]

43. Andrews, S. FastQC a Quality Control Tool for High Throughput Sequence Data. Available online: http://www.bioinformatics.babraham.ac.uk/projects/fastqc/ (accessed on 30 May 2018).

44. Bolger, A.M.; Usadel, B.; Lohse, M. Trimmomatic: A flexible trimmer for Illumina sequence data. *Bioinformatics* **2014**, *30*, 2114–2120. [CrossRef] [PubMed]

45. Bankevich, A.; Nurk, S.; Antipov, D.; Gurevich, A.A.; Dvorkin, M.; Kulikov, A.S.; Lesin, V.M.; Nikolenko, S.I.; Pham, S.; Prjibelski, A.D.; et al. SPAdes: A new genome assembly algorithm and its applications to single-cell sequencing. *J. Comput. Biol.* **2012**, *19*, 455–477. [CrossRef] [PubMed]

46. Gurevich, A.; Tesler, G.; Vyahhi, N.; Saveliev, V. QUAST: Quality assessment tool for genome assemblies. *Bioinformatics* **2013**, *29*, 1072–1075. [CrossRef] [PubMed]

47. Seemann, T. Prokka: Rapid prokaryotic genome annotation. *Bioinformatics* **2014**, *30*, 2068–2069. [CrossRef] [PubMed]

48. Bondi, A., Jr.; Spaulding, E.H.; Smith, D.E.; Dietz, C. A routine method for the rapid determination of susceptibility to penicillin and other antibiotics. *Am. J. Med. Sci.* **1947**, *213*, 221–225. [CrossRef] [PubMed]

Article

A Multi-screening Evaluation of the Nutritional and Nutraceutical Potential of the Mediterranean Jellyfish *Pelagia noctiluca*

Rosaria Costa [1], **Gioele Capillo** [2,*], **Ambrogina Albergamo** [1,3,*], **Rosalia Li Volsi** [2], **Giovanni Bartolomeo** [1,3], **Giuseppe Bua** [1,3], **Antonio Ferracane** [1,3], **Serena Savoca** [2], **Teresa Gervasi** [1,3], **Rossana Rando** [1], **Giacomo Dugo** [1,3] and **Nunziacarla Spanò** [1]

[1] Dipartimento di Scienze Biomediche, Odontoiatriche, e delle Immagini Morfologiche e Funzionali (Biomorf), University of Messina, Viale Annunziata, 98168 Messina, Italy; costar@unime.it (R.C.); gbartolomeo@unime.it (G.B.); gbua@unime.it (G.B.); aferracane@unime.it (A.F.); tgervasi@unime.it (T.G.); rrando@unime.it (R.R.); dugog@unime.it (G.D.); spano@unime.it (N.S.)

[2] Dipartimento di Scienze Chimiche, Biologiche, Farmaceutiche ed Ambientali (ChiBioFarAm), University of Messina, Viale Annunziata, 98168 Messina, Italy; linda89lv@gmail.com (R.L.V.); ssavoca@unime.it (S.S.)

[3] Science4Life s.r.l., a Spin-Off of the University of Messina, 98168 Messina, Italy

* Correspondence: gcapillo@unime.it (G.C.); aalbergamo@unime.it (A.A.); Tel.: +39-347-2236971 (G.C.); +39-090-3503996 (A.A.)

Received: 30 January 2019; Accepted: 13 March 2019; Published: 17 March 2019

Abstract: The phylum Cnidaria is one of the most important contributors in providing abundance of bio- and chemodiversity. In this study, a comprehensive chemical investigation on the nutritional and nutraceutical properties of Mediterranean jellyfish *Pelagia noctiluca* was carried out. Also, compositional differences between male and female organisms, as well as between their main anatomical parts, namely bell and oral arms, were explored in an attempt to select the best potential sources of nutrients and/or nutraceuticals from jellyfish. With the exception of higher energy densities and total phenolic contents observed in females than males, no statistically significant differences related to the specimen's sex were highlighted for the other compound classes. Rather, the distribution of the investigated chemical classes varied depending on the jellyfish's body parts. In fact, crude proteins were more abundant in oral arms than bells; saturated fatty acids were more concentrated in bells than oral arms, whereas polyunsaturated fatty acids were distributed in the exact opposite way. On the other hand, major elements and trace elements demonstrated an opposite behavior, being the latter most accumulated in oral arms than bells. Additionally, important nutraceuticals, such as eicosapentaenoic and docosahexaenoic acids, and antioxidant minerals, were determined. Overall, obtained data suggest the potential employment of the Mediterranean *P. noctiluca* for the development of natural aquafeed and food supplements.

Keywords: *Pelagia noctiluca*; Mediterranean jellyfish; chemical characterization; aquafeed and food supplements; sustainable fishing

---

1. Introduction

The marine environment and its inhabitants are today recognized as an enormous reservoir of bioactive substances to be exploited for pharmaceutical and aquaculture applications, and as nutraceuticals as well [1–4]. Nonetheless, a number of anthropogenic activities have negatively affected marine ecosystems, inhibiting the derived services, and their precious bioactive resources as well [5–8].

In this scenario, Cnidaria such as scypho-, cubo-, and hydromedusae, simply referred to as 'jellyfish', have become issues of public concern, due to the huge proliferations taking place in coastal waters. The triggers of jelly 'blooms' are typically identified in (i) global warming [9,10], (ii) overfishing [11,12], and (iii) eutrophication [13,14].

When they bloom, jellyfish are infamous for having a negative impact on the structure and function of marine ecosystems and, consequently, on the related economic and social activities [10]. They grow fast and feed mainly on zooplankton and fish larvae, thus depressing many fish catches (e.g., anchovy and sardines) [15,16]. Also, intense Cnidaria outbreaks have reported to damage mariculture, as small specimens and their oral arms can enter net-pens, inducing gill hemorrhage and subsequent fish suffocation [17,18], and can clog the cooling water pipelines of coastal plants, with consequent power reductions and shutdowns [19]. Last but not least, due to the presence of nematocysts causing painful stings, jellyfish are notoriously venomous, and, in some cases, may compromise the general health status of the wounded swimmers and bathers, affecting negatively the tourism of many coastal localities [20].

Despite being traditionally considered as indicators of perturbed ecosystems and trophic dead ends, Cnidaria are nowadays object of a 'paradigm shift' reconsidering their ecological role. Indeed, in marine ecosystems, jellyfish (i) represent relevant prey and predators [21,22]; (ii) provide a shelter for certain juveniles fish species [23] and a habitat for various invertebrate organisms [24–26]; (iii) serve as hosts with photosynthetic algae, such as zooxanthellae, which, in turn, are critical to the ephyrae metamorphosis and the survival of jellyfish, supplying much of the carbon requirements [27–29]; and (iv) contribute to the nutritional cycling of the trophic web [30,31]

Due to the high abundances and reproductive potential, paradoxically, jellyfish may be considered as value-added products, with various benefits for humans. Historically, jellyfish constitute a gourmet dish to be consumed in weddings and formal banquets, following a secret processing based on dehydration by water and alum. Umbrella are generally used for food consumption; however, oral arms from those species whose nematocyst toxin is relatively innocuous have been also considered [32,33]. According to recent estimates, at least 18 countries catch jellyfish for food, and a dozen or more countries are either exploring new fisheries or have been involved in jellyfish fisheries in the past [34]. Many countries do not report their catches of jellyfish explicitly to the Food and Agriculture Organization of the United Nations (FAO), as they include them either as 'other aquatic animals' (sea cucumbers, sea urchins, and edible jellyfish) or not at all. As a result, the average catch of 'other aquatic animals' reported by FAO in 2016, was approximately 938,500 tons (USD 6.8 billion) [35].

Species from the order Rhizostomeae, class Scyphozoa, are considered edible [36]. In this respect, *Rhopilema hyspidum* and *Rhopilema esculentum* represent relevant fishery and reared species in China [37], Malaysia [38], and Japan [39]; whereas *Stomolophus meleagris*, *Catostylus mosaicus*, and *Rhizostoma pulmo* are emerging fishery species respectively in the Gulf of Mexico [32], Australia [40], and Pakistan [41]. Overall, they are appreciated not only for the crunchy and crispy texture and taste, but also for their chemical composition, which ensures a low calorie intake, being low in carbohydrate, fat, and cholesterol. Although less desirable and not currently targeted at commercial scale, other Scyphozoa jellyfish (order Semaestomeae) such as Aurelia [42], Chrysaora [43], and Cyanea [44] have been consumed. Additionally, limited information suggests that cubozoans may be eaten in eastern regions, such as Taiwan [45]. Jellyfish have long been recognized also for their pharmaceutical value. *R. esculentum*, for example, is characterized by proteins from oral arms with a significant antioxidant activity [46]. Collagen from *S. meleagris* demonstrated to be an effective cure for rheumatoid arthritis in rats [47]; while collagen from *R. esculentum* showed antimelanogenic activity due to antioxidant properties and copper-chelating ability [48]. Jellyfish collagen might be also used in the biomedical area, for cartilage and bone reconstruction [49,50] and in the cosmetic field, for producing creams and lotions for skin care [51–53].

With this background, turning other yet unexplored jellyfish species from a nuisance into a sustainable resource becomes imperative for (i) controlling the size of jellyfish populations and (ii) maximizing the benefits related, but not limited, to their nutritional and nutraceutical potential.

Pelagia noctiluca (Scyphomedusae, Semaeostomae), also called "the mauve stinger", is a pelagic jellyfish characterized by wide distribution, abundance, and relevant ecological role as well [54]. Massive outbreaks of *P. noctiluca* have increasingly occurred in the Mediterranean area largely as a result of anthropogenic activity, as evidenced by numerous studies from the late 1970s and 1980s, [55–57]. *P. noctiluca* has been consequently explored also for its toxicological relevance. Indeed, nematocyst morphology, toxicity [58,59], and activation [60], as well as symptoms and epidemiology of stings [61,62], have been addressed.

Nonetheless, the number of studies reporting data on the chemical composition of *P. noctiluca*, and its exploitation as nutraceutical source also appears to be quite low. To the best knowledge of the authors, only Milisenda and coworkers recently investigated different compositional aspects of *P. noctiluca* from the Strait of Messina for studying (i) sexual reproduction [63], (ii) trophic relationships [64], and (iii) dynamics of fish predation [65].

Therefore, in the present study, the nutritional value and nutraceutical value of *P. noctiluca* were investigated through the application of a variety of techniques. Specifically, biometric properties, gross energy, total polyphenol, and protein contents of the mauve stinger were determined. Also, fatty acid composition and major and trace element profile were screened. Compositional differences related to sex, and to the main anatomical parts, namely bell and oral arms, were explored in an attempt to select the best source(s) of nutrients and/or nutraceuticals from such species.

Scope of the work was not only to provide a chemical fingerprint of the mauve stinger, but also to encourage the fishing of the natural populations in the Mediterranean area for supporting its potential employment as feed supplementation and/or nutraceutical.

2. Results

2.1. Biometrics

Biometrics of *P. noctiluca*, caught from March to June, are reported in Figure 1; while in Table S1, the summary output of the two-way ANOVA analysis performed on investigated parameters, such as length (cm), bell diameter (cm), and weight (g) of the wet mass, is reported.

For each dependent morphometric variable, the two-way ANOVA analysis pointed out statistically significant interactions ($p < 0.05$) between the two independent variables (namely, specimen's sex and sampling month) (Table S1).

Between March and April (2017), female organisms were characterized by length ranging between 10.6 and 13.0 cm ($p > 0.05$), and bell diameter between 8.8 and 9.9 cm ($p > 0.05$); while length and bell diameter of males were between 6.5 and 9.6 cm ($p > 0.05$) and 7.7 and 8.6 cm ($p > 0.05$), respectively. Weight oscillated between 39.5 and 58.2 g for females ($p < 0.05$) and 37.0 and 54.7 g for male specimens ($p < 0.05$) (Figure 1).

Between May and June, minor lengths and bell diameters were recorded both for females (9.0–7.7 cm, $p > 0.05$, and 6.5–6.7 cm, $p > 0.05$, respectively) and males (7.1–7.7 cm, $p > 0.05$, and 7.2–7.5 cm, $p > 0.05$, respectively). Also, females, on average, had a lower weight distribution than males, with weights ranging from 25.7 to 31.9 g ($p < 0.05$) and from 31.7 to 31.8 g ($p > 0.05$), respectively (Figure 1).

Considering the different sampling periods, greater fluctuations of length, bell diameter and weight, were observed between March and April than June and May, both in female and male specimens. Additionally, from March–April to June–May, the investigated parameters were significantly reduced ($p < 0.05$) in female jellyfish. A significant reduction ($p < 0.05$) of wet mass was also observed in male jellyfish, but similar lengths ($p > 0.05$) and bell diameters ($p > 0.05$) were recorded throughout the sampling period.

Considering the sex specimens, in March, female and male medusae showed similar bell diameters (8.8 and 7.7 cm, $p > 0.05$) and weights (39.5 and 37.0 cm, $p > 0.05$) but significantly different lengths (13.0 and 7.2 cm, $p < 0.05$). In April and May, the trend changed as bell diameters of females and males were similar (9.9 and 8.6 cm, $p > 0.05$, and 6.5 and 7.2 cm, $p > 0.05$, respectively); while significantly different lengths (10.6 and 9.0 cm, $p < 0.05$, 9.0 and 7.1 cm, $p < 0.05$, respectively) and weights (58.2 and 54.7 g, $p < 0.05$, and 25.7 and 31.7 g, $p < 0.05$, respectively) were observed. Finally in June both female and male organisms showed similar lengths, bell diameters and masses ($p > 0.05$) (Figure 1).

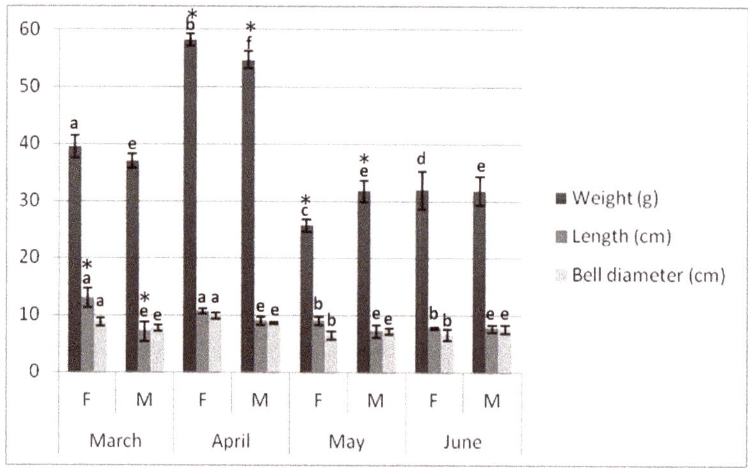

Figure 1. Fresh weight (g), bell diameter (cm), and length (cm) of jellyfish samples collected per sex and month. Data are reported as mean ± standard deviation ($n = 50$). According to the four sampling periods, female (or male) jellyfish marked by different letters for a given biometric characteristic, differ significantly ($p < 0.05$ by post hoc Tukey's Honestly Significant Difference (HSD) test). According to sex specimens, female and male jellyfish marked by the asterisk for a specific parameter in the same sampling month differ significantly ($p < 0.05$ by post hoc Tukey's HSD test). F = female. M = male.

2.2. Gross Energy Contents

The gross energy densities of female and male jellyfish bells and oral arms obtained by bomb calorimetry are outlined in terms of Kcal 100 g^{-1} on a dry weight (dw) basis (Figure 2). The two-way ANOVA analysis revealed that the interaction between specimen's sex and anatomical part (independent variables) was statistically significant ($p < 0.05$, Table S2). Overall, statistically significant differences were found independently of the considered body part ($p < 0.05$), as bell was characterized by higher energy content then oral arms, both in female and male medusae.

Additionally, considering the bell, statistically different energy values were detected between female and male specimens ($p < 0.05$). In fact, the sample from female bells showed the highest calorie value (621 Kcal 100 g^{-1}), followed by the one from male bells (357 Kcal 100 g^{-1}); whereas both male and female oral arms were characterized by inferior and nonsignificantly different calorie levels (174 and 151 Kcal 100 g^{-1}, respectively, $p > 0.05$).

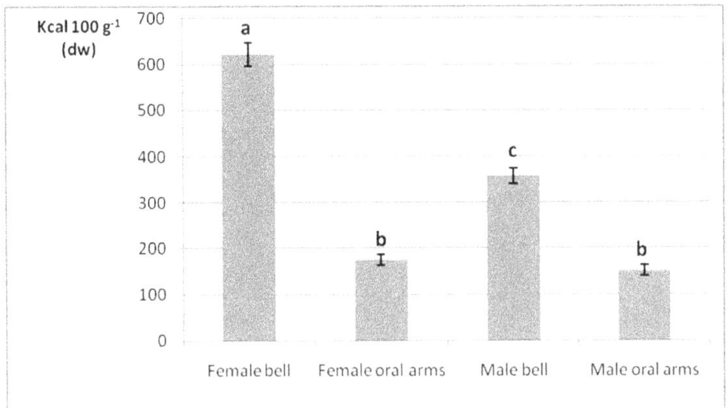

Figure 2. Gross energy densities of female and male jellyfish bells and oral arms. Data are reported as Kcal 100 g^{-1} on a dw basis, in terms of mean ± standard deviation (n = 3). Samples marked by different letters differ significantly ($p < 0.05$ by post hoc Tukey's HSD test).

2.3. Crude Protein

Total nitrogen and protein contents of female and male jellyfish bells and oral arms are reported on a % dw basis by the bar graph in Figure 3. A nonstatistically significant interaction between the two independent variables ($p > 0.05$) was found (Table S3). Overall, investigated bells showed lower protein contents than oral arms (12.09–15.7% vs. 23.5–26.4%, $p < 0.05$), and nonstatistically significant differences were reported between female and male organisms (12.09–23.53% vs. 15.7–26.45%, $p > 0.05$).

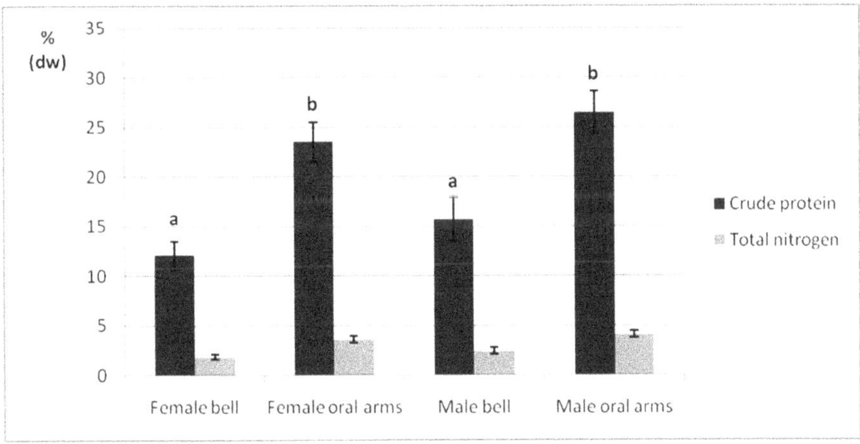

Figure 3. Total nitrogen and protein contents of female and male jellyfish bells and oral arms. Data are reported on a % dw basis, as mean ± standard deviation (n = 3). Samples marked by different letters differ significantly ($p < 0.05$ by post hoc Tukey's HSD test).

2.4. Phenolic Compounds

As mentioned in the Materials and Methods section, at an initial stage of analysis, samples underwent two different preparation procedures, based on the use of PBS and methanol, respectively. Once the method was optimized, the first procedure was preferred, due to better analytical performance in terms of speed and efficiency. The UV spectrophotometric analysis determined the total phenolic content in dried bells and oral arms of male and female organisms in terms of µg of gallic acid equivalents (GAE) g⁻ on a dw basis (Figure 4). According to the two-way ANOVA test, the interaction of specimen's sex and anatomical part was statistically significant ($p < 0.05$, Table S4). In general, phenolic compounds were more abundant in oral arms rather than in bells (1892–2126 µg GAE g^{-1} vs. 914–1303 µg GAE g^{-1}, respectively, $p < 0.05$). Considering the same anatomical part (bell or arms), a statistically significant difference ($p < 0.05$) was also determined between the phenolic contents of male and female specimens: 1303.32 µg GAE g^{-1} vs. 914.22 µg GAE g^{-1}; and 2126.56 µg GAE g^{-1} vs. 1892.69 µg GAE g^{-1}, in males and females, respectively.

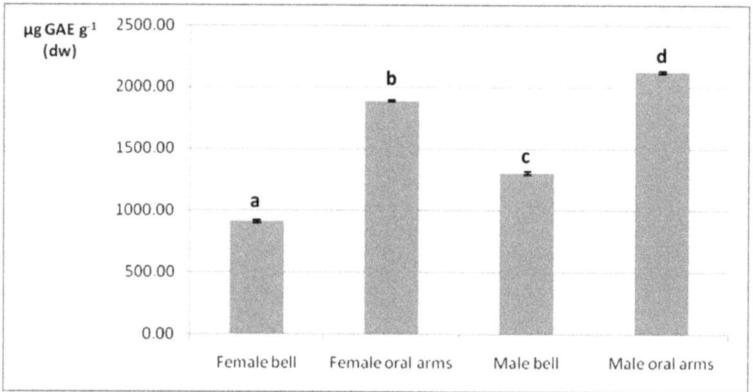

Figure 4. Total phenolic content (µg GAE g^{-1}) determined in male and female jellyfishes' bell and oral arms. Data are expressed as mean ± standard deviation (n = 3), on a dw basis. Samples marked by different letter differ significantly ($p < 0.05$ by post hoc Tukey's HSD test).

2.5. Fatty Acids

GC–MS analyses allowed us to determine around thirty fatty acid methyl esters (FAMEs), ranging from methyl hexanoate to methyl tetracosanoate, in dried samples from *P. noctiluca* (Table 1). With some exception, probably due to amounts below the detection limit, these FAMEs were common to each sample group, namely bells and oral arms of female and male specimens.

From a quantitative point of view, the main fatty acid classes varied depending on specimens' body part rather than sex. In fact, saturated fatty acids (SFAs), representing 70% ca. in bells and 65% ca. in oral arms, respectively; monounsaturated fatty acids (MUFAs), accounting for 15% ca. in all samples ($p > 0.05$); and polyunsaturated fatty acids (PUFAs), constituting 14% ca. in bells and 19% ca. in oral arms ($p < 0.05$). Only slight and nonstatistically significant differences ($p > 0.05$) could be attributed to the specimen's sex. For example, SFAs were slightly more abundant in male than female bells (70.6 vs. 69.5%, $p > 0.05$); MUFAs were slightly higher in female than male oral arms (15.8 vs. 14.6%, $p > 0.05$); and PUFAs content was slightly lower in male than female bells (13.8 vs. 15.4%, $p > 0.05$) and female than male oral arms (18.4 vs. 19.2%, $p > 0.05$).

It was not possible drawing conclusions on the distribution of single fatty acids according the organism's anatomical part or sex. The only exception was represented by PUFAs of the ω6 and ω3 series, which showed to be constantly more abundant in oral arms than bells, regardless of the sex (Table 1). However, fatty acid fingerprints pointed out, as predominant components of the mauve stinger, lauric acid (3.16–4.69%), palmitic acid (32.87–36.46%), stearic acid (19.01–25.25%), arachidic acid (2.24–3.09%), lignoceric acid (1.11–1.95%), palmitoleic acid (0.61–2.38%), oleic acid (12.26–12.95%), linoleic acid (1.25–1.68%), arachidonic acid (4.23–6.67%), eicosapentaenoic acid (EPA, 4.96–6.40%), and docosahexaenoic acid (DHA, 2.36–4.02%). As can be seen in Table 1, the ω6/ω3 ratios determined in all samples were close to 1:1 (Table 1).

Table 1. Fatty acid methyl esters determined in the bells and oral arms of female and male specimens from *Pelagia noctiluca*. Data are reported on a dw basis, as average Gas Chromatography-flame ionization detector (GC-FID) peak area percent $\pm$ standard deviation ($n = 3$).

Fatty Acid	Female Bell (%)	Female Oral Arms (%)	Male Bell (%)	Male Oral Arms (%)
C6:0	-	0.06 ± 0.01 [a]	-	0.04 ± 0.01 [b]
C8:0	0.98 ± 0.07 [a]	0.65 ± 0.03 [b]	0.79 ± 0.03 [c]	0.61 ± 0.03 [b,d]
C10:0	1.27 ± 0.16 [a]	0.85 ± 0.03 [b]	0.65 ± 0.05 [c]	0.88 ± 0.03 [b,d]
C12:0	4.69 ± 0.17 [a]	3.87 ± 0.09 [b]	3.16 ± 0.17 [c]	4.47 ± 0.09 [a,d]
C13:0	0.18 ± 0.02 [a]	0.21 ± 0.01 [b]	0.18 ± 0.02 [a]	0.15 ± 0.01 [a]
C14:0	0.98 ± 0.14 [a]	0.80 ± 0.06 [b]	0.94 ± 0.07 [a,c]	0.86 ± 0.03 [b,d]
C15:0	0.81 ± 0.03 [a]	0.63 ± 0.03 [b]	0.79 ± 0.04 [c]	0.69 ± 0.05 [b,d]
C16:0	36.46 ± 0.1 [a]	33.84 ± 0.19 [b]	32.87 ± 0.50 [c]	34.32 ± 0.18 [b,d]
C17:0	0.40 ± 0.04 [a]	0.65 ± 0.04 [b]	0.53 ± 0.06 [c]	0.94 ± 0.03 [d]
C18:0	19.60 ± 0.37 [a]	20.21 ± 0.21 [a]	25.25 ± 0.25 [b]	19.01 ± 0.06 [c]
C20:0	2.51 ± 0.07 [a]	2.42 ± 0.07 [a]	3.09 ± 0.06 [b]	2.24 ± 0.07 [c]
C22:0	0.31 ± 0.06 [a]	0.20 ± 0.02 [b]	0.21 ± 0.02 [b]	0.24 ± 0.01 [b]
C23:0	0.06 ± 0.01 [a]	0.15 ± 0.03 [b]	0.22 ± 0.03 [c]	0.18 ± 0.02 [b]
C24:0	1.27 ± 0.09 [a]	1.23 ± 0.04 [a]	1.95 ± 0.08 [b]	1.11 ± 0.03 [c]
SFAs	*69.53 ± 0.77* [a]	*65.76 ± 0.91* [b]	*70.64 ± 1.19* [a,c]	*65.76 ± 0.15* [b,d]
C14:1	0.70 ± 0.03 [a]	0.47 ± 0.07 [b]	0.68 ± 0.06 [a,c]	0.40 ± 0.02 [b,d]
C15:1	0.18 ± 0.04 [a]	0.16 ± 0.01 [a]	0.15 ± 0.01 [a]	0.09 ± 0.01 [b]
C16:1	1.32 ± 0.09 [a]	1.87 ± 0.09 [b]	2.38 ± 0.10 [c]	0.61 ± 0.03 [d]
C17:1	0.21 ± 0.03 [a]	0.18 ± 0.02 [a,b]	0.15 ± 0.02 [b,c]	0.09 ± 0.01 [d]
C18:1 n-9	12.26 ± 0.11 [a]	12.71 ± 0.22 [a]	12.48 ± 0.34 [a]	12.95 ± 0.10 [a]
C20:1 n-9	0.10 ± 0.01 [a]	0.06 ± 0.01 [b]	0.06 ± 0.01 [b]	0.10 ± 0.02 [a]
C22:1 n-9	-	0.06 ± 0.01 [a]	-	0.08 ± 0.01 [a]
C24:1 n-9	0.45 ± 0.04 [a]	0.32 ± 0.03 [b]	0.30 ± 0.03 [b]	0.32 ± 0.04 [b]
MUFAs	*15.22 ± 0.32* [a]	*15.83 ± 0.42* [a]	*16.19 ± 0.53* [a]	*14.64 ± 0.07* [b]
C18:2 n-6	1.42 ± 0.09 [a]	1.68 ± 0.04 [a]	1.25 ± 0.16 [b]	1.48 ± 0.05 [a]
C18:3 n-6	0.09 ± 0.02 [a]	0.11 ± 0.04 [a]	0.08 ± 0.01 [a]	0.10 ± 0.02 [a]
C18:3 n-3	0.24 ± 0.03 [a]	0.31 ± 0.05 [b]	0.13 ± 0.04 [c]	0.24 ± 0.02 [a,d]
C20:2 n-6	0.09 ± 0.01 [a]	0.17 ± 0.02 [b]	0.07 ± 0.03 [c]	0.18 ± 0.01 [b,d]
C20:3 n-6	-	0.04 ± 0.02 [a]	-	0.05 ± 0.01 [a]
C20:4 n-6	5.45 ± 0.22 [a]	6.61 ± 0.23 [b]	4.23 ± 0.31 [c]	6.67 ± 0.24 [b,d]
C20:5 n-3	5.79 ± 0.15 [a]	5.82 ± 0.13 [a]	4.96 ± 0.09 [b]	6.40 ± 0.10 [c]
C22:2 n-6	-	0.03 ± 0.01 [a]	-	0.04 ± 0.01 [a]
C22:6 n-3	2.36 ± 0.08 [a]	3.61 ± 0.08 [b]	3.08 ± 0.19 [c]	4.02 ± 0.06 [d]
PUFAs	*15.43 ± 0.56* [a]	*18.39 ± 0.55* [b]	*13.79 ± 0.76* [a,c]	*19.17 ± 0.37* [b,d]
n6/n3	*0.84*	*0.89*	*0.69*	*0.80*

[a–c]: Different superscript letters in the same row indicate significantly different values ($p < 0.05$ by post hoc Tukey's HSD test). Same superscript letters in the same row indicate not significantly different values ($p > 0.05$ by post hoc Tukey's HSD test).

2.6. Major and Trace Element Profiles

The contents of four major elements (Na, Mg, K, and Ca), five essential trace elements (Fe, Cu, Zn, Mn, and Se), and five nonessential/potentially toxic trace elements (Cr, Ni, As, Cd, and Pb), were evaluated by Inductively Coupled Plasma-Mass Spectrometry (ICP-MS) in male and female jellyfish bells and oral arms, on a dw basis (Table 2).

Table 2. Elemental signatures of male and female jellyfishes' bell and oral arms revealed by Inductively Coupled Plasma-Mass Spectrometry (ICP-MS). Contents of major elements (mg 100 g^{-1}) and trace elements (µg 100 g^{-1}) are expressed as mean $\pm$ SD (n = 3) on a dw basis.

	Female		Male	
	Bell	Oral Arms	Bell	Oral Arms
	Major elements (mg 100 g^{-1})			
Na	6544 $\pm$ 263 [a]	3887 $\pm$ 133 [b]	8079 $\pm$ 318 [c]	3740 $\pm$ 155 [b,d]
Mg	692 $\pm$ 36 [a]	427 $\pm$ 37 [b]	650 $\pm$ 26 [a,c]	440 $\pm$ 20 [b,d]
K	196 $\pm$ 13 [a]	126 $\pm$ 14 [b]	229 $\pm$ 21 [a,c]	143 $\pm$ 14 [b,d]
Ca	215 $\pm$ 22 [a]	143 $\pm$ 13 [b]	236 $\pm$ 12 [a,c]	133 $\pm$ 10 [b,d]
	Essential trace elements (µg 100 g^{-1})			
Fe	1465 $\pm$ 133 [a]	854 $\pm$ 72 [b]	1309 $\pm$ 140 [a,c]	1085 $\pm$ 114 [b,d]
Cu	699 $\pm$ 96 [a]	1095 $\pm$ 97 [b]	555 $\pm$ 83 [a,c]	1424 $\pm$ 180 [d]
Zn	570 $\pm$ 67 [a]	939 $\pm$ 62 [b]	695 $\pm$ 42 [a,c]	1106 $\pm$ 131 [b,d]
Mn	49.7 $\pm$ 5.99 [a]	77.8 $\pm$ 16 [b]	146 $\pm$ 20 [c]	212 $\pm$ 37 [d]
Se	46.2 $\pm$ 4.8 [a]	100 $\pm$ 12 [b]	31.2 $\pm$ 3.6 [a,c]	115 $\pm$ 39 [b,d]
	Nonessential trace elements (µg 100 g^{-1})			
Cr	401 $\pm$ 30 [a]	690 $\pm$ 45 [b]	573 $\pm$ 58 [c]	668 $\pm$ 93 [d]
Ni	142 $\pm$ 12 [a]	207 $\pm$ 19 [b]	178 $\pm$ 14 [b,c]	215 $\pm$ 9.5 [b,d]
As	412 $\pm$ 66 [a]	631 $\pm$ 94 [b]	528 $\pm$ 48 [b]	663 $\pm$ 90 [b,c]
Cd	42.6 $\pm$ 4.6 [a]	40.4 $\pm$ 2.8 [a]	46.2 $\pm$ 7.02 [a]	51.7 $\pm$ 10.6 [a]
Pb	140 $\pm$ 19 [a]	154 $\pm$ 15 [a]	117 $\pm$ 12 [a]	132 $\pm$ 16 [a]

[a-c]: Different superscript letters in the same row indicate significantly different values ($p < 0.05$ by post hoc Tukey's HSD test). Same superscript letters in the same row indicate not significantly different values ($p > 0.05$ by post hoc Tukey's HSD test).

Table S5 reports data of the validation procedure carried out by means of reference standards. The ICP-MS method showed good linearity for all the elements, with coefficients of correlation between 0.994 and 0.999. Acceptable recoveries between 93.49% (Ni) and 103.31% (Cr) were obtained. Evaluated in terms of RSD%, precision (intraday repeatability) and intermediate precision (interday repeatability), resulted to be within the range of 2.02 to 6.53%, and below 8.52%, respectively.

Major element signatures varied most in dependence of the organ rather than sex (Table 2). Overall, these metals appeared to bioaccumulate mainly in bells than the respective oral arms ($p < 0.05$) by the decreasing order of Na > Mg > Ca $\approx$ K, regardless of female and male sex ($p > 0.05$). As a result, Na was characterized by the highest levels (6544–8079 mg 100 g^{-1} for bells; 3877–3740 mg 100 g^{-1} for oral arms); while K was the less abundant one, showing contents inferior by one order of magnitude (196–229 mg 100 g^{-1} for bells; 126–143 mg 100 g^{-1} for oral arms) (Table 2).

Dealing with essential trace elements, they were found in the decreasing order of Fe > Cu > Zn > Mn > Se, in both male and female bells and oral arms. Fe was characterized by a behavior similar to major elements, since it was more abundant in bells (1309–1465 µg 100 g^{-1}) than oral arms (854–1085 µg 100 g^{-1}) ($p < 0.05$), regardless of specimens' sex. However, Cu, Zn, Mn, and Se completely inverted such trend, showing to be more abundant in oral arms than bells ($p < 0.05$). Also, Zn and Mn contents showed a slight dependence on the specimens' sex, though in a nonstatistically significant manner ($p > 0.05$), since males showed to most bioaccumulate such elements than females in both bell and oral arms. Being not yet determined in any jellyfish species, Se levels ranged from 31.2 to 46.2 µg 100 g^{-1} in bells and from 100 to 115 µg 100 g^{-1} in oral arms (Table 2).

Nonessential/toxic trace elements were reported in the decreasing order of Cr $\approx$ As > Ni > Pb > Cd. Such elements were slightly higher in oral arms than in bells ($p > 0.05$ in almost all cases), and, with the exception of Pb, marginally more bioaccumulated in male organisms than female ones ($p > 0.05$) as well. Cr and As varied respectively between 401 and 573 µg 100 g^{-1} and 412 and 528 µg 100 g^{-1} in female and male bells; while female and male oral arms reported Cr and As levels of 668 to 631 µg·100 g^{-1} and 690 to 663 µg·100 g^{-1}, respectively. Contrary to Cr and As, Pb exhibited the lowest contents in both bells (140–117 µg 100 g^{-1}) and oral arms (154–132 µg 100 g^{-1}), with comparable values between female and male specimens (Table 2).

3. Discussion

Despite of the short period (four months) of sample collection, morphometric parameters, namely body length, bell diameter, and wet mass were measured in order to integrate the knowledge about *P. noctiluca* in the Mediterranean Sea, particularly in the peculiar environment of the Strait of Messina.

As already reported by Sandrini et al. [66], the metabolism of *P. noctiluca* is directly proportional to the sea temperature, so that a temperature increase results in an increased metabolism rate and food requirement and accelerates growth. Also, enhanced food availability, in terms of zooplankton and ichthyoplankton blooms, typically occurring in the Strait of Messina during spring [67–69], may be responsible of the higher weights recorded in female and male organisms, in agreement with the findings reported by Rosa et al. for *P. noctiluca* caught in the same seawaters [70]. Clearly, during March and April female organisms had the highest weight as they reached full gonads' maturity, resulting to be also sparingly bigger than males. In fact, biometrical and reproduction analyses performed by Rosa and coworkers [70] suggest that in the Strait of Messina, *P. noctiluca* reproduces with a maximum peak in late autumn/early winter. In this season, medusae diameter rapidly increases and most of the females are mature. They continue to grow and spawn until March–April, and in May they can reach the end of their life cycle, bringing empty, small, and slightly colored gonads. This could explain our observations on the decrease of weight during May and June in parallel to bell diameter reduction, especially in the female organisms.

As concerns bell diameter, findings of the present study are also in accordance with those reported by Rosa et al. [70]; in that case, bell diameter and water temperature were negatively correlated, providing an explanation for bell's reduction as the weather approaches the warm climate (May–June). Additionally, the mean bell's reduction recorded in May and June may be related to the presence of different age jellyfish groups, including younger (and consequently smaller) individuals, as suggested by Milisenda and colleagues [63]. Even small variations of bell size had a remarkable effect on body weight, as previously reported by Lilley et al. [71], where bell diameter accounted for 98% of the variation in wet mass.

As a basic nutritional aspect, the gross energy content of *P. noctiluca* was first evaluated. The greater calorie values of bells compared to oral arms, both in female and male specimens, was presumably due to the presence of the high-calorie gonads. In particular, energy density of female bell exceeded the male counterpart, due to higher carbohydrate, lipid, and protein contents of eggs next to reproduction maxima period [63]. The energetic value of jellyfish was barely addressed with respect to other common planktonic preys. Milisenda and coworkers carried out the energy count of *P. noctiluca*'s gonads and oral arms, the latter resulting in similar values in male and female specimens (~52.3 Kcal 100 g^{-1}) [65]. Doyle et al. [72] assessed the gross energy densities among several jellyfish species from Ireland (*Cyanea capillata*, *Rhizostoma octopus* and *Chrysaora hysoscella*) and between different tissues (bell and oral arms) within the same species as well. Besides the fact that each species was characterized by its own energy value, it was pointed out that oral arms were coherently characterized by higher gross energy contents than bells in any case. Indeed, gonads were not considered in these calorie counts, being excised from the respective bells, and separately assessed. Recently, higher gross energy densities in edible Malaysian jellyfish, such as *Rhopilema esculentum*, *Acromitus hardenbergi*, and *Rhopilema hispidum*, have been reported, with calorie values within in the range of 211 to 97 kcal·100 g^{-1} for bells and 282 to 200 kcal·100 g^{-1} for oral arms [38]. However, as contemplated by the experimental plan, gonad-provided bells were apparently investigated in such study.

Proteins are known to constitute the largest portion of the organic matter present in jellyfish dried samples, which typically contain only lower amounts of either lipids or carbohydrates [70,71,73]. Being presumably attributable to the dominant structural collagen distributed throughout the jellyfish body, the peculiar protein fraction is an appealing source of essential nutrients, thus, making such marine invertebrates potential suitable for food/feed purposes [46,74,75]. Looking at *P. noctiluca* samples, oral arms reported higher crude protein content than bells, regardless of specimen's sex, probably not only because of the abundant structural collagen, but also due to the labile protein toxins present in the numerous nematocysts typically present in oral arms. Significant variations between the different body parts were confirmed also in *C. capillata*, *R. octopus*, and *C. hysoscella*, with oral arms contents (13.1–34.1%) higher than the ones detected in gonad excised bells (5.2–11.2%) [72]; and in *R. esculentum*, *A. hardenbergi*, and *R. hispidum* species, where bell and oral arms proteins were within the range of 19.95 and 38.12% and 33.6 and 53.8%, respectively [38]. However, the qualitative and quantitative distribution of proteins both in bells and oral arms of the Mediterranean *P. noctiluca* should be further explored by appropriate proteomic techniques.

Polyphenols are widely acknowledged antioxidant compounds, widespread in nature and particularly concentrated in plants. When referring to polyphenols from the marine environment, literature indicates micro- and macroalgae as the main sources for such compounds [76–78]. Nevertheless, in consideration of the scope of the present study, namely to investigate the nutritional and nutraceutical properties of the mauve stinger, the determination of the total polyphenol content was reasonably carried out. Previous reports on total polyphenols in jellyfish species are very few. Leone et al. [74,75] reported the total phenolic contents of three jellyfish species, none belonging to *Pelagia* genus. In two of them, ~1800–2000 µg GAE g^{-1} (dw) were determined. Such findings were quite in accord with the present report. The total phenolic content determined in *P. noctiluca* is only a preliminary and rough screening that should be followed by a targeted qualitative analysis for the identification of the specific polyphenolic structures. In fact, it has been speculated that amino acidic residues from the protein fraction could contribute to the value of the total phenolic content [74].

The results obtained from the analyses of fatty acids can be hardly compared with previous studies, since *P. noctiluca* results to be underinvestigated with respect to this issue. Mastronicolis et al. reported 0.19% *ww* of total lipids, 41.7% of which were free fatty acids [79]; whereas Cardona et al. analyzed the fresh mass of *P. noctiluca* obtaining a fatty acid composition roughly represented by 70% palmitic acid, 15% pentadecanoic acid, 6% EPA, and 2% DHA [80]. Approximately, the fatty acid profiles here presented are in accordance with those reported for other jellyfish species, collected in the Mediterranean Sea and investigated with the same analytical procedure here applied [74,80,81].

Notoriously, essential fatty acids belonging to the ω-6 and ω-3 groups, and are significantly present in a variety of vertebrate and invertebrate marine species, and, when correctly balanced, confer a relevant nutraceutical value to the source, which they come from [82]. As can be observed in Table 2, essential ω-3 fatty acids, such as EPA and DHA, and essential ω-6 fatty acids, such as linoleic acid, were determined also in the Mediterranean *P. noctiluca*, being more abundant in oral arms than bells. Since pioneer works on the positive correlation between essential fatty acids and human health, different schools of thought have arisen during the years, attributing a higher healthy power one time to ω6 acids, another to ω3 acids. However, in the last decades, scientists concluded that it is the balance between ω6 and ω3 that promotes normal development and preserves homeostasis. Ideally, this ratio should be 1:1, due to the observation of chronic diseases especially in those individuals having a diet deficient of ω3- fatty acids and, consequently, with an unbalanced ω6/ω3 ratio [83]. In this respect, all samples of *P. noctiluca* reported a correct balance of the two classes of fatty acids, thus confirming its nutraceutical value. In particular, bell and oral arms from female organisms showed the ratios closest to 1:1 (respectively, 0.89 and 0.80).

The screening of major and trace elements provided further insights on the nutritional and nutraceutical value of *P. noctiluca*. Jellyfish bioaccumulates and transfers essential minerals and trace elements from lower trophic levels to high-order fish predators, having a key role in balancing any potential nutritional shortfall of the food chain. The same applies to nonessential and potentially toxic elements, which could instead represent a threat for the health of marine ecosystems, and, not least, human consumers [84,85]. To date, few published data dealt with inorganic elements in this gelatinous zooplankton, mainly to test the environmental quality of coastal systems [81,84–87], rather than for food/feed purposes [38].

According to obtained data, jellyfish bell exhibited higher Na, Mg, K, and Ca levels than oral arms, probably because of the buffering activities carried out to maintain the osmotic balance and, thus, the floating capacity of the bell [38]. Although such a mineral distribution was confirmed also by previous works on different jellyfish species from Portuguese (*Catostylus tagi*) and Australian (*Cassiopea* sp.) coasts [81,84], not comparable contents were detected, because of clear taxonomic and ecological reasons. For example, Mg in bell and oral arms belonging to *C. stagi* and *Cassiopea* sp. was equal to 328 mg 100 g^{-1} and 240 mg 100 g^{-1} [81], and 74.4–129.7 mg 100 g^{-1} and 73.4–125.7 mg 100 g^{-1}, respectively [86]. Ca levels distributed between bells and oral arms within the ranges 25.8 to 46.6 mg 100 g^{-1} and 27.2 to 44.6 mg 100 g^{-1} [84] in the Australian *Cassiopea* sp.; whereas, in the Portoguese *C. stagi*, bell and oral arms showed Ca contents equal to 1026 mg 100 g^{-1} and 736 mg 100 g^{-1} [81].

Trace elements, such as Fe, Zn, Cu Mn, and Se, are well known to be both limiting nutrients and toxicants. In fact, they are essential for the metabolism of vertebrate and invertebrate marine organisms, as they constitute a variety of metalloproteins and antioxidant enzymes and play a key role in cellular detoxification activity, but, at the same time, they become toxic at high concentrations, leading to damaging oxidative processes [88]. In the same way as for major elements, the comparison of trace element levels from *P. noctiluca* with those from other jellyfish species addressed in previous works [81,84,85] becomes challenging. In fact, an increasing discrepancy seems to appear dealing thoroughly with getting smaller contents of essential trace elements; however, similarly to *P. noctiluca*, also other species, such as *C. stagi* [81], *Cassiopea* sp. [84], and *Cotylorhiza tuberculata* [85], showed to more bioaccumulate such trace elements in oral arms than bells. These distribution patterns could be related to the metal uptake via food, occurring by oral arms directly involved in suctorial feeding, and fine filtering functions [89]. Assuming that the elemental requirements of an organism, including the human being, are driven almost entirely by utility (i.e., cellular function, with shifts in biological requirements decoupled from corresponding environmental abundances), and that they should be assessed in depth case-by-case, it is reasonable to hypothesize that both essential major and trace elements found in the Mediterranean *P. noctiluca*, may be exploited for developing natural food and aquafeed supplements. In the latter case, other jellyfish species, such as *A. aurita* and *Chrysaora pacifica*, have already been demonstrated to support the growth and survival of specific farmed species, thanks also to a peculiar metal profile [90].

The main source of nonessential and potentially toxic elements, such as Cr, As, Ni, Pb, and Cd, typically comes from anthropogenic activities, negatively impacting the health status of marine environments. Previous works have already pointed out that different jellyfish species can bioconcentrate these harmful elements and reflect a time-integrated measured of their levels in the water, demonstrating to be useful bioindicators of coastal environments [84,85]. The uptake and accumulation of nonessential/toxic elements in *P. noctiluca* varied, although in nonsignificant manner, between selected tissues and sex specimens. However, as the present study does not rely on an ecotoxicological basis, and no literature on the pollutant accumulation capacity of the Mediterranean *P. noctiluca* has been yet produced, no conclusion will be drawn about the health status of such coastal areas. Nonetheless, obtained data could be useful for stressing on the bioaccumulating capacity of jellyfish, other than constituting an input for future environmental studies on such species.

Concerning the toxicological value of *P. noctiluca* as aquafeed and food supplement, it is useful to mention the Commission Regulation (EU) N. 744/2012 [91] setting the limits of heavy metals in animal feeds, and the Commission Regulation (EC) N. 629/2008 [92], amending the Regulation (EC) N. 1881/2006 and fixing the maximum levels of heavy metals in food supplements. Indeed, according the EU Regulation N. 744/2012, the investigated samples were characterized by means concentrations of As and Pb (5.58 and 1.35 mg kg^{-1}) well within the maximum contents set at 10 and 5 mg kg^{-1} for a complete feed, respectively, and at 10 and 10 mg kg^{-1} for a complementary feed, respectively, in every case with a moisture content of 12%. On the other hand, according the EC Regulation (EC) N.

629/2008, *P. noctiluca* showed mean levels of Pb and Cd (1.35 and 0.45 mg kg^{-1}), well below the limits set respectively at 3.0 and 1.0 mg kg^{-1}.

As a result, potential aquafeed and food supplements based on *P. noctiluca* from the Strait of Messina would be safe for farmed fish and human consumers, in terms of toxic heavy metals.

The pioneering chemical composition data of *P. noctiluca* herein discussed may be a valuable starting point for acknowledging its nutritional and nutraceutical properties for food and aquafeed purposes.

Concerning food applications, it has been widely discussed the relevant value of certain jellyfish species as food for human consumption in Asian countries. Additionally, a specific protein fraction of the edible *R. esculentum* with high antioxidant power [93] and a new bioactive polysaccharide isolated from edible jellyfish [94] were proposed for developing novel and natural food supplements.

Regarding aquafeed field, Marques et al. recently revealed that all developmental stages of *Aurelia* sp. were accepted as a feed source by *Sparus aurata* [73]; *Aurelia aurita* was also demonstrated to be a valid feed source for *Thamnaconus modestus*, when other common preys were not visible [95], and for rearing commercial fish such as *Pampus argenteus* [96]. Furthermore, Milisenda and colleagues proposed *P. noctiluca* as valuable food source for fish predators, such as Boops boops, in the Strait of Messina, due to an increased energy content during the period of gonad maturation, and to the high available biomass present during blooms [65].

Clearly, before getting into any practical application of *P. noctiluca*, other in vitro studies will be necessary for firstly developing the most suitable procedures of biomass processing and venom neutralization. Then, if the applications turned out to be economically and realistically feasible, further research will be required to investigate (i) the in vivo toxicity and effectiveness of food/aquafeed supplements and (ii) the potential food web cascading effects due to the harvesting of wild populations.

4. Materials and Methods

4.1. Chemicals

All materials and reagents employed for the histological analyses were supplied by Bio-Optica Milano S.p.a. (Milan, Italy). The following reagents were all provided by Sigma-Aldrich (Milan, Italy): sulfuric acid (95–98% purity), sodium hydroxide, boric acid, hydrogen chloride, benzoic acid, and phosphate-buffered saline (PBS). The Kjeldahl catalyst was supplied by Carlo Erba (Milan, Italy). n-Heptane was purchased from PanReac AppliChem (Barcelona, Spain). Potassium hydroxide, chloroform, methanol, sodium carbonate anhydrous, Folin Ciocalteau's reagent, stock standard solution of Re (1000 mg L^{-1} in 2% HNO$_3$), gallic acid, and stock standard solutions of Sc, Ge, In, and Bi (1000 mg L^{-1} in 2% HNO$_3$) were from Fluka (St. Gallen, Switzerland). Nitric acid (65%, v/v) was of Suprapur grade (Mallinckrodt Baker, Milan, Italy). Ultrapure water (<5 mg L^{-1} TOC) was obtained from a Barnstead Smart2Pure 12 water purification system (Thermo Scientific, Milan, Italy).

4.2. Sample Collection

Specimens of *P. noctiluca* were sampled from the Strait of Messina, South Italy. In this peculiar environment, the mauve stinger can be found in the Strait of Messina during the whole year, reaching the highest diffusion in the period of March to June. Approximately 400 adult specimens were collected from March to June 2017 (~100 organisms per month), in the coastal waters of Capo Peloro (Messina). Adult jellyfish samples were taken by means of a small net from water's edge, put in tanks filled with seawater, and immediately transported to laboratory for morphometric measurements and sex and chemical determinations. The experimental protocol was developed in accord with the ethical standards reported in the European Directive 2010/63/EU on the protection of animals used for scientific purposes [97].

4.3. Biometrics and Sex Determination

Maximum length, bell diameter, and total weight (fresh mass) of each jellyfish sample were measured with the support of a ruler and of a lab-made apparatus, shown in Figure 5. Each specimen was placed inside the most appropriate circular hole, chosen on the basis of bell diameter (Figure 5A). Oral arms were left suspended on the other side of the flat surface. Length was measured from the top of the bell to the end of the longest oral arm, along the oral–aboral axis (Figure 5B). Bell diameter was considered as the maximum distance between distal tips of opposed interradial rhopalia. Biomass (w/w) was measured after washing medusae with distilled water for salt removal. For each specimen, the sex was determined by inspection of the gonads through a stereomicroscope (Carl Zeiss Stemi SV11): male individuals presented purple follicles, while female subjects were characteristically pink-red colored, and crowded by Milisenda et al. [65]. When identification of sex resulted to be challenging or uncertain, histological assessments were carried out following a standard protocol [98].

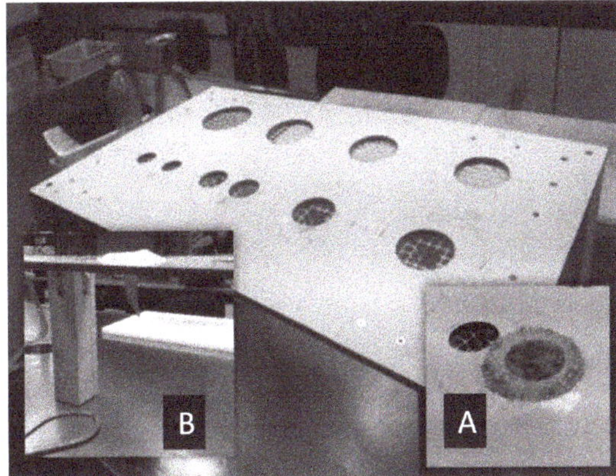

Figure 5. Lab-made apparatus for morphometric measurements. (**A**) Jellyfish were placed inside the most appropriate circular hole, chosen among different predefined measures. (**B**) Length was measured from the top of the bell to the end of the longest oral arm.

4.4. Sample Lyophilization

After biometrical measurements and sex determinations were carried out, the organisms were sacrificed separating the bells by the manubrium bearing the oral arms, with the help of a Teflon knife, avoiding metal cross-contamination. Bells and oral arms were grouped to provide four different pools (n = 200 each), as reported in Table 3. Each pool was weighed, cleansed thoroughly with distilled water, minced, and lyophilized by an Alpha 1-2/LD Plus freeze dryer (Martin Christ, Osterode, Germany), for 72 h at $-55\,^{\circ}$C using a chamber pressure of 0.110 mbar. Then, the freeze dried pools were weighed, and stored at $-20\,^{\circ}$C until use. For each pool, a moisture content of ~95% w/w and a yield of 5% w/w were assessed.

Table 3. Sample pools considered for the present study.

Collected Specimens	Pool	n
200 males	Bells	200
	Oral arms	200
200 females	Bells	200
	Oral arms	200

4.5. Gross Energy Assessments

To measure the gross energy content, a benchtop isoperibol calorimeter (Parr® 6200 Oxygen Bomb Calorimeter, Parr Instrument Company, Moline, IL, USA) was employed.

Approximately 1 g of each powdered pool was placed in a 1108 model oxygen bomb. To determine the gross energy densities of bombed samples, the calorimeter chamber was previously calibrated for the heat of combustion of 1 g of benzoic acid (26.46 kJ g^{-1}), under controlled and reproducible operating conditions. In fact, the known amount of heat produced by the combustion of the calibration standard determined the energy equivalent (W) per change in water temperature between initial and postcombustion of the sample (ΔT). Therefore, the energy content of each sample (ES) was calculated as follows

$$ES = W \times \Delta T / exact\ sample\ weight$$

Each sample pool was run in triplicate. Results were expressed as Kcal 100 g^{-1}, dw.

4.6. Crude Protein

The crude protein content was determined using the AOAC Official Method 976.05 (automated Kjeldahl method) [99]. Approximately 1 g of each powdered pool was separately digested by the SpeedDigester K-439 (Büchi, Switzerland) and then analyzed by the KjelMaster System K- 375 (Büchi, Switzerland) and equipped with a scrubber of gases and vapors (Scrubber K-415, Büchi, Switzerland). For the calculation of the % protein in a sample, the obtained % nitrogen was multiplied by a conversion factor of 6.5. Each determination was conducted in triplicate.

4.7. Total Phenolic Content

As a preliminary part of method development, an initial group of samples was subjected to two different extraction procedures: (i) with PBS (phosphate-buffered saline), pH 3.5 and (ii) with 80% methanol [74]. Aliquots (1.0 g) of each lyophilized sample pool were added with 16 mL of PBS and shaken for 2 h at 4 °C, in one case; with 16 mL of methanol and shaken for 16 h at 4 °C, in the other case. Successively, samples were homogenized at 9000 rpm and 4 °C for 30 min. To 1.0 mL of supernatant, 5.0 mL of Folin–Ciocalteau reagent and 5.0 mL of sodium carbonate (20%) were added. The resulting solution was kept in the dark for 2 h, and later analyzed at the UV–Vis spectrophotometer, model UV-2401PC (Shimadzu, Milan, Italy). The wavelength of absorbance was set at 760 nm. A 5-point calibration plot was built up by using solutions of gallic acid in methanol in the range 50 to 2000 µg mL^{-1}. Each point corresponded to three replicates.

4.8. Fatty Acids

The extraction of fatty acids was carried out following the procedure reported by Bligh & Dyer [100]. An aliquot (10.0 mg) of each lyophilized pool was homogenized with 1.0 mL of chloroform ($CHCl_3$) and 3.0 mL of methanol (MeOH). This mixture was then added with 1.0 mL of MeOH and 1.0 mL of water, homogenized, let to settle, and filtered through paper. After settling, the filtrate was centrifuged at 3000 rpm for 15 min. Two layers were formed: the bottom one ($CHCl_3$), containing the isolated lipids, was transferred to a rotating evaporator, model P/N Hei-VAP Precision ML/G3 (Heidolph Instruments GmbH & Co., Schwabach, Germany). The upper layer (MeOH/H_2O) was subjected again to all the steps above described in order to reach an exhaustive extraction. For the fatty acid methylation, the dried lipidic extract was recovered through addition of 1 mL hexane, then added with reagent (CH_3OH/H_2SO_4, 9:1), and heated at 100 °C for 1 h. The hydrocarbon layer was collected and injected into the GC instrumentation. Qualitative analyses were carried out in a GCMS-TQ8030 (Shimadzu, Kyoto, Japan) system equipped with an AOC-20i autosampler and a capillary column Supelco SLB-IL100 (60 m × 0.25 mm, film thickness 0.20 µm). GC conditions were set as follows: injector, 280 °C; injection volume: 1.0 mL; head pressure: 26.7 kPa; carrier gas: He, at a linear velocity of 30.0 cm/s (constant); split ratio 1:100; oven temperature program: 50–280 °C at 3 °C/min, held 10 min.

MS conditions: operation mode was in full scan; ion source and interface temperatures were 220 °C and 250 °C, respectively; and scan mass range was 40 to 400 m/z. For FAMEs identification, a triple means methodology was used: (i) spectral matching with Wiley and NIST databases; (ii) co-injection with standards (Supelco 37 component FAME mix, Supelco, St. Louis, MO, USA); and (iii) comparison with literature data [101]. Data handling was performed by GCMS solution software. Quantitative analyses were carried out on a Master GC-DANI system, equipped with a capillary column Supelco SLB-IL100 (60 m × 0.25 mm, film thickness 0.20 μm). Oven temperature program: 120–200 °C at 1 °C min^{-1} (10 min). Injector and FID temperatures were respectively set at 220 and 240 °C. Carrier gas was He, at a constant linear velocity of 30.0 cm s^{-1}. FID conditions: sampling frequency: 25 Hz; gases: makeup (He), 25 mL min^{-1}; H$_2$, 40 mL min^{-1}; air, 280 mL min^{-1}. Data were processed through the Clarity software (Dani). All determinations were run in triplicate.

4.9. Elemental Analysis

Approximately 500 mg of each lyophilized pool were digested with 10 mL of HNO$_3$, exploiting a closed-vessel microwave digestion system Ethos 1 (Milestone, Bergamo, Italy) equipped with PTFE vessels. The mineralization was carried out by setting the following temperature program; 0–200 °C in 10 min (step 1), 200 °C held for 5 min (step 2), and 200–220 °C in 5 min (step 3), with a constant microwave power of 1000 W. Due to the expected very high salt concentrations, each digested pool was diluted by ultrapure water with a dilution factor of 1000, and stored at 4 °C until ICP-MS analysis. A quadrupole ICP-MS iCAP Q (ThermoScientific, Waltham, MA, USA), equipped with an ASX-520 autosampler (Cetac Technologies Inc., Omaha, NE, USA), was employed for the analytical determinations. The ICP-MS operating conditions are shown in Table 4. All samples were analyzed in triplicate, along with blanks to check for any loss or cross contamination. For quantitative purposes, the external calibration procedure was carried out with the help of multielemental standard solutions. For building up six point calibration plots, six different concentrations were evaluated per analyte, and each point resulted from triplicate extractions and analyses [102]. For method validation, a linear least-square regression analysis of the calibration graphs was performed to check for the linearity between the instrumental response and the nominal concentration of each elemental standard. The intra-assay and interassay variabilities were determined by quantifying three replicates on the same day and six consecutive days, respectively.

Table 4. ICP-MS operating conditions applied to elemental analysis.

Parameter	Value
Forward power	1550 W
Plasma gas flow rate (Ar)	14 L min^{-1}
Auxiliary gas flow rate (Ar)	0.89 L min^{-1}
Carrier gas flow rate (Ar)	0.91 L min^{-1}
Collision gas flow rate (He)	4.5 mL min^{-1}
Spray chamber temperature	2.70 °C
Sample uptake/wash time	45 s
Injection volume	200 μL
Sample introduction flow rate	0.93 mL min^{-1}
Scan mode	Full scan
Dwell time	Optimized for each analyte

4.10. Statistical Analysis

With the exception of biometrics, all data are reported as mean ± standard deviation of triplicate measurements. Statistical analysis was conducted by the SPSS Statistics Software v. 21.0 (SPSS Inc., Chicago, IL, USA), performing for each investigated variable a two-way analysis of variance (ANOVA) followed by the Tukey' s honestly significant difference (HSD) post-hoc test. Level of significance was set at $p < 0.05$

5. Conclusions

In this study, male and female specimens of *Pelagia noctiluca* from the Strait of Messina were chemically characterized with respect to their nutritional and nutraceutical properties, taking into consideration both their bells and oral arms. Overall, the gross energy content was higher in bells rather than oral arms, while proteins and total phenolics were inversely concentrated. The fatty acid composition included important MUFAs, such as oleic acid, and PUFAs, such as EPA and DHA. Also, antioxidant inorganic elements were revealed. Both chemical classes varied in dependence of the investigated jellyfish part. In fact, SFAs were more concentrated in bells than oral arms, whereas PUFAs were distributed in the exact opposite way. MUFAs appeared to be equally present in both organs. On the other hand, major elements and trace elements demonstrated an opposite behavior, being the latter most accumulated in oral arms than bells.

With the exception of the significantly high energy contents and total phenolic contents observed in female organisms, no remarkable differences that could be ascribed to the variable "sex" were highlighted for the other compound classes. With the appropriate precautions, obtained results may support the potential employment of *P. noctiluca* as aquafeed or food supplement.

Supplementary Materials: The following are available online at http://www.mdpi.com/1660-3397/17/3/172/s1, Table S1: Summary output of the two-way ANOVA with replication analysis performed on biometric data from *P. noctiluca*. SS: sum-of-squares; df: degrees of freedom; MS: mean squares. Table S2. Summary output of the two-way ANOVA with replication analysis performed on gross energy contents from *P. noctiluca*. SS: sum-of-squares; df: degrees of freedom; MS: mean squares. Table S3. Summary output of the two-way ANOVA with replication analysis performed on crude protein contents derived from *P. noctiluca*. SS: sum-of-squares; df: degrees of freedom; MS: mean squares. Table S4. Summary output of the two-way ANOVA with replication analysis performed on total polyphenol levels obtained from *P. noctiluca*. SS: sum-of-squares; df: degrees of freedom; MS: mean squares. Table S5: Performance of the ICP-MS method in terms of linearity, LOD, LOQ, intra- and interday repeatability (*n* = 3), and accuracy.

Author Contributions: R.C., G.C., and A.A.: Project Design and Writing; R.L.V.: Sample Collection and Biometric Data Collection; G.B. (Giovanni Bartolomeo): Phenolics Analysis; G.B. (Giuseppe Bua): Elemental Analysis; A.F.: Calorimetry Analysis; S.S.: Biometric Data Collection; T.G.: Protein Analysis; R.R.: Lipid Analysis; G.D. and N.S.: Project Supervision.

Funding: This research received no external funding.

Conflicts of Interest: The authors declare no conflict of interest.

References

1. Alasalvar, C.; Shahidi, F.; Quantick, P. Food and health applications of marine nutraceuticals: A review. In *Seafoods—Quality, Technology and Nutraceutical Applications*; Alasalvar, C., Taylor, T., Eds.; Springer: Berlin/Heidelberg, Germany, 2002; pp. 175–204.

2. Suleria, H.; Osborne, S.; Masci, P.; Gobe, G. Marine-based nutraceuticals: An innovative trend in the food and supplement industries. *Mar. Drugs* **2015**, *13*, 6336–6351. [CrossRef] [PubMed]

3. Freitas, A.C.; Rodrigues, D.; Rocha-Santos, T.A.; Gomes, A.M.; Duarte, A.C. Marine biotechnology advances towards applications in new functional foods. *Biotechnol. Adv.* **2012**, *30*, 1506–1515. [CrossRef] [PubMed]

4. Malve, H. Exploring the ocean for new drug developments: Marine pharmacology. *J. Pharm. Bioall. Sci.* **2016**, *8*, 83–91. [CrossRef]

5. Gomes, T.; Albergamo, A.; Costa, R.; Mondello, L.; Dugo, G. Potential use of proteomics in shellfish aquaculture: From assessment of environmental toxicity to evaluation of seafood quality and safety. *Curr. Org. Chem.* **2017**, *21*, 402–425. [CrossRef]

6. Albergamo, A.; Rigano, F.; Purcaro, G.; Mauceri, A.; Fasulo, S.; Mondello, L. Free fatty acid profiling of marine sentinels by nanoLC-EI-MS for the assessment of environmental pollution effects. *Sci. Total Environ.* **2016**, *571*, 955–962. [CrossRef]

7. Hoegh-Guldberg, O.; Bruno, J.F. The impact of climate change on the world's marine ecosystems. *Science* **2010**, *328*, 1523–1528. [CrossRef]

8. Scheffer, M.; Carpenter, S.; De Young, B. Cascading effects of overfishing marine systems. *Trends Ecol. Evol.* **2005**, *20*, 579–581. [CrossRef] [PubMed]

9.   Graham, W.M. Numerical increases and distributional shifts of *Chrysaora quinquecirrha* (Desor) and *Aurelia aurita* (Linné) (Cnidaria: Scyphozoa) in the northern Gulf of Mexico. *Hydrobiologia* **2001**, *451*, 97–111. [CrossRef]

10.   Brodeur, R.D.; Sugisaki, H.; Hunt, G.L. Increases in jellyfish biomass in the Bering Sea: Implications for the ecosystem. *Mar. Ecol. Prog. Ser.* **2002**, *233*, 89–103. [CrossRef]

11.   Arai, M.N. Predation on pelagic coelenterates: A review. *J. Mar. Biol. Assoc.* **2005**, *85*, 523–536. [CrossRef]

12.   Purcell, J.E.; Sturdevant, M.V. Prey selection and dietary overlap among zooplanktivorous jellyfish and juvenile fishes in Prince William Sound, Alaska. *Mar. Ecol. Prog. Ser.* **2001**, *210*, 67–83. [CrossRef]

13.   Han, C.-H.; Uye, S. Combined effects of food supply and temperature on asexual reproduction and somatic growth of polyps of the common jellyfish *Aurelia aurita* S.L. *Plankt. Benthos Res.* **2010**, *5*, 98–105. [CrossRef]

14.   Rutherford, L.D.; Thuesen, E.V. Metabolic performance and survival of medusae in estuarine hypoxia. *Mar. Ecol. Prog. Ser.* **2005**, *294*, 189–200. [CrossRef]

15.   Cheng, J.H.; Ding, F.Y.; Li, S.F.; Yan, L.P.; Ling, J.Z.; Li, J.S. A study on the quantity distribution of macro-jellyfish and its relationship to seawater temperature and salinity in the East China Sea Region. *Acta Ecol. Sin.* **2005**, *25*, 440–445.

16.   Lynam, C.P.; Gibbons, M.J.; Axelsen, B.E.; Sparks, C.A.J.; Coetzee, J.; Heywood, B.G.; Brierley, A.S. Jellyfish overtake fish in a heavily fished ecosystem. *Curr. Biol.* **2006**, *16*, R492–R493. [CrossRef] [PubMed]

17.   Becerra-Amezcua, M.P.; Guerrero-Legarreta, I.; González-Márquez, H.; Guzmán-García, X. In vivo analysis of effects of venom from the jellyfish *Chrysaora* sp. in zebrafish (*Danio rerio*). *Toxicon* **2016**, *113*, 49–54. [CrossRef]

18.   Bosch-Belmar, M.; Azzurro, E.; Pulis, K.; Milisenda, G.; Fuentes, V.; Kéfi-Daly Yahia, O.; Micallef, A.; Deidun, A.; Piraino, S. Jellyfish blooms perception in Mediterranean finfish aquaculture. *Mar. Policy* **2017**, *76*, 1–7. [CrossRef]

19.   Matsumura, K.; Kamiya, K.; Yamashita, K.; Hayashi, F.; Watanabe, I.; Murao, Y.; Miyasaka, H.; Kamimura, N.; Nogami, M. Genetic polymorphism of the adult medusae invading an electric power station and wild polyps of *Aurelia aurita* in Wakasa Bay, Japan. *J. Mar. Biol. Assoc.* **2005**, *85*, 563–568. [CrossRef]

20.   Killi, N.; Mariottini, G.L. Cnidarian Jellyfish: Ecological aspects, nematocyst isolation, and treatment methods of sting. In *Marine Organisms as Model Systems in Biology and Medicine*; Kubiak, J.Z., Kloc, M., Eds.; Springer International Publishing: Cham, Germany, 2018; pp. 477–513.

21.   Hays, G.C.; Doyle, T.K.; Houghton, J.D.R. A Paradigm shift in the trophic importance of jellyfish? *Trends Ecol. Evol.* **2018**, *33*, 874–884. [CrossRef]

22.   Boero, F.; Bouillon, J.; Gravili, C.; Miglietta, M.P.; Parsons, T.; Piraino, S. Gelatinous plankton: Irregularities rule the world (sometimes). *Mar. Ecol. Prog. Ser.* **2008**, *356*, 299–310. [CrossRef]

23.   Lynam, C.P.; Brierley, A.S. Enhanced survival of 0-group gadoid fish under jellyfish umbrellas. *Mar. Biol.* **2007**, *150*, 1397–1401. [CrossRef]

24.   Pagès, F. Biological associations between barnacles and jellyfish with emphasis on the ectoparasitism of *Alepas pacifica* (lepadomorpha) on *Diplulmaris malayensis* (scyphozoa). *J. Nat. Hist.* **2000**, *34*, 2045–2056. [CrossRef]

25.   Pages, F.; Corbera, J.; Lindsay, D. Piggybacking pycnogonids and parasitic narcomedusae on *Pandea rubra* (Anthomedusae, Pandeidae). *Plankt. Benthos Res.* **2007**, *2*, 83–90.

26.   Martorelli, S.R. Digenea parasites of jellyfish and ctenophores of the southern Atlantic. *Hydrobiologia* **2001**, *451*, 305–310. [CrossRef]

27.   Stoner, E.W.; Layman, C.A.; Yeager, L.A.; Hassett, H.M. Effects of anthropogenic disturbance on the abundance and size of epibenthic jellyfish *Cassiopea* spp. *Mar. Pollut. Bull.* **2011**, *62*, 1109–1114. [CrossRef] [PubMed]

28.   Welsh, D.T.; Dunn, R.J.K.; Meziane, T. Oxygen and nutrient dynamics of the upside down jellyfish (*Cassiopea* sp.) and its influence on benthic nutrient exchanges and primary production. *Hydrobiologia* **2009**, *635*, 351–362. [CrossRef]

29.   Purcell, J.E.; Clarkin, E.; Doyle, T.K. Foods of *Velella velella* (Cnidaria: Hydrozoa) in algal rafts and its distribution in Irish seas. *Hydrobiologia* **2012**, *690*, 47–55. [CrossRef]

30.   Pitt, K.A.; Welsh, D.T.; Condon, R.H. Influence of jellyfish blooms on carbon, nitrogen and phosphorus cycling and plankton production. *Hydrobiologia* **2009**, *616*, 133–149. [CrossRef]

31. Condon, R.H.; Steinberg, D.K.; del Giorgio, P.A.; Bouvier, T.C.; Bronk, D.A.; Graham, W.M.; Ducklow, H.W. Jellyfish blooms result in a major microbial respiratory sink of carbon in marine systems. *Proc. Natl. Acad. Sci. USA* **2011**, *108*, 10225–10230. [CrossRef]

32. Hsieh, Y.P.; Leong, F.M.; Rudloe, J. Jellyfish as food. In *Jellyfish Blooms: Ecological and Societal Importance Proceedings of the International Conference on Jellyfish Blooms*; Purcell, J.E., Graham, W.M., Dumont, H.J., Eds.; Springer Science & Business Media: Dordrecht, The Netherlands, 2001; pp. 11–17.

33. Omori, M.; Nakano, E. Jellyfish fisheries in southeast Asia. *Hydrobiologia* **2001**, *451*, 19–26. [CrossRef]

34. Brotz, L. Jellyfish fisheries—A global assessment. In *So Long, and Thanks for All the Fish: The Sea Around Us, 1999–2014, A Fifteen Year Retrospective*; Pauly, D., Zeller, D., Eds.; The University of British Columbia: Vancouver, BC, Canada, 2016; pp. 77–82.

35. FAO. The State of World Fisheries and Aquaculture 2018. Available online: http://www.fao.org/3/i9540en/I9540EN.pdf (accessed on 21 February 2019).

36. Purcell, J.E.; Baxter, E.J.; Fuentes, V.L. Jellyfish as products and problems of aquaculture. In *Advances in Aquaculture Hatchery Technology*; Allan, G., Burnell, G., Eds.; Woodhead Publishing: Cambridge, UK, 2013; pp. 404–430.

37. Dong, J.; Jiang, L.X.; Tan, K.F.; Liu, H.Y.; Purcell, J.E.; Li, P.J.; Ye, C.C. Stock enhancement of the edible jellyfish (*Rhopilema esculentum* Kishinouye) in Liaodong Bay, China: A review. *Hydrobiologia* **2009**, *616*, 113–118. [CrossRef]

38. Khong, N.M.H.; Yusoff, F.M.; Jamilah, B.; Basri, M.; Maznah, I.; Chan, K.W.; Nishikawa, J. Nutritional composition and total collagen content of three commercially important edible jellyfish. *Food Chem.* **2016**, *196*, 953–960. [CrossRef] [PubMed]

39. Omori, M.; Kitamura, M. Taxonomic review of three Japanese species of edible jellyfish (Scyphozoa: Rhizostomeae). *Plankt. Biol. Ecol.* **2004**, *51*, 36–51.

40. Pitt, K.A.; Kingsford, M.J. Geographic separation of stocks of the edible jellyfish *Catostylus mosaicus* (Rhizostomeae) in New South Wales, Australia. *Mar. Ecol. Prog. Ser.* **2000**, *196*, 143–155. [CrossRef]

41. Muhammed, F.; Sultana, R. New record of edible jellyfish, *Rhizostoma pulmo* (Cnidaria: Scyphozoa: Rhizostomitidae) from Pakistani waters. *Mar. Biodivers. Rec.* **2008**, *1*, e67. [CrossRef]

42. Sloan, N.A.; Gunn, C.R. Fishing, processing, and marketing of the jellyfish *Aurelia aurita* (L.), from Southern British Columbia. *Can. Ind. Rep. Fish. Aquat. Sci.* **1985**, *157*, 1–29.

43. Morikawa, T. Jellyfish. *FAO Infofish Mark. Dig.* **1984**, *1*, 37–39.

44. Dong, Z.; Liu, D.; Keesing, J.K. Jellyfish blooms in China: Dominant species, causes and consequences. *Mar. Pollut. Bull.* **2010**, *60*, 954–963. [CrossRef]

45. Purcell, J.E.; Uye, S.I.; Lo, W.T. Anthropogenic causes of jellyfish blooms and their direct consequences for humans: A review. *Mar. Ecol. Prog. Ser.* **2007**, *350*, 153–174. [CrossRef]

46. Yu, H.H.; Liu, X.G.; Xing, R.E.; Liu, S.; Guo, Z.Y.; Wang, P.B.; Li, C.P.; Li, P.C. In vitro determination of antioxidant activity of proteins from jellyfish *Rhopilema esculentum*. *Food Chem.* **2006**, *95*, 123–130. [CrossRef]

47. Hsieh, Y.H.P. Use of Jellyfish Collagen (Type II) in the Treatment of Rheumatoid Arthritis. U.S. Patent No. 6,894,029, 13 November 2001.

48. Zhuang, Y.; Sun, L.; Zhao, X.; Wang, J.; Hou, H.; Li, B. Antioxidant and melanogenesis-inhibitory activities of collagen peptide from jellyfish (*Rhopilema esculentum*). *J. Sci. Food Agric.* **2009**, *89*, 1722–1727. [CrossRef]

49. Hoyer, B.; Bernhardt, A.; Lode, A.; Heinemann, S.; Sewing, J.; Klinger, M.; Notbohm, H.; Gelinsky, M. Jellyfish collagen scaffolds for cartilage tissue engineering. *Acta Biomater.* **2014**, *10*, 883–892. [CrossRef]

50. Addad, S.; Exposito, J.Y.; Faye, C.; Ricard-Blum, S.; Lethias, C. Isolation, characterization and biological evaluation of jellyfish collagen for use in biomedical applications. *Mar. Drugs* **2011**, *9*, 967–983. [CrossRef] [PubMed]

51. Wolfinbarger, J.R.L. Method and Process for the Production of Collagen Preparations from Invertebrate Marine Animals and Compositions Thereof. U.S. Patent No. 6,337,389, 28 November 2001.

52. Sionkowska, A.; Skrzyński, S.; Śmiechowski, K.; Kołodziejczak, A. The review of versatile application of collagen. *Polym. Adv. Technol.* **2017**, *28*, 4–9. [CrossRef]

53. Alves, A.; Marques, A.; Martins, E.; Silva, T.; Reis, R. Cosmetic potential of marine fish skin collagen. *Cosmetic* **2017**, *4*, 39. [CrossRef]

54. Mariottini, G.L.; Giacco, E.; Pane, L. The mauve stinger *Pelagia noctiluca* (Forsskål, 1775). Distribution, ecology, toxicity and epidemiology of stings. A review. *Mar. Drugs.* **2008**, *6*, 496–513. [CrossRef] [PubMed]

55. Benović, A. Appearance of the jellyfish *Pelagia noctiluca* in the Adriatic Sea during the summer season of 1983. In Proceedings of the Workshop on Jellyfish Blooms in the Mediterranean, Athens, Greece, 31 October–4 November 1983; pp. 3–8.

56. Axiak, V.; Galea, C.; Schembri, P.J. Coastal aggregations of the jellyfish *Pelagia noctiluca* (Scyphozoa) in the Maltese coastal waters during 1980–1986. In Proceedings of the II Workshop on Jellyfish in the Mediterranean Sea, Trieste, Italy, 2–5 September 1987.

57. Zavodnik, D. Spatial aggregations of the swarming jellyfish *Pelagia noctiluca* (Scyphozoa). *Mar. Biol.* **1987**, *94*, 265–269. [CrossRef]

58. Avian, M.; Del Negro, P.; Sandrini, L.R. A comparative analysis of nematocysts in *Pelagia noctiluca* and *Rhizostoma pulmo* from the North Adriatic Sea. *Hydrobiologia* **1991**, *216–217*, 615–621. [CrossRef]

59. Burnett, J.W.; Calton, G.J. Venomous pelagic coelenterates: Chemistry, toxicology, immunology and treatment of their stings. *Toxicon* **1987**, *25*, 581–602. [CrossRef]

60. Salleo, A.; La Spada, G.; Barbera, R. Gadolinium is a powerful blocker of the activation of nematocytes of *Pelagia noctiluca*. *J. Exp. Biol.* **1994**, *187*, 201–206.

61. Kokelj, F.; Burnett, J.W. Treatment of a pigmented lesion induced by a *Pelagia noctiluca* sting. *Cutis* **1990**, *46*, 62–64. [PubMed]

62. Vlachos, P.; Kontos, P. Epidemiology and therapeutic methods of "jellyfish" poisoning in Greece. In Proceedings of the II Workshop on Jellyfish in the Mediterranean Sea, Trieste, Italy, 2–5 September 1987.

63. Milisenda, G.; Martinez-Quintana, A.; Fuentes, V.L.; Bosch-Belmar, M.; Aglieri, G.; Boero, F.; Piraino, S. Reproductive and bloom patterns of *Pelagia noctiluca* in the Strait of Messina, Italy. *Estuar. Coast. Shelf Sci.* **2018**, *201*, 29–39. [CrossRef]

64. Milisenda, G.; Rossi, S.; Vizzini, S.; Fuentes, V.L.; Purcell, J.E.; Tilves, U.; Piraino, S. Seasonal variability of diet and trophic level of the gelatinous predator *Pelagia noctiluca* (Scyphozoa). *Sci. Rep.* **2018**, *8*, 1–13. [CrossRef] [PubMed]

65. Milisenda, G.; Rosa, S.; Fuentes, V.L.; Boero, F.; Guglielmo, L.; Purcell, J.E.; Piraino, S. Jellyfish as prey: Frequency of predation and selective foraging of *Boops boops* (Vertebrata, Actinopterygii) on the mauve stinger *Pelagia noctiluca* (Cnidaria, Scyphozoa). *PLoS ONE* **2014**, *9*, e94600. [CrossRef] [PubMed]

66. Rottini Sandrini, L.; Avian, M. Reproduction of *Pelagia noctiluca* in the central and northern Adriatic Sea. *Hydrobiologia* **1991**, *216–217*, 197–202. [CrossRef]

67. Malej, A.; Malej, M. Population dynamics of the jellyfish Pelagia noctiluca (Forsskal, 1775). In *Marine Eutrophication and Population Dynamics, Proceedings of the 25th European Marine Biology Symposium (EMBS)*; Colombo, G., Ceccherelli, V.U., Rossi, R., Eds.; Olsen & Olsen: Fredensborg, Denmark, 1992.

68. Canepa, A.; Fuentes, V.; Sabatés, A.; Piraino, S.; Boero, F.; Gili, J.M. Pelagia noctiluca in the Mediterranean sea. In *Jellyfish Blooms*; Pitt, K.A., Lucas, C.H., Eds.; Springer Science & Business Media: Dordrecht, The Netherlands, 2014; pp. 237–266.

69. Guglielmo, L.; Minutoli, R.; Bergamasco, A.; Granata, A.; Zagami, G.; Brugnano, C.; Boero, F. The Strait of Messina: A aey area for *Pelagia noctiluca* (Cnidaria, Scyphozoa). In *Jellyfish—Ecology, Distribution Patterns and Human Interactions*; Mariottini, G.L., Ed.; Nova Publishers: New York, NY, USA, 2017; pp. 71–90.

70. Rosa, S.; Pansera, M.; Granata, A.; Guglielmo, L. Interannual variability, growth, reproduction and feeding of *Pelagia noctiluca* (Cnidaria: Scyphozoa) in the Strait of Messina (Central Mediterranean Sea): Linkages with temperature and diet. *J. Mar. Syst.* **2013**, *111*, 97–107. [CrossRef]

71. Lilley, M.K.S.; Elineau, A.; Ferraris, M.; Thiéry, A.; Stemmann, L.; Gorsky, G.; Lombard, F. Individual shrinking to enhance population survival: Quantifying the reproductive and metabolic expenditures of a starving jellyfish, *Pelagia Noctiluca*. *J. Plankton Res.* **2014**, *36*, 1585–1597. [CrossRef]

72. Doyle, T.K.; Houghton, J.D.R.; McDevitt, R.; Davenport, J.; Hays, G.C. The energy density of jellyfish: Estimates from bomb-calorimetry and proximate-composition. *J. Exp. Mar. Bio. Ecol.* **2007**, *343*, 239–252. [CrossRef]

73. Marques, R.; Bouvier, C.; Darnaude, A.M.; Molinero, J.C.; Przybyla, C.; Soriano, S.; Tomasini, J.A.; Bonnet, D. Jellyfish as an alternative source of food for opportunistic fishes. *J. Exp. Mar. Bio. Ecol.* **2016**, *485*, 1–7. [CrossRef]

74. Leone, A.; Lecci, R.M.; Durante, M.; Meli, F.; Piraino, S. The bright side of gelatinous blooms: Nutraceutical value and antioxidant properties of three mediterranean jellyfish (Scyphozoa). *Mar. Drugs* **2015**, *13*, 4654–4681. [CrossRef]

75. Leone, A.; Lecci, R.M.; Durante, M.; Piraino, S. Extract from the zooxanthellate jellyfish Cotylorhiza tuberculata modulates gap junction intercellular communication in human cell cultures. *Mar. Drugs* **2013**, *11*, 1728–1762. [CrossRef] [PubMed]

76. Goiris, K.; Muylaert, K.; Fraeye, I.; Foubert, I.; De Brabanter, J.; De Cooman, L. Antioxidant potential of microalgae in relation to their phenolic and carotenoid content. *J. Appl. Phycol.* **2012**, *24*, 1477–1486. [CrossRef]

77. Gall, E.A.; Lelchat, F.; Hupel, M.; Jegou, C.; Stiger-Pouvreau, V. Extraction and purification of phlorotannins from brown algae. *Methods Mol. Biol.* **2015**, *1308*, 131–143.

78. Capillo, G.; Savoca, S.; Costa, R.; Sanfilippo, M.; Rizzo, C.; Lo Giudice, A.; Albergamo, A.; Rando, R.; Bartolomeo, G.; Spanò, N.; et al. New insights into the culture method and antibacterial potential of *Gracilaria Gracilis*. *Mar. Drugs* **2018**, *16*, 492. [CrossRef] [PubMed]

79. Mastronicolis, S.; Miniadis, S.; Nakhel, I.; Smirniotopoulou, A. Biochemical and ecological research on jellyfish and other organisms in the Mediterranean Sea. In Proceedings of the II Workshop on Jellyfish in the Mediterranean Sea, Trieste, Italy, 2–5 September 1987; pp. 268–282.

80. Cardona, L.; Martínez-Iñigo, L.; Mateo, R.; González-Solís, J. The role of sardine as prey for pelagic predators in the western Mediterranean Sea assessed using stable isotopes and fatty acids. *Mar. Ecol. Prog. Ser.* **2015**, *531*, 1–14. [CrossRef]

81. Morais, Z.B.; Pintao, A.M.; Costa, I.M.; Calejo, M.T.; Bandarra, N.M.; Abreu, P. Composition and in vitro antioxidant effects of jellyfish *Catostylus tagi* from Sado Estuary (SW Portugal). *J. Aquat. Food Prod. Technol.* **2009**, *18*, 90–107. [CrossRef]

82. Bergé, J.P.; Barnathan, G. Fatty acids from lipids of marine organisms: Molecular biodiversity, roles as biomarkers, biologically active compounds, and economical aspects. In *Marine biotechnology I*; Le Gal, Y., Ulber, R., Eds.; Springer: Berlin/Heidelberg, Germany, 2005; Volume 96, pp. 49–125.

83. Simopoulos, A.P. Evolutionary aspects of diet, the omega-6/omega-3 ratio and genetic variation: Nutritional implications for chronic diseases. *Biomed. Pharmacother.* **2006**, *60*, 502–507. [CrossRef]

84. Templeman, M.A.; Kingsford, M.J. Trace element accumulation in *Cassiopea* sp. (Scyphozoa) from urban marine environments in Australia. *Mar. Environ. Res.* **2010**, *69*, 63–72. [CrossRef]

85. Muñoz-Vera, A.; García, G.; García-Sánchez, A. Metal bioaccumulation pattern by *Cotylorhiza tuberculata* (Cnidaria, Scyphozoa) in the Mar Menor coastal lagoon (SE Spain). *Environ. Sci. Pollut. Res.* **2015**, *22*, 19157–19169. [CrossRef]

86. Fowler, S.W.; Teyssi, J.L.; Cotret, O.; Danis, B.; Rouleau, C.; Warnau, M. Applied radiotracer techniques for studying pollutant bioaccumulation in selected marine organisms (jellyfish, crabs and sea stars). *Nukleonika* **2004**, *49*, 97–100.

87. Hay, S. Marine ecology: Gelatinous bells may ring change in marine ecosystems. *Curr. Biol.* **2006**, *16*, R679–R682. [CrossRef] [PubMed]

88. Sunda, W.G.; Huntsman, S.A. Processes regulating cellular metal accumulation and physiological effects: Phytoplankton as model systems. *Sci. Total Environ.* **1998**, *219*, 165–181. [CrossRef]

89. Hamner, W.M.; Dawson, M.N. A review and synthesis on the systematics and evolution of jellyfish blooms: Advantageous aggregations and adaptive assemblages. *Hydrobiologia* **2009**, *616*, 161–191. [CrossRef]

90. Wakabayashi, K.; Sato, H.; Yoshie-Stark, Y.; Ogushi, M.; Tanaka, Y. Differences in the biochemical compositions of two dietary jellyfish species and their effects on the growth and survival of Ibacus novemdentatus phyllosomas. *Aquac. Nutr.* **2016**, *22*, 25–33. [CrossRef]

91. Commission Regulation (EU), N. 744/2012 of 16 August 2012 amending Annexes I and II to Directive 2002/32/EC of the European Parliament and of the Council as regards maximum levels for arsenic, fluorine, lead, mercury, endosulfan, dioxins, Ambrosia spp., diclazuril and lasalocid A sodium and action thresholds for dioxins Text with EEA relevance. *Off. J. Eur. Union* **2012**, *219*, 5–12.

92. Commission Regulation (EC), N. 629/2008 of 2 July 2008 amending Regulation (EC) No 1881/2006 setting maximum levels for certain contaminants in foodstuffs. *Off. J. Eur. Union* **2008**, *173*, 6–9.

93. Liu, X.; Zhang, M.; Shi, Y.; Qiao, R.; Tang, W.; Sun, Z. Production of the angiotensin I converting enzyme inhibitory peptides and isolation of four novel peptides from jellyfish (*Rhopilema esculentum*) protein hydrolysate. *J. Sci. Food Agric.* **2016**, *96*, 3240–3248. [CrossRef] [PubMed]

94. Li, Q.-M.; Wang, J.-F.; Zha, X.-Q.; Pan, L.-H.; Zhang, H.-L.; Luo, J.-P. Structural characterization and immunomodulatory activity of a new polysaccharide from jellyfish. *Carbohydr. Polym.* **2017**, *159*, 188–194. [CrossRef]

95. Miyajima, Y.; Masuda, R.; Yamashita, Y. Feeding preference of threadsail filefish *Stephanolepis cirrhifer* on moon jellyfish and lobworm in the laboratory. *Plankt. Benthos Res.* **2011**, *6*, 12–17. [CrossRef]

96. Liu, C.S.; Chen, S.Q.; Zhuang, Z.M.; Yan, J.P.; Liu, C.L.; Cui, H.T. Potential of utilizing jellyfish as food in culturing *Pampus argenteus* juveniles. *Hydrobiologia* **2015**, *754*, 189–200. [CrossRef]

97. Directive 2010/63/EU Of The European Parliament and of the Council of 22 September 2010 on the protection of animals used for scientific purposes. *Off. J. Eur. Union.* **2010**, *276*, 33–79.

98. Lauriano, E.R.; Faggio, C.; Capillo, G.; Spanò, N.; Kuciel, M.; Aragona, M.; Pergolizzi, S. Immunohistochemical characterization of epidermal dendritic-like cells in giant mudskipper, *Periophthalmodon Schlosseri*. *Fish Shellfish Immunol.* **2018**, *74*, 380–385. [CrossRef]

99. Padmore, J.M. Animal feed—AOAC official method 976.05—Protein (crude) in animal feed, automated Kjeldhal method. In *Official Methods of Analysis*, 15th ed.; Helrich, K., Ed.; AOAC International: Arlington, WV, USA, 1990; Volume 1, p. 72.

100. Bligh, E.G.; Dyer, W.J. A rapid method for total lipid extraction. *Biochem. Cell Biol.* **1959**, *37*, 911–917.

101. Costa, R.; Beccaria, M.; Grasso, E.; Albergamo, A.; Oteri, M.; Dugo, P.; Fasulo, S.; Mondello, L. Sample preparation techniques coupled to advanced chromatographic methods for marine organisms investigation. *Anal. Chim. Acta* **2015**, *875*, 41–53. [CrossRef]

102. Bua, G.D.; Albergamo, A.; Annuario, G.; Zammuto, V.; Costa, R.; Dugo, G. High-throughput icp-ms and chemometrics for exploring the major and trace element profile of the Mediterranean sepia ink. *Food Anal. Methods* **2017**, *10*, 1181–1190. [CrossRef]

Review

Cnidarian Interaction with Microbial Communities: From Aid to Animal's Health to Rejection Responses

Loredana Stabili [1,2,*], **Maria Giovanna Parisi** [3], **Daniela Parrinello** [3] and **Matteo Cammarata** [3,*]

1   Istituto per l'Ambiente Marino Costiero, U.O.S. di Taranto, CNR, Via Roma 3, 74123 Taranto, Italy
2   Dipartimento di Scienze e Tecnologie Biologiche ed Ambientali, Università del Salento, via Prov.le Lecce Monteroni, 73100 Lecce, Italy
3   Laboratory of Marine Immunobiology, Dipartimento delle Scienze della Terra e del Mare, Università di Palermo, Viale delle Scienze Ed. 16, 90128 Palermo, Italy; mariagiovanna.parisi@unipa.it (M.G.P.); daniela.parrinello@unipa.it (D.P.)
*   Correspondence: loredana.stabili@iamc.cnr.it (L.S.); matteo.cammarata@unipa.it (M.C.)

Received: 27 July 2018; Accepted: 16 August 2018; Published: 23 August 2018

Abstract: The phylum Cnidaria is an ancient branch in the tree of metazoans. Several species exert a remarkable longevity, suggesting the existence of a developed and consistent defense mechanism of the innate immunity capable to overcome the potential repeated exposure to microbial pathogenic agents. Increasing evidence indicates that the innate immune system in Cnidarians is not only involved in the disruption of harmful microorganisms, but also is crucial in structuring tissue-associated microbial communities that are essential components of the Cnidarian holobiont and useful to the animal's health for several functions, including metabolism, immune defense, development, and behavior. Sometimes, the shifts in the normal microbiota may be used as "early" bio-indicators of both environmental changes and/or animal disease. Here the Cnidarians relationships with microbial communities and the potential biotechnological applications are summarized and discussed.

Keywords: cnidarian; anthozoa; microbial communities; cnidarian holobiont; zooxanthellae; bleaching; antibacterial activity

---

1. Introduction to Cnidarian

Cnidarian are a group made up of more than 9,000 living species, exclusively aquatic, getting their name from the presence of cnidocysts connected to supporting cells and neurons. These in turn form a unique chemosensor and mechanoreceptor neuronal cell complex that releases highly-ordered secretion products upon stimulation. The phylum Cnidaria includes the corals, hydras, jellyfish, Portuguese men-of-war, sea anemones, sea pens, sea whips, and sea fans. Cnidaria are taxonomically subdivided into: Anthozoa (Hexacorallia and Octocorallia) with the absence of a medusa stage, and the Medusozoa, that usually exhibit a medusa stage in their life cycle and includes the classes Cubozoa, Hydrozoa, Scyphozoa, and Staurozoa [1] (Figure 1).

The Cnidaria are one of the earliest branches in the animal tree, with tissue layers, muscles, and sense organs. They are diploblastic, have a radial symmetry, do not possess a real brain having only two cell layers; the epithelial cells are involved in all the innate immune responses. The endodermal epithelium functions as a chemical barrier using antimicrobial peptides, while the ectodermal epithelium represents a physicochemical barrier. Furthermore, Cnidaria are present in the fossil record since the Precambrian, when the other animals similar to the present ones were absent [1,2].

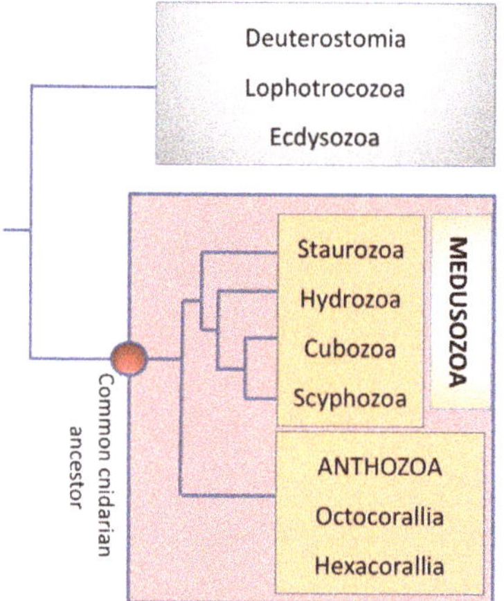

Figure 1. The phylogenetic relationships of Medusozoa (Staurozoa, Hydrozoa, Cubozoa, and Scyphozoa) and Anthozoa as reported by Boero et al. [1]. Molecular data sustain the separation in two class of the Anthozoa, which are common distinguished by tentacles morphology. Octocorallia is a group of hard coral species living at a depth of more than 100 m. This is a very slow growing species, formed by polyps with eight tentacles which capture floating materials of up to several hundred microns and included soft corals. Hexacorallia is a group of several hundred reef-building coral species including stony coral and sea anemones. The polyps of this coral have tentacles in groups of six, instead of eight.

Several important issues related to immunity can be inferred from the diversity in cnidarian life histories and habitats. In particular, in some cases their life cycles are very long and they may be subjected to repeated exposure to pathogenic agents [2]. Consequently, in the absence of specific immune cells, cnidarians must have effective mechanisms to defend against microbial pathogens. Furthermore, colonial forms, in order to save tissue integrity, rely on their capacity of self/nonself discrimination to rapidly recognize approaching allogeneic cells as foreign and to remove them [3]. Finally, successful growth for cnidarians is related to their capacity to differentiate between beneficial symbionts and pathogenic intruders [4,5], since they are colonized by complex bacterial communities and in several cases constitute home to algal symbionts. On account of these considerations it is of interest to understand how animal's longevity have modified the defense to innate components of immunity, leading us to consider Cnidaria as good candidates at the crossroad of metazoan evolution. Several molecular "omics" studies on the hydrozoans [6,7] sea anemones [8,9] and corals [8,10,11] demonstrated that some genes, associated with the immune responses, resulted conserved from cnidarians to vertebrates.

2. Cnidarians Associated Microbial Communities

Marine microorganisms are present at high density representing a major component in terms of the biomass on Earth. By the advent of the powerful tools of the molecular biology, remote sensing, and deep sea exploration, amazing discoveries on the abundance and diversity of marine microbial life and its function in global ecology have been made. In particular, researches on the relationships of microbial components with other organisms have furnished new information on the phenomena of food networks, symbiosis and pathogenicity [12,13].

Recently, there has been increasing interest in microbes as a relevant portion of the animal phenotype, responsible for the fitness as well as the ecological features of their hosts [14,15]. Several studies accomplished by genetic and genomic approaches have provided evidence for several animal–bacteria interactions in invertebrates and vertebrates revealing that bacteria play a crucial role in facilitating animals' origin and evolution [16,17]. Moreover, these findings clarified that animals and bacteria mutually influence their genomes [18] and that the homeostasis between animals and their symbionts is maintained by complex mechanisms [19,20]. Considering that microbial communities colonize all epithelia in animals, each animal with its associated microbes can be can be treated as a metaorganism (Figure 2) composed of the macroscopic host and the mutual symbiotic association with bacteria, archaea, fungi, and other microbial and eukaryotic species [21]. In such a community, membership is often influenced by interactions among species and properties [22].

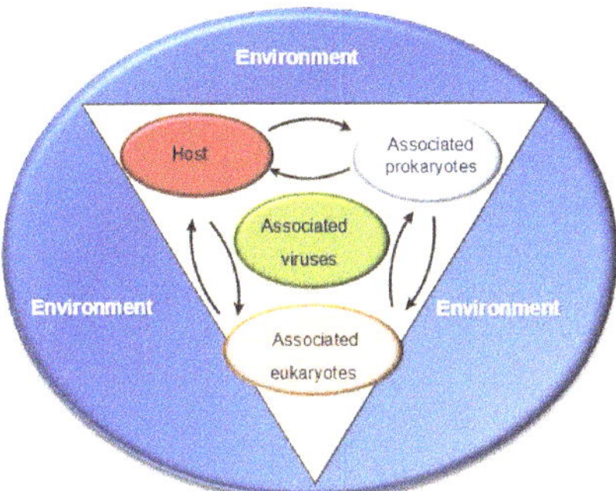

Figure 2. Multicellular organisms as "metaorganism" including the macroscopic host and the synergistic associated bacteria, archaea, fungi, and numerous other microbial and eukaryotic species. Modified from Bosch T.C. and McFall-Ngai M.J. 2011.

In the case of Cnidarians, associated protists, bacteria, archea, and viruses are essential components of the Cnidarian holobiont, capable of influencing, for example, the health of corals and coral reef ecosystems [21,23,24]. The associated bacteria perform several potential roles, such as nitrogen fixation, antibiotics synthesis [25,26], organic compounds decomposition [27], and space utilization; avoiding pathogen colonization [28]. Complexity and diversity are peculiar characteristics of coral-associated bacteria which reveal host species-specificity [29,30] and differ when compared to the bacterial communities recorded in the surrounding seawater [29,31,32]. Coral-associated microbial communities are influenced in their composition by several ecological parameters. When changes in environmental parameters are recorded, e.g., increases of seawater temperature, microbial species change in their density, making the coral holobiont capable of adapting to the new condition. Most studies have been carried out conclude that bacteria are directly involved in coral diseases [23,24,30,33,34]. Microbial communities associated with corals constitute a key factor useful to understand the coral reef health. Changes in bacterial composition over time may influence coral health and consequently their sensitivity to disease. Some researchers [35] have shown that when a small portion of the colony exerts signs of disease, the bacterial community associated with the colony is affected and modified. As a consequence, these data indicate that the evaluation of the shifts in the normal microbiota may be employed as "early" bio-indicators of both environmental changes and coral disease. Stress due to anthropogenic activities as well as environmental impacts may result in changes in the coral-associated microbial communities reflected as negative effects on the entire coral [28]. Climate change has been indicated as one of the foremost threats to Indo-Pacific reefs strictly related to coral bleaching. In some instances, at high temperatures, certain bacterial species increase their virulence and have been considered involved, for example, in bleaching [36]. It is therefore possible that the disappearance of key bacterial associates (by biotic or abiotic disturbances) amongst these communities provide an entry niche for opportunistic species that can further interfere with the microbial community structure and health status of the coral holobiont [35].

In particular, studies on the coral *Oculina patagonica* in the Mediterranean Sea have shown that the causative agent of the bleaching disease (resulting in the expulsion of the endosymbiotic zooxanthellae) is *Vibrio shiloi* [37]. In association with the onset of bleaching this bacterium exerts some virulence factors when high temperature values are recorded [38,39]. Moreover, *Vibrio coralliilyticus* was responsible for the bleaching of the coral *Pocillopora damicornis* on the coral reefs in the Indian Ocean and Red Sea [40]. In some coral diseases, such as black band, white pox, and white plague, bacteria are involved [33,36,41] and more than twenty coral diseases have been described (Table 1: from Rosenberg et al. [42]).

These diseases and their etiological mechanisms have been widely investigated over several years [43]. However, Reshef et al. [44] indicated that *O. patagonica* has become resistant against the infection supported by *V. shiloi*; thus, this bacterial species can no longer be isolated on the corals, and this *Vibrio* species, previously infecting corals is unable to produce disease on the existing corals. In order to explicate these results Reshef et al. [44] proposed the "Coral Probiotic Hypothesis". The term 'probiotic' means 'for life' and is referred to live microorganisms capable of determining a benefit in terms of health on their host [45,46]. Thus, these invertebrates seem not only to tolerate, but also to need the colonization by beneficial microorganisms for several functions including metabolism, immune defense, development, and behavior [47–49].

Table 1. Coral microbial pathogens.

Disease	Pathogen	Coral Host
Black band	*Roseofilum reptotaenium, Desulfovibrio, Beggiatoa* sp.	Several
White band I	Gram (-) bacterium	Several
White band II *	*Vibrio carchariae*	*Acropora* sp.
Aspergillosis *	*Aspergillus sidowii*	Gorgonians (sea fans)
White pox *	*Serratia marcescens*	*Acropora palmata*
Bleaching *	*Vibrio shiloi*	*Oculina patagonica*
Bleaching and lysis *	*Vibrio corallilyticus*	*Pocillopora damicornis*
Yellow blotch	*Vibrio alginolyticus*	*Monastraea* sp.
Red band	*Oscillatoria* sp. and other cyanobacteria	Several
Dark spots I	*Vibrio* sp. ?.	Several
Dark bands	?	Several
White plague (Eilat)	*Thalassomonas loyana*	Several
White plague	*Aurantimonas coralicida*	Several
White plague I	Gram (-) bacterium	Several
Porites ulcerative white spots	*Vibrio* sp.	Several

* Koch's postulates fulfilled.

3. Tissue-Associated Microbial Communities

As suggested by the increasing evidence the innate immune system in Cnidarians is not only involved in the disruption of harmful microorganisms, but also plays a crucial role in maintaining those tissue-associated microbial communities, useful to the host's health [50,51]. This is also the case with Hydra. Bacteria are indeed an important component of the Hydra holobiont in which 36 bacterial phylotypes were identified belonging to three different bacterial divisions and are dominated by the phyla Proteobacteria and Bacteroidetes [52]. The health of the whole animal can be compromised by disturbances or shifts in any of these partners [53]. In laboratory studies Hydra have been cultivated under standard conditions (constant temperature and identical food), and surprisingly, it was observed that in different Hydra species, maintained in the laboratory for more than 20 years, a complex microbial community colonized the epithelium which greatly differed in individuals from different species, but was cultured under identical conditions. On account of these evidences it was concluded that the microbiota in Hydra is specific for each species [12].

When closely related Hydra species were examined, the associated microbial community that resulted was similar. This, for instance, was the case of *H. vulgaris* and *H. magnipapillata*. Moreover, the early branching linage of Hydra species *H. oligactis*, examined so far [54], results associated with the most distinct microbial community in comparison with the other Hydra species. These observations lead to conclude that on and within the Hydra epithelium distinct selective pressures are imposed [50]: the colonization of a certain epithelium by bacteria is related to several ecological factors, such as host immune responses, the availability of nutrients, and the space competition between bacterial strains. Thus, it can be suggested that the colonizing microbial composition is shaped by both host factors such as components of Hydra's innate immune system and frequency-dependent bacteria–bacteria interactions [51].

In contrast to microbiomes of tropical corals, characterized by high diversity, Mediterranean octocorals harbor structured bacterial assemblages in which only a few species prevail for >90% [55,56]. This makes them ideal model organisms to investigate cnidarian microbe interactions. In the Mediterranean Sea, gorgonians contribute significantly to the structural complexity, biomass, and biodiversity of these ecosystems constituting the most significant habitat-forming species of benthic communities [19]. The success of gorgonians is in part due to the specific symbioses with bacteria which are relatively stable across spatial scales. Bacteria belonging to the genus Endozoicomonas are the most prevalent [57,58] in certain species from the Gorgoniidae family. Endozoicomonas have been recorded in several marine invertebrates and have been recognized as crucial to the health of corals [59], with a loss of these microorganisms producing a conspicuous negative impact on the holobiont functioning. Spirochaetes represent one of the (co-)dominant microbial associates [60], presumably involved in nitrogen and carbon fixation in tropical octocorals and deep-sea gorgonians, particularly the precious red coral *Corallium rubrum* [61]. The spatial stability of these bacteria–host associations, which may exist in in the same habitat and location, lead to hypothesize strong selection mechanisms used by the holobiont of Mediterranean gorgonians.

Therefore, exploring the structure and functioning of the microbiome is a major challenge of current research in Cnidarians, also taking into account that their tissue and mucus support a diverse microbial community [24,34,62].

4. Mucus-Associated Microbial Communities

Mucus adhesion and colonization by bacteria represents one of the best characterized symbiosis in the marine environment. Specifically, mucus released by certain marine invertebrates furnishes a habitat for several bacteria [63,64]. During mucus production, marine organisms consume a significant portion of energy. For instance, in corals mucus release consumes up to 50% of the assimilated energy [65]. Some marine invertebrates are coated by a layer of mucus to prevent bacteria and debris from accumulating on the body surface [66]. This matrix is involved in a number of defense mechanisms [67–72], to cope with the rich mixture of microorganisms in the surrounding water. However, for many species, including corals, as reported by Coffroth [73], the mucus also represents a home site [63,64] and may function as a potential food source [74]. Mucus contains primarily polysaccharides and proteins with C:N ratios of 1:5, on account of this composition it was suggested that this matrix, released by some marine invertebrates, is readily degradable by microbes, thus supporting microbial growth. As regards Cnidarian, mucus contains many microorganisms [24,75,76], and in particular, in the coral mucus, the mean concentration of colony-forming bacteria is about 0.2% of the total counts determined microscopically by using SYBR Gold Staining and ranges between 10^5 and 10^6 mL^{-1} [77,78]. Furthermore, in several other corals, including the Caribbean coral *Monastrea franks* and *Oculina patagonica* [28,79], the mucus layer was shown to contain a high bacterial density (3×10^8 mL^{-1}). Species belonging to the three primary domains Archaea, Eubacteria, and Eukarya have been recorded in coral mucus [80]. In *H. vulgaris* mucus *V. splendidus* was the most-abundant species attaining 68% and 50% in the winter and summer, respectively [78]. Moreover, vibrios prevailed in the culturable bacterial isolates from the mucus of *Acropora palmata*. Most of the studies have conducted on the microbiota associated with corals and other marine invertebrates' mucus [76,81], by contrast, zoanthids have received little attention in that respect, although only few studies by Chimetto et al. [75,82,83] investigated the diversity of bacteria on zoanthids and found 16S rDNA sequences belonging to the *Vibrio* genus. As already proposed by Calow [74], differences in biochemical composition may render the mucus more or less susceptible to microbial attack. The mucus rich in proteins released by some invertebrates is rapidly used by microbes possessing exoenzymes potentially capable to degrade mucoid polymers [73,84]. On account of this activity, microbial communities harbored in such mucus may use mucus-derived, dissolved, and particulate matter transforming them into living biomass [84]. As a consequence, the mucus can represent the foundation from which microbial organisms are then preyed upon by other organisms [85–88]. In contrast to this, some mucus bacteria are involved in the defense of their hosts by the production of antibacterial compounds. In turn, mucus bacteria capable of producing antibacterial molecules have an advantage over other microorganisms, assisting in competition over space and nutrition. Several bacteria with antimicrobial activity against presumed coral pathogens have been isolated from corals [26,89,90], and the antimicrobial activity of coral mucus appears decreased in corals displaying signs of coral bleaching or disease [26]. These observations suggest intriguing relationships between different coral-associated bacteria and between bacterial associates and the coral host. It is currently unknown whether bacterial communities are selected by extrinsically mediated factors or whether the holobiont itself selects for beneficial associates [90]. Analogous studies of sponges hint at the latter hypothesis as possible, where it is suggested that the species *Mycale adhaerens* may selectively sequester bacterial epibionts with antimicrobial activities [91]. One major group of coral-associated bacteria exerting antibacterial activity is *Pseudoalteromonas* sp. Several *Pseudoalteromonas* produce antibacterial compounds, toxins, bacteriolytic substances, and enzymes, all of which may aid the bacterial cells in their competition for space, nutrients, and in the security from predation [92]. It is also plausible that bacteria such as *Pseudoalteromonas* sp. can affect the microbial community by releasing active compounds into the coral mucus. This is in accordance with the "Coral Probiotic Hypothesis" [44], whereby active *Pseudoalteromonas* sp. can be considered as "probiotic" to corals, taking part in coral holobiont defense against bacteria.

Extraordinary recent progress in sequencing technologies and the ability to culture simple but genetically accessible model organisms for some time under germ-free conditions are revealing details of host–microbe interactions that highlight the value of an evolutionary perspective thus undermining prior concepts. However, in spite of these insights, the factors involved in microbial colonization of mucosal surfaces are still unknown. Moreover, the accumulated data are not still coherently integrated in order to obtain a truly mechanistic understanding of host–microbe interactions on host mucosal surfaces.

5. Innate Immune System as a Regulator in Maintaining Homeostasis between Animals and Their Resident Microbiota?

In a recent review, Bosch [50] has reviewed the pre-existing idea that immune systems evolve exclusively to control invading pathogens furnishing evidence that host-specific microbiota is established by the crucial role played by major factors of the immunological system. The involvement of components of the innate immunity systems, such as antimicrobial peptides, in shaping the microbiota is now undeniable. His thesis, based mainly on Hydra examples, is that the need to control of the resident beneficial microbes induced the evolution of the immune systems. He suggested that it is reasonable to assume that the inferences drawn apply to both invertebrates and vertebrates. Stem cell proliferation, microbiota composition, and innate immunity seem to have a mutual direct link. Particularly, he highlighted that recent discoveries in Hydra show that homeostasis between animals and the resident microbiota is assured by the action of innate immune system factors and transcriptional regulators of stem cells. He stated that, in early-branching metazoans, the evolution of the innate immune system and its host-specific components is due to the need to control the resident beneficial microbes, rather than the action of invasive pathogens. In this framework, disease onset is considered as the result of a complex network of interactions among different associated partners capable of affecting the fitness of the entire metaorganism [93].

This is also the case of microbial hypothesis of coral bleaching. According to this hypothesis, several physical and biological factors, including variation in sea surface temperatures [94,95], UV irradiation [96], low salinity and pollution [97], and bacterial infection [13] are responsible for bleaching, which is a symptom of stress. These different kinds of stress act on both the coral microorganisms and the coral host, determining a change in the microbial community that in some cases is reflected directly or indirectly on bleaching. This also [98] emerges from research conducted over the last decade, which has supported that the coral host, its endosymbiotic zooxanthellae, and a large number and variety of accompanying microorganisms form a complex and dynamic symbiosis represented by the coral holobiont. In a healthy coral, the growth, reproduction, and disease resistance of the holobiont is due to the metabolic activities of each organism interacting with the other ones. Thus, the coral host, by capturing and feeding on prey, through their digestion, provides nutrients for its associated microorganisms. The associated microorganisms may be also used directly by the coral [99] with the production of carbon dioxide and water as byproducts of cellular respiration. The zooxanthellae, in turn, employing the carbon dioxide and water, accomplishes photosynthesis. In particular, *Symbiodinium* reside in host tissues at millions of cells per square centimeter and provide the energy required by reef building corals to grow, calcify, and reproduce. The zooxanthellae cells produce, as products of photosynthesis, sugars, lipids (fats), and oxygen; major components needed for animal and bacterial respiration. Thus, the driving force behind the growth and productivity of coral reefs is represented by the tight recycling of products between the polyp cells, bacteria, and the zooxanthellae. Under stress, the components of the symbiosis separate and the associated endosymbionts may be digested [100]. In this framework, the roles of bacteria in contributing to the holobiont health is a matter of current interest and debate. It has been shown that coral bacteria can fix nitrogen, degrade complex polysaccharides, and produce antibiotics useful in helping to prevent infection by pathogens. Rosenberg et al. [98] suggested that coral bleaching happens when the equilibrium between the different components of the coral holobiont is destroyed and results in a decrease in the endosymbiotic zooxanthellae. Bleaching is now considered a host innate internal defense response to compromised symbionts and, in particular, Cnidarian bleaching is due to a breakdown in the symbiotic relationship between host cnidarians and photosynthetic dinoflagellates belonging to the genus *Symbiodinium*. The symbiosis between anthozoan polyps and zooxanthellae are considered nonharmful infections, where the unicellular organisms are able to control the host defense response until the environmental conditions are optimal for survival of autotrophic and heterotrophic organisms. The oxygen reactive species (ROS) and the reactive nitrogen species nitric oxide (NO) are involved in host–pathogen interactions and bleaching events. The stress triggered by alterations of physical factors, pathogenic infections, or injuries indeed involves the increase of ROS and NO by the symbionts. These molecules activate the cascade mechanisms in internal defense systems and eliminate the zooxanthellae. The loss of the symbionts unable to perform the photosynthesis process occurs through traditional mechanisms of the innate immune system including exocytosis, host cell detachment, and apoptosis [101]. Although bleaching is induced by a variety of environmental stressors like global climate change and high solar radiation, the temperature increase in the superficial seawater and anthropogenic stress also caused an enhancement in diseases of species of the genus Anthozoa responsible for the bleaching or tissue death [102]. However, most works have concentrated on the innate immune repertoire of anthozoans, the immune effector mechanisms mediated corals adaptation to such events remain almost unknown.

6. Antimicrobial Peptides, Multifunctionality, and Biotechnological Implications

Humoral response is realized by the synthesis and release of an array of chemical compounds, including melanin, reactive oxygen species (ROS), antimicrobial peptides (AMP), and secondary metabolites whose most important purpose is to destroy microbes by: (1) opsonizing and agglutinating invaders; (2) permeabilizing the invader's cell membrane, causing lysis; or (3) disruption of their metabolism [103]. Antimicrobial agents do not possess functional specificity since exert a broad spectrum of activity against Gram-positive and Gram-negative bacteria, fungi, viruses, and protists [103,104]. However, all antibacterial agents are not equally effective toward bacteria [105]. The size of the antimicrobials may impact structural microbial components differently. For instance, small peptides (<23 amino acids long) mainly destroy the cell membrane integrity of the invaders, while larger peptides pose lytic properties, or may be proteins with specific domains that sequester essential nutrients from microbes [106]. Several studies have been attained to evidence the presence of antimicrobial compounds in cnidarians. The antimicrobial activity of the eight species of gorgonian corals *Plexaura homomalla*, *Pseudoplexaura flagellosa*, *Plexaurella fusifera*, *Eunicea clavigera*, *Eunicea tourneforti*, *Eunicea laciniata*, *Eunicea calyculata* (Plexauridae), and *Pseudopterogorgia americana* were assayed against five species of bacteria including marine bacteria as well as human pathogenic species (*Vibrio harveyi*, *Pseudomonas aeruginosa*, *Serratia marcescens*, *Staphylococcus aureus*, *Bacillus megaterium*, and *Escherichia coli*). Antimicrobial activity was evaluated on polar and nonpolar extracts and was most apparent in the nonpolar fractions. In general, marine bacteria were not as sensitive to the extracts as the nonmarine species [107]. Subsequently, from the West Indian gorgonian coral *Pseudopterogorgia elisabethae*, by using NMR spectroscopy, the structure of two compounds capable to inhibit the growth of *Mycobacterium tuberculosis* H37Rv t. was determined [108]. The activity was ascribed to two diterpenoid alkaloids, namely pseudopteroxazole, producing a 97% growth inhibition, and seco-pseudopteroxazole, responsible for a 66% inhibition at 12.5 µg/mL. From the same octocoral *P. elisabethae* of San Andrés and Providencia Islands (Southwest Caribbean Sea) [109] the cytotoxic and antimicrobial activity of pseudopterosins and secopseudopterosins, active against *Staphylococcus aureus* and *Enterococcus faecalis* but inactive against *Pseudomonas aeruginosa* and *Candida albicans*, was investigated. Shapo et al. [110] reported that crude extracts from the gorgonian coral *Leptogorgia virgulata*, likely containing homarine, showed inhibitory activity against *Escherichia coli* and *Vibrio harveyi* as well as other bacteria. Uncharacterized antimicrobial agents have been also documented in over a dozen members of the Plexauridae, Gorgonidae, and Ellisellidae families [107,111,112]. Chen et al. [113] reported that fifteen guaiazulene-based terpenoids (anthogorgienes A–O) and eight analogues, isolated from the lipophilic extract of *Anthogorgia* sp., were effective against *S. aureus* and *Streptococcus pneumoniae* and three fungi (*Aspergillus fumigatus*, *Aspergillus flavus*, and *Fusarium oxysporum*). In the sea-whip *Dichotella gemmacea* Li et al. [113] isolated six briarane diterpenoids and two analogs showing a weak antimicrobial action against the growth of *E. coli*. Scleractinian corals also possess secondary compounds with antimicrobial properties, even though they have been less well investigated than those derived from the gorgonians [114]. Gochfeld and Aeby [114] reported that crude extracts from three Hawaiian corals i.e., *Montipora capitata*, *Porites lobata*, and *Pocillopora meandrina*, exerted antibacterial activity against coral pathogens such as *Serratia marcescens*, *Vibrio coralliityticus*, and *V. shilo*. However, the antibacterial activity of extracts varied among species and as a function of the state of health of the host. The antimicrobial properties of extracts from Red Sea soft corals (alcyonaceans) *Litophyton arboreum*, *Rythisma fulvum*, *Heteroxenia fuscescens*, *Sarcophyton glaucum*, *Dendronephthya hemprichi*, and *Xenia macrospiculata*, were compared with those of stony (scleractinian) corals, *Acropora variabilis*, *Fungia scutaria*, *Fungia granulosa*, *Turbinaria* sp., *Stylophora pistillata*, and *Favia favus* and the majority of soft corals (83%) resulted to affect remarkably the growth of the marine bacteria *Arthrobacter* sp. and scarcely the growth of *Vibrio* sp., while stony corals showed very little or no activity [115].

In recent studies, thermostable proteases and antimicrobial peptides have been characterized from the body and tentacles of the sea anemones *Actinia equina* and *Anemonia sulcata* with application for biocleaning and as antifungals [115]. In particular, bioactive molecules (BMs) isolated from the sea anemone *Actinia equina*, were proven to hydrolyse aged/altered protein layers or coatings as well as to control bacteria/fungi growth [116]. On account of these features these molecules represent an innovative tool in conservative restoration procedures. Particularly, the BMs molecules with proteolytic activity were tested in order to remove protein layers or to control microbial colonizations.

The removal of undesired layers under "room temperature" (19–25 °C) conditions, without heating the enzyme solution or the artwork surface on which it was applied was tested. Agreeable results were obtained after application of gelled enzymatic solution, in removing coherent protein layer both from the surface of polychrome wood or wax sculpture. In both cases the complete removal of the protein layer, without producing whitening phenomena was observed. The best advantage of these molecules is their temperature of action (<30 °C) which is different from that of the commercial proteases active at higher temperature (37 °C). Moreover, the antimicrobial activity of BMs was assayed to inhibit the growth of some bacteria such as *Enterobacter* spp. and *Micrococcus luteus*, and fungi as *Aspergillus niger* and *Penicillium chrysogenum*. Thus, the employment of these molecules for biocleaning represents an innovative procedure that minimizes the exposure to harmful solvents and chemicals compounds for both the workers and the environment [116]. Furthermore, these molecules are totally safe for works of art, restores, and the environment, requiring a short time of application. Consequently, we hypothesize that these bioactive molecules represent a valid alternative to the traditional procedures in sustainable restoration projects [116].

Moreover, recently it was established that anthozoans could also benefit of the multifunctionality of some of their bioactive molecules [117]. *Actinia viridis* and *Actinia equina* possess a toxin with bifunctional characteristics: Neurotoxin ATX II, isolated from *A. viridis*, is a sodium channel type 1 toxin constituted of 47aa, characterized by the presence of three disulfide bridges capable of binding to the sodium voltage ionic channel, delaying the inactivation phase during the transmission of action potential and exerting antimicrobial activity towards *Micrococcus lysodeikticus*. ATX II can be considered as a neurotoxin with an additional antimicrobial peptide property. Thus, anemones could adopt the multifunctionality of toxins as an evolutionary strategy in order to amplify their predation capacity. Moreover, the antimicrobial molecules would assure the polyps to survive avoiding bacterial infections [117]. *Actinia equina* lives in the temperate coastal area and this intertidal species is a suitable and exemplary model for the study of bioactive molecules and their evolution. Hemolytic molecules such as equinotoxin [118,119] and proteins for potassium and sodium voltage dependent channels [120] have been characterized. The mucus of this sea anemone contains a complex mixture of proteins and polysaccharides with differential biological activity implicated in the immune defense. This matrix plays a crucial role in a series of biological processes including structural support, locomotion, food particle trapping, and defense against multiple environmental stresses, predators, parasites, and pathogens. In this mucosal matrix hemolytic activity versus rabbit erythrocytes, cytotoxic activity against human erythromyeloblastoid leukemia T cell line (K562) and lysozyme-like activity was observed [71]. Lysozyme is involved in internal innate defense and acts as an antimicrobial enzyme system and in particular as a glycoside hydrolase, catalyzing the hydrolysis of 1,4-beta-linkages between *N*-acetylmuramic acid and *N*-acetyl-D-glucosamine in peptoglycans component of bacterial cell wall. As a consequence, the integrity of bacterial pathogens through the lysis of their cell wall results compromised. The presence in *A. equina* mucus of an antibacterial activity in association with a hemolytic and cytotoxic activity indicates its participation in the defense system against pathogenic invaders suggesting that the humoral effectors of the internal defense system can be released in mucus layer. The activity against *Micrococcus lysodeikticus* as well as the satisfactory results obtained at 37 °C lead to consider *A. equina* mucus an interesting prospect for future biotechnological applications of pharmaceutical and marine technology interest. With regards to pharmaceuticals, the increasing development of bacteria resistant to traditional antibiotics has reached alarming levels, and thus there is the need to develop new antimicrobial agents. In this framework, lysozyme was recently selected as a model protein to develop more potent bactericidal agents thus introducing, a new conceptual employment of lysozyme [69]. Lastly, the antibacterial proteins of *A. equina* mucus could be used to deter the settlement of bacteria representing the primary colonizers in the development of marine biofouling thus constituting an alternative natural antisettlement agents compared to the banned paints and organic biocides [69]. In this framework, it is intriguing that lysozyme-like proteins have also already been evidenced in other cnidarians [121,122].

7. Conclusions

Comparative immunobiology studies have led to the abandonment of the idea that invertebrates do not possess immune capacity. Cnidarians possess components of the main routes of immunity of invertebrates. The receptors and pathways already identified indicate that these basal invertebrates are far from being "simple" in the range of methods they have to deal with potential germs and pathogens.

Cnidarian-associated microbial communities are probably a result of a functional cross-talking because cnidarian need to control the resident beneficial microbes, not as a response to invasive pathogens, but because, just as black can exist only if white is visible, so too the use of the same thrifty ways for distinguishing pathogens could be considered the possible origin of the first immunity arms.

In Cnidarians, the crucial activities in structuring tissue-associated microbial communities, useful to the animal's health, are related to the increasing evidence of the existing innate immune responses involved in the disruption of harmful microorganisms. The present review represents a contribution to reduce the gaps in the current knowledge, regarding the complex relationships established between cnidarians and microorganisms, as well as to provide an overview of the potential biotechnological applications of the defensive compounds present in these invertebrates.

Author Contributions: The first two authors contributed equally. L.S. reviewed all the aspects related to the Cnidarian-microorganisms relationships. L.S., M.G.P., D.P. and M.C. examined all the immunological aspects. L.S., M.G.P. and M.C. wrote the manuscript.

Acknowledgments: Financial support was provided by PJ_RIC_FFABR_2017_004312 incentivi alle attività base di Ricerca ANVUR and the Life Project REMEDIA Life "REmediation of Marine Environment and Development of Innovative Aquaculture: exploitaiton of edible/not edible biomasss" (LIFE Environment and Resource Efficiency 2016. LIFE16 ENV/IT/000343) funded by EU and by FFR-Cammarata from Scientific Research University of Palermo (2014) and MC RITMARE Project (CNR and CONISMA).

Conflicts of Interest: The authors declare no conflict of interest.

References

1. Boero, F.; Bouillon, J.; Piraino, S. The role of Cnidaria in evolution and ecology. *Ital. J. Zool.* **2005**, *72*, 65–71. [CrossRef]

2. Petralia, R.S.; Mattson, M.P.; Yao, J. Aging and longevity in the simplest animals and the quest for immortality. *Ageing Res. Rev.* **2014**, *16*, 66–82. [CrossRef] [PubMed]

3. Hidaka, M. Tissue compatibility between colonies and between newly settled larvae of *Pocillopora damicornis*. *Coral Reefs* **1985**, *4*, 111–116. [CrossRef]

4. Campbell, R.D.; Bibb, C. Transplantation in coelenterates. *Transplant. Proc.* **1970**, *2*, 202–211. [PubMed]

5. Lubbock, R. Clone-specific cellular recognition in a sea anemone. *Proc. Natl. Acad. Sci. USA* **1980**, *77*, 6667–6669. [CrossRef] [PubMed]

6. Soza-Ried, J.; Hotz-Wagenblatt, A.; Glatting, K.; del Val, C.; Fellenberg, K.; Bode, H.R.; Franck, U.; Hoheisel, J.; Frohme, M. The transcriptome of the colonial marine hydroid *Hydractinia echinata*. *FEBS J.* **2010**, *277*, 197–209. [CrossRef] [PubMed]

7. Wenger, Y.; Galliot, B. RNAseq versus genome-predicted transcriptomes: A large population of novel transcripts identified in an Illumina-454 Hydra transcriptome. *BMC Genomics* **2013**, *14*, 204–221. [CrossRef] [PubMed]

8. Miller, D.; Hemmrich, G.; Ball, E.E.; Hayward, D.; Khalturin, K.; Funayama, N.; Agata, K.; Bosch, T.C. The innate immune repertoire in cnidaria-ancestral complexity and stochastic gene loss. *Genome Biol.* **2007**, *8*, R59. [CrossRef] [PubMed]

9. Putnam, N.H.; Srivastava, M.; Hellsten, U.; Dirks, B.; Chapman, J.; Salamov, A.; Terry, A.; Shapiro, H.; Lindquist, E.; Kapitonov, V.V.; et al. Sea anemone genome reveals ancestral eumetazoan gene repertoire and genomic organization. *Science* **2007**, *317*, 86–94. [CrossRef] [PubMed]

10. Schwarz, J.A.; Brokstei, P.B.; Voolstra, C.; Terry, A.Y.; Miller, D.J.; Szmant, A.M.; Coffroth, M.A.; Medina, M. Coral life history and symbiosis: Functional genomic resources for two reef building Caribbean corals, *Acropora palmata* and *Montastraea faveolata*. *BMC Genomics* **2010**, *97*, 1–16.

11. Vidal-Dupiol, J.; Adjeroud, M.; Roger, E.; Foure, L.; Duval, D.; Mone, Y.; Ferrier, C.; Tambutte, E.; Tmabutte, S.; Zoccola, D.; et al. Coral bleaching under thermal stress: putative involvement of host/symbiont recognition mechanisms. *BMC Physiol.* **2009**, *14*, 1–16. [CrossRef] [PubMed]

12. Munn, C. *Marine Microbiology: Ecology and Applications*, 2nd ed.; CRC Press: New York, NY, USA, 2011; p. 364.

13. Bevins, C.L.; Salzman, N. The potter's wheel: The host's role in sculpting its microbiota. *Cell Mol. Life Sci.* **2011**, *68*, 3675–3685. [CrossRef] [PubMed]

14. McFall-Ngai, M. Adaptive immunity: Care for the community. *Nature* **2007**, *7124*, 153. [CrossRef] [PubMed]

15. Nyholm, S.V.; Graf, J. Knowing your friends: Invertebrate innate immunity fosters beneficial bacterial symbioses. *Nat. Rev. Microbiol.* **2012**, *10*, 815–827. [CrossRef] [PubMed]

16. Nichols, S.A.; Dayel, M.J.; King, N. Genomic, phylogenetic, and cell biological insights into metazoan origins. In *Animal Evolution: Genes, Genomes, Fossils and Trees*; Telford, M.J., Littlewood, D.T.J., Eds.; Oxford University Press: New York, NY, USA, 2009; pp. 24–32.

17. Alegado, R.A.; Brown, L.W.; Cao, S.; Dermenjian, R.K.; Zuzow, R.; Fairclough, S.R.; Clardy, J.; King, N. A bacterial sulfonolipid triggers multicellular development in the closest living relatives of animals. *eLife* **2009**, *15*, 1:e00013. [CrossRef] [PubMed]

18. Hughes, D.T.; Sperandio, V. Inter-kingdom signalling: Communication between bacteria and their hosts. *Nat. Rev. Microbiol.* **2008**, *6*, 111–120. [CrossRef] [PubMed]

19. McFall-Ngai, M.; Hadfield, M.G.; Bosch, T.C.; Carey, H.V.; Domazet-Loso, T.; Douglas, A.E.; Dubilier, N.; Eberl, G.; Fukami, T.; Gilbert, S.F.; et al. Animals in a bacterial world: a new imperative for the life sciences. *Proc. Natl. Acad. Sci. USA* **2013**, *110*, 3229–3236. [CrossRef] [PubMed]

20. Pradeu, T.A. Mixed self: The role of symbiosis in development. *Biol. Theory* **2011**, *6*, 80–88. [CrossRef]

21. Bosch, T.C.; McFall-Ngai, M.J. Metaorganisms as the new frontier. *Zoology* **2011**, *114*, 185–190. [CrossRef] [PubMed]

22. Drake, J.A. The mechanics of community assembly and succession. *J. Theor. Biol.* **1990**, *47*, 213–233. [CrossRef]

23. Rosenberg, E.; Koren, O.; Reshef, L.; Efrony, R.; Zilber-Rosenberg, I. The role of microorganisms in coral health, disease and evolution. *Nat. Rev. Microbiol.* **2007**, *5*, 355–362. [CrossRef] [PubMed]

24. Kimes, N.E.; Johnson, W.R.; Torralba, M.; Nelson, K.E.; Weil, E.; Morris, P.J. The *Montastraea faveolata* microbiome: Ecological and temporal influences on a Caribbean reef-building coral in decline. *Environ. Microbiol.* **2013**, *15*, 2082–2094. [CrossRef] [PubMed]

25. Kelman, D.; Kashman, Y.; Rosenberg, E.; Kushmar, A.; Loya, Y. Antimicrobial activity of Red Sea corals. *Mar. Biol.* **2006**, *149*, 357–363. [CrossRef]

26. Ritchie, K.B. Regulation of microbial populations by coral surface mucus and mucus-associated bacteria. *Mar. Ecol. Prog. Ser.* **2006**, *32*, 1–14. [CrossRef]

27. Di Salvo, L.H. Isolation of bacteria from the corallum of *Porites lobata* (Vaughn) and its possible significance. *Am. Zool.* **1969**, *9*, 735–740. [CrossRef]

28. Ritchie, K.B.; Smith, G.W. Microbial communities of coral surface mucopolysaccharide layers. In *Coral Health and Disease*; Rosenberg, E., Loya, Y., Eds.; Springer: Berlin/Heidelberg, Germany, 2004; pp. 259–263.

29. Rohwer, F.; Seguritan, V.; Azam, F.; Knowlton, N. Diversity and distribution of coral-associated bacteria. *Mar. Ecol. Prog. Ser.* **2002**, *243*, 1–10. [CrossRef]

30. Sunagawa, S.; Woodley, C.M.; Medina, M. Threatened corals provide underexplored microbial habitats. *PLoS ONE* **2005**, *5*, 1–7. [CrossRef] [PubMed]

31. Frias-Lopez, J.; Zerkle, A.L.; Bonheyo, G.T.; Fouke, B.W. Partitioning of bacterial communities between sea-water and healthy, black band diseased, and dead coral surfaces. *Appl. Environ. Microbiol.* **2002**, *68*, 2214–2228. [CrossRef] [PubMed]

32. Bourne, D.G.; Munn, C.B. Diversity of bacteria associated with the coral *Pocillopora damicornis* from the Great Barrier Reef. *Environ. Microbiol.* **2005**, *7*, 1162–1174. [CrossRef] [PubMed]

33. Denner, E.B.M.; Smith, G.; Busse, H.J.; Shumann, P.; Narzt, T.; Polson, S.W.; Lubitz, W.; Richardson, L.L. *Aurantimonas coralicida* sp. nov, the causative agent of white plague type II on Caribbean scleractinian corals. *Int. J. Syst. Evol. Microbiol.* **2003**, *53*, 1115–1122. [CrossRef] [PubMed]

34. Roder, C.; Arif, C.; Bayer, T.; Aranda, M.; Daniels, C.; Shibl, A.; Chavanich, S.; Voolstra, C.R. Bacterial profiling of white plague disease in a comparative coral species framework. *ISME J.* **2014**, *8*, 31–39. [CrossRef] [PubMed]

35. Pantos, O.; Cooney, R.P.; Le Tissier, M.D.A.; Barer, M.R.; O'Donnell, A.G.; Bythell, J.C. The bacterial ecology of a plague-like disease affecting the Caribbean coral *Montastrea annularis*. *Environ. Microbiol.* **2003**, *5*, 370–382. [CrossRef] [PubMed]

36. Rosenberg, E.; Ben-Haim, Y. Microbial diseases of corals and global warming. *Environ. Microbiol.* **2002**, *4*, 318–326. [CrossRef] [PubMed]

37. Kushmaro, A.; Loya, Y.; Fine, M.; Rosenberg, E. Bacterial infection and coral bleaching. *Nature* **1996**, *380*, 396. [CrossRef]

38. Toren, A.; Landau, L.; Kushmaro, A.; Loya, Y.; Rosenberg, E. Effect of temperature on adhesion of Vibrio strain AK-1 to *Oculina patagonica* and on coral bleaching. *Appl. Environ. Microbiol.* **1998**, *64*, 1379–1384. [PubMed]

39. Banin, E.; Israely, T.; Fine, M.; Loya, Y.; Rosenberg, E. Role of endosymbiotic zooxanthellae and coral mucus in the adhesion of the coral-bleaching pathogen *Vibrio shiloi* to its host. *FEMS Microbiol. Lett.* **2001**, *199*, 33–37. [CrossRef] [PubMed]

40. Ben-Haim, Y.; Thompson, F.L.; Thompson, C.C.; Cnockaert, M.C.; Hoste, B.; Swings, J.; Rosenberg, E. *Vibrio coralliilyticus* sp nov, a temperature dependent pathogen of the coral *Pocillopora damicornis*. *Int. J. Syst. Evol. Microbiol.* **2003**, *53*, 309–315. [CrossRef] [PubMed]

41. Richardson, L.L. Coral diseases: What is really known? *Trends Ecol. Evol.* **1998**, *13*, 438–443. [CrossRef]

42. Rosenberg, E.; Kellog, C.A.; Rohwer, R. Coral Microbiology. *Oceanography* **2007**, *20*, 146–154. [CrossRef]

43. Rosenberg, E.; Falkovitz, L. The *Vibrio shiloi/Oculina patagonica* model system of coral bleaching. *Annu. Rev. Microbiol.* **2004**, *58*, 143–159. [CrossRef] [PubMed]

44. Reshef, L.; Kore, O.; Loya, Y.; Zilber-Rosenberg, I.; Rosenberg, E. The coral probiotic hypothesis. *Environ. Microbiol.* **2006**, *8*, 2068–2073. [CrossRef] [PubMed]

45. Schrezenmeir, J.; de Vrese, M. Probiotics, prebiotics, and symbiotics–approaching a definition. *Am. J. Clin. Nutr.* **2001**, *73*, 361–364. [CrossRef] [PubMed]

46. Mao-Jones, J.; Ritchie, K.B.; Jones, L.E.; Ellner, S.P. How microbial community composition regulates coral disease development. *PLoS Biol.* **2010**, *8*, 1–16. [CrossRef] [PubMed]

47. Nyholm, S.V.; McFall-Ngai, M.J. The winnowing: Establishing the squid-vibrio symbiosis. *Nat. Rev. Microbiol.* **2004**, *2*, 632–642. [CrossRef] [PubMed]

48. Mazmanian, S.K.; Liu, C.; Tzianabos, A.O.; Kasper, D.L. An immunomodulatory molecule of symbiotic bacteria directs maturation of the host immune system. *Cell* **2005**, *122*, 107–118. [CrossRef] [PubMed]

49. Chow, J.; Lee, S.M.; Shen, Y.; Khosravi, A.; Mazmanian, S.K. Host-bacterial symbiosis in health and disease. *Adv. Immunol.* **2010**, *107*, 243–274. [PubMed]

50. Bosch, T.C. Cnidarian-microbe interactions and the origin of innate immunity in metazoans. *Annu. Rev. Microbiol.* **2013**, *67*, 499–518. [CrossRef] [PubMed]

51. Franzenburg, S.; Fraune, S.; Altrock, P.M.; Kunzel, S.; Baines, J.F.; Traulsen, A.; Bosch, T. Bacterial colonization of Hydra hatchlings follows a robust temporal pattern. *ISME J.* **2013**, *7*, 781–790. [CrossRef] [PubMed]

52. Fraune, S.; Bosch, T.C.G. Why bacteria matter in animal development and evolution. *BioEssays* **2010**, *32*, 571–580. [CrossRef] [PubMed]

53. Fraune, S.; Abe, Y.; Bosch, T. Disturbing epithelial homeostasis in the metazoan Hydra leads to drastic changes in associated microbiota. *Environ. Microbiol.* **2009**, *11*, 2361–2369. [CrossRef] [PubMed]

54. Hemmrich, G.; Anokhin, B.; Zacharias, H.; Bosc, T.C.G. Molecular phylogenetics in Hydra, a classical model in evolutionary developmental biology. *Mol. Phylogenet. Evol.* **2007**, *44*, 281–290. [CrossRef] [PubMed]

55. van de Water, J.; Lamb, J.B.; van Oppen, M.J.H.; Willis, B.L.; Bourne, D.G. Comparative immune responses of corals to stressors associated with offshore reef-based tourist platforms. *Conserv. Physiol.* **2015**, *3*, cov032. [CrossRef] [PubMed]

56. van de Water, J.A.J.M.; Melkonian, R.; Junca, H.; Voolstra, C.R.; Reynaud, S.; Allemand, D.; Ferrier-Pagès, C. Spirochaetes dominate the microbial community associated with the red coral *Corallium rubrum* on a broad geographic scale. *Sci. Rep.* **2016**, *6*, 27277. [CrossRef] [PubMed]

57. Bayer, T.; Arif, C.; Ferrier-Pagès, C.; Zoccola, D.; Aranda, M.; Voolstra, C. Bacteria of the genus *Endozoicomonas* dominate the microbiome of the Mediterranean gorgonian coral *Eunicella cavolini*. *Mar. Ecol. Prog. Ser.* **2013**, *479*, 75–84. [CrossRef]

58. La Rivière, M.; Garrabou, J.; Bally, M. Evidence for host specificity among dominant bacterial symbionts in temperate gorgonian corals. *Coral Reefs* **2015**, *34*, 1087–1098. [CrossRef]

59. Neave, M.J.; Apprill, A.; Ferrier-Pagès, C.; Voolstra, C.R. Diversity and function of prevalent symbiotic marine bacteria in the genus *Endozoicomonas*. *Appl. Microbiol. Biotechnol.* **2016**, *100*, 8315–8324. [CrossRef] [PubMed]

60. Wessels, W.; Sprungala, S.; Watson, S.A.; Miller, D.J.; Bourne, D.G. The microbiome of the octocoral *Lobophytum pauciflorum*: Minor differences between sexes and resilience to short-term stress. *FEMS Microbiol. Ecol.* **2017**, *93*, fix013. [CrossRef] [PubMed]

61. van de Water, J.A.J.M.; Voolstra, C.R.; Rottier, C.; Cocito, S.; Peirano, A.; Allemand, D.; Ferrier-Pagès, C. Seasonal stability in the microbiomes of temperate gorgonians and the red coral *Corallium rubrum* across the Mediterranean Sea. *Microb. Ecol.* **2018**, *75*, 274–288. [CrossRef] [PubMed]

62. Ainsworth, T.D.; Thurber, R.V.; Gates, R.D. The future of coral reefs: a microbial perspective. *Trends Ecol. Evol.* **2010**, *25*, 233–240. [CrossRef] [PubMed]

63. Cook, A.; Bamford, O.S.; Freeman, J.B.; Teidman, D.J. A study on the homing habit of the limpet. *Anim. Behav.* **1969**, *117*, 330–339. [CrossRef]

64. McFarlane, I.D. Trail-following and trail-searching behaviour in homing of the intertidal gastropod mollusc, *Onchidium verruculatum*. *Mar. Behav. Physiol.* **1980**, *7*, 95–108.

65. Wild, C.; Huettel, M.; Klueter, A.; Kremb, S.G.; Rasheed, M.Y.; Jørgensen, B.B. Coral mucus functions as an energy carrier and particle trap in the reef ecosystem. *Nature* **2004**, *428*, 66–70. [CrossRef] [PubMed]

66. Baier, R.E.; Gucinski, H.; Meenaghan, M.A.; Wirth, J.; Glantz, P.Q. Biophysical studies of mucosal surfaces. In *Oral Interfacial Reactions of Bone, Soft Tissue and Saliva, Proceedings of a workshop, Marstrand, Sweden, 9–11 November 1984*; IRL Press: Oxford, UK, 1984; pp. 83–95.

67. Clare, A.S. Marine natural product antifoulants: Status and potential. *Biofouling* **1995**, *9*, 211–229. [CrossRef]

68. Stabili, L.; Schirosi, R.; Licciano, M.; Giangrande, A. The mucus of *Sabella spallanzanii* (Annelida, Polychaeta): Its involvement in chemical defence and fertilization success. *J. Exp. Mar. Biol. Ecol.* **2009**, *374*, 144–149. [CrossRef]

69. Stabili, L.; Schirosi, R.; Di Benedetto, A.; Merendino, A.; Villanova, L.; Giangrande, A. First insights into the biochemistry of *Sabella spallanzanii* (Annelida: Polychaeta) mucus: A potentially unexplored resource for applicative purposes. *J. Mar. Biol. Assoc. UK* **2011**, *91*, 199–208. [CrossRef]

70. Stabili, L.; Schirosi, R.; Licciano, M.; Giangrande, A. Role of *Myxicola infundibulum* (Polychaeta, Annelida) mucus: From bacterial control to nutritional home site. *J. Exp. Mar. Biol. Ecol.* **2014**, *46*, 344–349. [CrossRef]

71. Stabili, L.; Schirosi, R.; Parisi, M.G.; Piraino, S.; Cammarata, M. The mucus of *Actinia equina* (Anthozoa, Cnidaria): An unexplored resource for potential applicative purposes. *Mar. Drugs* **2015**, *13*, 5276–5296. [CrossRef] [PubMed]

72. Suzuki, Y.; Tasumi, S.; Tsutsui, S.; Okamoto, M.; Suetake, H. Molecular diversity of skin mucus lectins in fish. *Comp. Biochem. Physiol. B* **2003**, *136*, 723–730. [CrossRef]

73. Coffroth, M.A. Mucus sheet formation on poritid corals: An evaluation of coral mucus as a nutrient source on reefs. *Mar. Biol.* **1990**, *105*, 39–49. [CrossRef]

74. Calow, P. Why some metazoan mucus secretions are more susceptible to microbial attack than others. *Am. Nat.* **1979**, *114*, 149–152. [CrossRef]

75. Chimetto, L.A.; Brocchi, M.; Gondo, M.; Thompson, C.C.; Gomez-Gil, B.; Thompson, F.L. Genomic diversity of vibrios associated with the Brazilian coral *Mussismilia hispida* and its sympatric zoanthids (*Palythoa caribaeorum*, *Palythoa variabilis* and *Zoanthus solanderi*). *J. Appl. Microbiol.* **2009**, *106*, 1818–1826. [CrossRef] [PubMed]

76. Castro, A.P.D.; Araujo, S.D.; Reis, A.M.; Moura, R.L.; Francini-Filho, R.B.; Pappas, G.; Rodrigues, T.B.; Thompson, F.L.; Kruger, R.H. Bacterial community associated with healthy and diseased reef coral *Mussismilia hispida* from Eastern Brazil. *Microb. Ecol.* **2010**, *59*, 658–667. [CrossRef] [PubMed]

77. Ducklow, H.W.; Mitchell, R. Bacterial populations and adaptations in the mucus layers of living corals. *Limnol. Oceanogr.* **1979**, *24*, 715–725. [CrossRef]

78. Koren, O.; Rosenberg, E. Bacteria Associated with mucus and tissues of the coral *Oculina patagonica* in summer and winter. *Appl. Environ. Microbiol.* **2006**, *72*, 5254–5259. [CrossRef] [PubMed]

79. Rohwer, F.; Breitbart, M.; Jara, J.; Azam, F.; Knowlton, N. Diversity of bacteria associated with the Caribbean coral *Montastraea franski*. *Coral Reefs* **2001**, *20*, 85–95.

80. Wegley, L.; Edwards, R.; Rodriguez-Brito, B.; Liu, H.; Rohwer, F. Metagenomic analysis of the microbial community associated with the coral *Porites astreoides*. *Environ. Microbiol.* **2007**, *29*, 2707–2719. [CrossRef] [PubMed]

81. Menezes, C.B.; Bonugli-Santos, R.C.; Miqueletto, P.B.; Passarini, M.R.Z.; Silva, C.H.D.; Justo, M.R.; Leal, R.R.; Fantinatti-Garboggini, F.; Oliveira, V.M.; Berlinck, R.G.S.; et al. Microbial diversity associated with algae, ascidians and sponges from the north coast of São Paulo state. *Brazil Microbiol. Res.* **2010**, *165*, 466–482. [CrossRef] [PubMed]

82. Chimetto, L.A.; Brocchi, M.; Thompson, C.C.; Martins, R.C.; Ramos, H.R.; Thompson, F.L. Vibrios dominate as culturable nitrogen-fixing bacteria of the Brazilian coral *Mussismilia hispida*. *Syst. Appl. Microbiol.* **2008**, *31*, 312–319. [CrossRef] [PubMed]

83. Chimetto, L.A.; Cleenwerck, I.; Alves, N.; Silva, B.S.; Brocchi, M.; Willems, A.; De Vos, P.; Thompson, F.L. *Vibrio communis* sp nov, isolated from the marine animals *Mussismilia hispida*, *Phyllogorgia dilatata*, *Palythoa caribaeorum*, *Palythoa variabilis* and *Litopenaeus vannamei*. *Int. J. Syst. Evol. Microbiol.* **2011**, *61*, 362–368. [CrossRef] [PubMed]

84. Azam, F.; Smith, D.C.; Steward, G.F.; Hagström, A. Bacteria–organic matter coupling and its significance for oceanic carbon cycling. *Microb. Ecol.* **1993**, *28*, 167–179. [CrossRef] [PubMed]

85. Davies, M.S.; Hawkins, S.J.; Jones, H.D. Pedal mucus and its influence on the microbial food supply of two intertidal gastropods, *Patella vulgata* L and *Littorina littorea* (L). *J. Exp. Mar. Biol. Ecol.* **1992**, *161*, 57–77. [CrossRef]

86. Herndl, G.J.; Peduzzi, P. Potential microbial utilisation rates of sublittoral gastropod mucus trails. *Limnol. Oceanogr.* **1989**, *34*, 780–784. [CrossRef]

87. Peduzzi, P.; Herndl, G.J. Mucus trails in the rocky intertidal: A highly active microenvironment. *Mar. Ecol. Prog. Ser.* **1991**, *75*, 267–274. [CrossRef]

88. Imrie, D.W. The role of pedal mucus in the feeding behaviour of *Littorina littorea* (L). In Proceedings of the 3rd International Symposium on Littorinid Biology, Dale Fort Field Center, Wales, UK, 5–12 September 1990; Grahame, J., Mill, P.J., Reid, D.G., Eds.; The Malacological Society of London: London, UK, 1992; p. 221.

89. Nissimov, J.R.E.; Munn, C.B. Antimicrobial properties of resident coral mucus bacteria of *Oculina patagonica*. *FEMS Microb. Lett.* **2009**, *292*, 210–215. [CrossRef] [PubMed]

90. Charlotte, E.; Kvennefors, E.; Sampayo, E.; Kerr, C.; Vieira, G.; Roff, G.; Barnes, A. Regulation of bacterial communities through antimicrobial activity by the coral holobiont. *Microb. Ecol.* **2012**, *63*, 605–618.

91. Lee, O.O.; Qian, P.Y. Potential control of bacterial epibiosis on the surface of the sponge *Mycale adhaerens*. *Aquat. Microb. Ecol.* **2004**, *34*, 11–21. [CrossRef]

92. Holmström, C.; Kjelleberg, S. Marine *Pseudoalteromonas* species are associated with higher organisms and produce biologically active extracellular agents. *FEMS Microbiol. Ecol.* **1999**, *30*, 285. [CrossRef]

93. Rosenstiel, P.; Philipp, E.E.; Schreiber, S.; Bosch, T.C. Evolution and function of innate immune receptors insights from marine invertebrates. *J. Innate. Immun.* **2009**, *1*, 291–300. [CrossRef] [PubMed]

94. Hoegh-Guldberg, O.; Fine, M. Coral bleaching following wintry weather. *Limnol. Oceanogr.* **2005**, *50*, 265–271. [CrossRef]

95. Brown, B.E. Coral bleaching: Causes and consequences. *Coral Reefs* **1997**, *16*, 129–138. [CrossRef]

96. Glynn, P.W. Coral-reef bleaching-ecological perspectives. *Coral Reefs* **1993**, *12*, 1–17. [CrossRef]

97. Rosenberg, E.; Kushmaro, A.; Kramarsky-Winter, E.; Banin, E.; Yossi, L. The role of microorganisms in coral bleaching. *ISME J.* **2009**, *3*, 139–146. [CrossRef] [PubMed]

98. Kushmaro, A.; Kramarsky-Winter, E. Bacteria as a source of coral nutrition. In *Coral Health and Disease*; Rosenberg, E., Loya, Y., Eds.; Springer: New York, NY, USA, 2004; pp. 231–241.

99. Titlyanov, E.A.; Titlyanova, T.V.; Leletkin, V.A.; Tsukahara, J.; van Woesik, R.; Yamazato, K. Degradation of zooxanthellae and regulation of their density in hermatypic corals. *Mar. Ecol. Progr. Ser.* **1996**, *139*, 167–178. [CrossRef]

100. Davy, S.; Allemand, D.; Weis, V. Cell biology of cnidarian-dinoflagellate symbiosis. *Microbiol. Mol. Biol. Rev.* **2012**, *76*, 229–261. [CrossRef] [PubMed]

101. Sutherland, K.P.; Shaba, S.; Joyner, J.L.; Porter, J.W.; Lipp, E.K. Human pathogen shown to cause disease in the threatened eklhorn coral *Acropora palmata*. *PLoS ONE* **2011**, *6*, 1–7. [CrossRef] [PubMed]

102. Ellis, R.P.; Parry, H.; Spicer, J.I.; Hutchinso, T.H.; Pipe, R.K.; Widdicombe, S. Immunological function in marine invertebrates: Responses to environmental perturbation. *Fish Shellfish Immunol.* **2011**, *30*, 1209–1222. [CrossRef] [PubMed]

103. Otero-González, A.J.; Magalhães, B.S.; Garcia-Villarino, M.; López-Abarrategui, C.; Sousa, D.A.; Dias, S.C.; Franco, O.L. Antimicrobial peptides from marine invertebrates as a new frontier for microbial infection control. *FASEB J.* **2010**, *24*, 1320–1334. [CrossRef] [PubMed]

104. Gochfeld, J.; Aeby, G.S. Antibacterial chemical defenses in Hawaiian corals provide possible protection from disease. *Mar. Ecol. Prog. Ser.* **2008**, *362*, 119–128. [CrossRef]

105. Ganz, T. The role of antimicrobial peptides in innate immunity. *Intergr. Comp. Biol.* **2003**, *43*, 300–304. [CrossRef] [PubMed]

106. Kim, K. Antimicrobial activity in gorgonian corals (Coelenterate, Octocorallia). *Coral Reefs* **1994**, *13*, 75–80. [CrossRef]

107. Rodríguez, A.D.; Ramírez, C.; Rodríguez, I.I.; González, E. Novel antimycobacterial benzoxazole alkaloids, from the west Indian Sea whip *Pseudopterogorgia elisabethae*. *Org. Lett.* **1999**, *121*, 527–530. [CrossRef]

108. Correa, H.; Haltli, B.; Duque, C.; Kerr, R. Bacterial communities of the gorgonian Octocoral *Pseudopterogorgia elisabethae*. *Microb. Ecol.* **2013**, *66*, 972–985. [CrossRef] [PubMed]

109. Shapo, J.L.; Moeller, P.D.; Galloway, S.B. Antimicrobial activity in the common seawhip, *Leptogorgia virgulata* (Cnidaria: Gorgonaceae). *Comp. Biochem. Physiol. B Biochem. Mol. Biol.* **2007**, *148*, 65–73. [CrossRef] [PubMed]

110. Kim, K.; Harvell, C.D.; Kim, P.D.; Smith, G.W.; Merkel, S.M. Fungal disease resistance of Caribbean sea fan corals (*Gorgonia* spp.). *Mar. Biol.* **2000**, *136*, 256–267. [CrossRef]

111. Kim, K.; Kim, P.D.; Alker, A.P.; Harvell, C.D. Chemical resistance of gorgonian corals against fungal infection. *Mar. Biol.* **2000**, *137*, 393–401. [CrossRef]

112. Chen, G.; Swem, L.R.; Swem, D.L.; Stauff, D.L.; O'Loughlin, C.T.; Jeffrey, P.D.; Bassler, B.L.; Hughson, F.M. A strategy for antagonizing quorum sensing. *Mol. Cell* **2011**, *42*, 199–209. [CrossRef] [PubMed]

113. Li, C.; La, M.P.; Tang, H.; Pan, W.H.; Sun, P.; Krohn, K.; Yi, Y.H.; Li, L.; Zhang, W. Bioactive briarane diterpenoids from the South China Sea gorgonian *Dichotella gemmacea*. *Bioorg. Med. Chem. Lett.* **2012**, *22*, 4368–4372. [CrossRef] [PubMed]

114. Kelman, D.; Kushmaro, A.; Loya, Y.; Kashman, Y.; Benayahu, Y. Antimicrobial activity of a Red Sea soft coral, *Parerythropo-dium fulvum fulvum*: Reproductive and developmental considerations. *Mar. Ecol. Prog. Ser.* **1998**, *169*, 87–89. [CrossRef]

115. Saxby, T.; Dennison, W.C.; Hoegh-Guldberg, O. Photosynthetic responses of the coral *Montipora digitata* to cold temperature stress. *Mar. Ecol. Prog. Ser.* **2003**, *248*, 85–97. [CrossRef]

116. Palla, F.; Barresi, G.; Giordano, A.; Schiavone, S.; Trapani, M.R.; Rotolo, V.; Parisi, M.G.; Cammarata, M. Cold-active molecules for a sustainable preservation and restoration of historic artistic manufacts. *IJCS* **2016**, *7*, 239–246.

117. Trapani, M.R.; Parisi, M.G.; Parrinello, D.; Sanfratello, M.A.; Benenati, G.; Palla, F.; Cammarata, M. Specific inflammatory response of *Anemonia sulcata* (Cnidaria) after bacterial injection causes tissue reaction and enzymatic activity alteration. *J. Invertebr. Pathol.* **2016**, *135*, 15–21. [CrossRef] [PubMed]

118. Anderluh, G.; Macek, P. Cytolytic peptide and protein toxins from sea anemones (Anthozoa: Actiniaria). *Toxicon* **2002**, *40*, 111–124. [CrossRef]

119. Parisi, M.G.; Trapani, M.R.; Cammarata, M. Granulocytes of sea anemone *Actinia equina* (Linnaeus, 1758) body fluid contain and release cytolysins forming plaques of lysis. *Invert. Surviv. J.* **2014**, *11*, 39–46.

120. Minagawa, S.; Ishida, M.; Nagashima, Y. Primary structure of a potassium channel toxin from the sea anemone *Actinia equina*. *FEBS Lett.* **1998**, *427*, 149–151. [CrossRef]

121. Lesser, M.P.; Stochaj, W.R.; Tapley, D.W.; Shick, J.M. Bleaching in coral reef anthozoans: Effects of irradiance, ultraviolet radiation, and temperature on the activities of protective enzymes against active oxygen. *Coral Reefs* **1995**, *8*, 225–232. [CrossRef]

122. Leclerc, M. Humoral factors in marine invertebrates. In *Molecular and Subcellular Biology: Invertebrate Immunology*; Rinkevich, B., Muller, W.E.G., Eds.; Springer: Berlin, Germany, 1996; pp. 1–9.

MDPI

St. Alban-Anlage 66

4052 Basel

Switzerland

Tel. +41 61 683 77 34

Fax +41 61 302 89 18

www.mdpi.com

Marine Drugs Editorial Office

E-mail: marinedrugs@mdpi.com

www.mdpi.com/journal/marinedrugs